AF251473

BIOTECHNOLOGY IN AGRICULTURE SERIES

General Editor: Gabrielle J. Persley, Biotechnology Adviser, Environmentally Sustainable Development, The World Bank, Washington, DC, USA.

For a number of years, biotechnology has held out the prospect for major advances in agricultural production, but only recently have the results of this new revolution started to reach application in the field. The potential for further rapid developments is, however, immense.

The aim of this book series is to review advances and current knowledge in key areas of biotechnology as applied to crop and animal production, forestry and food science. Some titles focus on individual crop species, others on specific goals such as plant protection or animal health, with yet others addressing particular methodologies such as tissue culture, transformation or immunoassay. In some cases, relevant molecular and cell biology and genetics are also covered. Issues of relevance to both industrialized and developing countries are addressed and social, economic and legal implications are also considered. Most titles are written for research workers in the biological sciences and agriculture, but some are also useful as textbooks for senior-level students in these disciplines.

Editorial Advisory Board:
E.P. Cunningham, Trinity College, University of Dublin, Ireland.
P. Day, Rutgers University, New Jersey, USA.
J.H. Dodds, Attorney at Law/Patent Attorney, Washington, DC, USA.
S.L. Krugman, United States Department of Agriculture, Forest Service.
I. Morrison, Institute for Animal Health, Compton, UK.
W.J. Peacock, CSIRO, Division of Plant Industry, Australia.

BIOTECHNOLOGY IN AGRICULTURE SERIES

Titles Available:

*Out of print

The Biotechnology Revolution in Global Agriculture: Innovation, Invention and Investment in the Canola Industry

Edited by

Peter W.B. Phillips

Professor of Agricultural Economics, NSERC/SSHRC Chair in Managing Knowledge-based Agri-food Development

and

George G. Khachatourians

Professor of Applied Microbiology and Food Sciences, University of Saskatchewan, Canada

CABI *Publishing*

CABI *Publishing* is a division of CAB *International*

CABI Publishing
CAB International
Wallingford
Oxon OX10 8DE
UK

Tel: +44 (0)1491 832111
Fax: +44 (0)1491 833508
Email: cabi@cabi.org
Web site: http://www.cabi.org

CABI Publishing
10 E 40th Street,
Suite 3203
New York, NY 10016
USA

Tel: +1 212 481 7018
Fax: +1 212 686 7993
Email: cabi-nao@cabi.org

A catalogue record for this book is available from the British Library, London, UK.

Library of Congress Cataloging-in-Publication Data

The biotechnology revolution in global agriculture : innovation, invention, and investment in the canola industry / edited by W.B. Phillips and G.G. Khachatourians.
   p. cm. -- (Biotechnology in agriculture series; #24)
Includes bibliographical references (p.   ).
ISBN 0-85199-513-6 (alk. paper)
   1. Canola--Biotechnology. I. Phillips, Peter W.B. II. Khachatourians, George G., 1940–   III. Biotechnology in agriculture series; 24.

SB299.R2 B56 2001
633.8′53--dc21   00-045427

ISBN 0 85199 513 6

Typeset by Columns Design Ltd, Reading.
Printed and bound in the UK by Cromwell Press, Trowbridge, UK.

Contents

Contributors

Murray E. Fulton is Professor of Agricultural Economics and Head, Department of Agricultural Economics, University of Saskatchewan, Canada

Richard S. Gray is Professor of Agricultural Economics, University of Saskatchewan, Canada

Grant E. Isaac is Associate Professor of Management and Marketing, College of Commerce, University of Saskatchewan, Canada

Lynette Keyowski is an M.Sc. Agricultural Economics student, Department of Agricultural Economics, University of Saskatchewan, Canada

George G. Khachatourians is Professor of Applied Microbiology and Food Sciences, University of Saskatchewan, Canada

Stavroula T. Malla is an SSHRC Post-doctoral Fellow, Department of Agricultural Economics, University of Saskatchewan, Canada

Peter W.B. Phillips is Professor of Agricultural Economics, NSERC/SSHRC Chair in Managing Knowledge-based Agri-food Development, University of Saskatchewan, Canada

Arthur K. Sumner is Professor Emeritus of Applied Microbiology and Food Science, University of Saskatchewan, Canada

Preface

This volume is the result of a single chance meeting. Shortly after assuming the Van Vliet Professorship in Agricultural Economics at the University of Saskatchewan in 1997 Peter Phillips gave a talk at the Canadian Wheat Board Grain World event. He used that event to float a few ideas about how knowledge-based development would affect the agricultural trade negotiations scheduled to begin in 1999. He received little feedback at the event but had an opportunity to replay the talk as a guest at the Joel, a small group in Saskatoon that for 65 years has brought together persons from the campus of the university with those from the town to debate topics of mutual interest. George Khachatourians was in the group and we engaged in an excited and far-ranging discussion about the implications of knowledge change. We agreed that night to begin to examine canola as an example of this phenomenon.

It became clear early on that neither of us had the time nor skills to examine all aspects of the story. Over the intervening 3 years, we found collaborators in our faculty and among the graduate students. Each has made a significant contribution to the scope and conclusions of the study.

Peter W.B. Phillips and George G. Khachatourians
University of Saskatchewan
Saskatoon, Canada
September 2000

Acknowledgements

A number of agencies directly or indirectly provided funds that assisted with gathering data or supporting research. The key support came from the Van Vliet Research Fund at the University of Saskatchewan. The final changes to the book were in part financed by NSERC and SSHRC. In addition, data were gathered and specific studies undertaken with support from the Canadian Food Inspection Agency, Ag-West Biotech Inc., the International Food Policy Research Institute and federal and provincial summer student job subsidy programmes.

Over the past 2 years six student researchers assisted with this work: Cameron McCormick, Lynette Keyowski, Leif Carlson, Brian Perillat, Grant Kuntz and Monica Wilson. Their contributions were important in many unseen ways.

Finally, we would like to thank all our colleagues, friends and family who have put up with our endless expositions on this topic. Your questions and observations have contributed enormously to the quality of this work.

The Setting

Introduction and Overview 1

Peter W.B. Phillips and
George G. Khachatourians

Background

Some 40% of the world's market economy is based upon biological products and processes (Gadbow and Richards, 1990). Innovation, knowledge and technology are increasingly affecting the competitive base for much of that industry. Although biotechnology applications have been with us for centuries – one of the oldest large-scale applications of biotechnologies by industrial societies was the purification of waste water through microbial treatment in the 19th century – modern, Mendelian plant breeding has, since 1973, been increasingly influenced and driven by new molecular biology techniques (OECD, 1999). This transformation, which is influencing the structure and location of global agricultural activities, has not been studied in any comprehensive way.

This transformation is clearly visible in western Canada, where plant, animal and microbial products and processes are the base of the modern regional economy. In the past, western Canada's competitive position in agri-food production was based on high-quality land and capital-intensive production processes. That now appears to be changing, with knowledge becoming the defining factor in much of the food industry.

This book examines the canola sector to illustrate this phenomenon. Innovation has been the defining feature of the canola sector for more than 40 years. Government research in the 1960s bred a new type of rapeseed with only a small amount of two undesirable traits – erucic acid and glucosinolates – and named it canola, thereby creating the base for a knowledge-based, innovation-driven industry centred around Saskatoon, Canada. This precipitated a myth that Saskatoon and Canada were the centre of the global canola industry. To a point, the myth reflects reality. Initially a large portion of the research, all of the commercial varieties and an increasing proportion of the production of canola were produced in Saskatoon and the surrounding farming areas in western Canada. Nevertheless, after the first breakthrough, the research into and production of canola began to disperse to other locations.

With the establishment of private intellectual property rights and the development of new biotechnology processes in the 1980s and 1990s, private seed and agrochemical companies began to invest in and to undertake substantial research and development in the canola sector around the world. Economic theory suggests that innovation-driven industries like this are inherently imperfectly competitive because large up-front research and development costs and low marginal costs yield rapidly increasing returns to scale in production. When combined with the presence of spillovers that are localized, the theory suggests that over time the research, commercialization and even production activities of an innovative industry will converge on fewer locations, or even a single location. Thus, the 'myth' of Saskatoon and Saskatchewan as the centre of the industry may be actually becoming a reality.

This study examines relevant economic theories, reviews the scientific and historical base for the industry, uses scholarly citations to investigate the evolution of canola research across both time and geography, analyses the commercialization and adoption of canola in western Canada and the world, and estimates the costs and benefits of innovation in the industry. This work is then used to examine prospective trends and to investigate the role of public policy in supporting and encouraging commercial success in the worldwide canola sector.

The Research Context

Knowledge-based growth and development theory has been articulated, debated and taught for more than 15 years but has remained for the most part simply a theoretical concept that has been applied in only a limited way. The few cases where it has been used, such as examining Silicon Valley and other industrial agglomerations, have not included any agri-food examples. This may be partly understood given the prevailing view that agri-food sectors are low tech and not focal points for innovation.

Before beginning this research, the authors undertook a literature search to determine what economic or policy work, if any, had been done on canola. A search of the ISI Social Sciences Citations Index showed that only 53 social science journal articles written by about 35 researchers had been produced between 1980 and 1996 relating to canola. Of those written by economists, many were simply market assessments produced for annual outlook conferences and then republished as part of proceedings. The other major type of research undertaken focused on market issues, such as the impacts of tariffs and exchange rate variability on trade (e.g. Griffith and Meikle, 1993). On further investigation, a number of papers undertaken in the early period (e.g. Nagy and Furtan, 1978; Ulrich *et al.*, 1984) estimated the gains from research into new canola varieties. All of these papers were completed before canola was granted GRAS status in the US and ultimately became the third largest source of edible oil in the world, planted by hundreds of thousands of farmers worldwide. The fact that these papers were addressing a marginal oil that had only limited market access at least partly explains why the research was seldom cited by others. The 53 papers identified in the citations search produced only 18 citations between them; an average group of papers of this type would have been cited 57 times. Since then there has been little work done on the nature and impact of innovation in the canola sector. In the past few years, interest has risen. A number of graduate students at the University of Saskatchewan (Malla, 1995; Mayer, 1997; Keyowsky, 1998) have begun to investigate the research benefits from the introduction of new varieties of canola. More recently, Carew (2000) undertook a partial analysis of the impact of intellectual property rights on canola. Elsewhere, a group of sociologists led by Lawrence Busch at Michigan State University has used a sociological approach to examine the research institutions and processes in the public breeding programmes. Apart from that, the only major canola-related publication was the polemic by Brewster Kneen (1992), entitled *The Rape of Canola*.

Given the major changes that have occurred in the agri-food sector, and more particularly in the canola industry, it is a subject ready and amenable for analysis. Canola exhibits some highly relevant features that made it a logical choice for investigation. First, the industry has undergone two large innovation periods, first in the 1970s as rapeseed was converted to canola and more recently as biotechnology has enabled more targeted trait introduction. Secondly, the two transformations were managed by different lead actors. Unlike maize, cotton and soybeans, where private activity has been dominant for decades, canola started out as a publicly managed sector and now is predominantly privately managed. When biotechnology is introduced into the traditionally publicly led breeding programmes for cereals, pulses and small crops, they may face similar circumstances as canola. Thirdly, although much of the industry has been privatized in the past 15 years, it remains relatively open to investigation. Many of the key scientists and business leaders in the sector began their careers in the public sector and still appreciate the value of exchanging information about what they are doing. One notable example is the annual industry research committee meetings chaired by Keith Downey of Agriculture Canada (renamed Agriculture and AgriFood Canada in 1985), the acknowledged 'father' of canola, where firms and public agencies share information about what they are doing in their laboratories and greenhouses. For all these reasons, and probably at least partly just because 'it was there', this book investigates the canola sector as a case study of how the agri-food sector is being transformed due to increased private innovation, invention and investment.

The Characteristics of Knowledge-based Growth

Professor Peter Drucker (1993) has argued that 'the basic economic resource – "the means of production", to use the economist's term – is no longer capital, or natural resources (the economist's "land"), nor "labour". It is and will be knowledge.' Western Canada has been labelled the 'breadbasket' of the world because of the inherent competitive position of its soils and the accumulation of labour and capital in the farm industry. USDA studies have shown that on that basis, Canada has a comparative advantage in producing wheat, canola and some red meats. The knowledge explosion, however, is challenging western Canada's comparative advantage for agri-food production. It appears, as Grossman and Helpman (1991) argue, that comparative advantage is endogenously generated and evolving over time. As the rate of innovation accelerates, the possibility of firms, sectors or areas losing existing or gaining new comparative advantages increases.

In the industrial economy, land, labour and capital were the key assets for growth. In the knowledge economy, the key asset is innovation – the ability to develop new ideas, products and organizational structures by combining existing ideas, products and structures in new ways.

Agricultural policy has traditionally been modelled on the assumption that agricultural markets are perfectly competitive. Research, production and marketing analyses all tend to take as given that the agri-food sector produces 'commodities' which are sold in markets characterized by perfectly competitive features. When there is a choice in specifying a model, economists inevitably choose agriculture or food to be the competitive product. This model, however, does not explain recent agri-food development, which is characterized by increased innovation, more tightly integrated production systems and two-way trade in differentiated products. Douglass North (1991), in his recent Nobel lecture, concludes that 'neo-classical theory is simply an inappropriate tool to analyse and prescribe policies that induce development. It is concerned with the operation of markets, not with how markets develop'.

The challenge is to find an appropriate theoretical specification for agriculture, which explains what has been happening in the agricultural and food sectors. The purpose of the following exposition is not to theorize for its own sake but to find the threads of economic theory from other investigations and to weave them into an explanatory framework that will help policy makers to understand the dynamics in the sector and examine and compare alternative policy options.

This book examines the hypothesis that the agri-food sector is being transformed into an innovation-driven, vertically coordinated business, exporting differentiated products. Innovation is much more than invention. While a prototype fax machine is an invention, the millionth fax machine in use marks a transformative innovation. Innovation most frequently occurs within organizations whose aim is to transform creations into socially valued products, and whose success is marked by the ease in which creations are absorbed into and persist in society. Innovation is characterized by the fact that society always reshapes what it uses;

in turn, the ability to renew innovation is dependent on understanding the changing context in which successive innovation occurs. Innovation is thus a creative activity that takes place within an organizational and a social context and has organizational and social consequences. Three aspects of innovation – a creative activity, an organizational and social context, and organizational and social consequences – tend to concentrate innovations in business, organizations and the economy in clusters in which new knowledge and skills complement imaginative industry leadership, all of which are supported by active partners, including communities and governments. This pattern is frequently seen in the innovation corridors of Silicon Valley, Boston, Austin, Cambridge and Bangalore.

Agri-food systems, in particular, are increasingly driven to innovate to improve cost competitiveness and to differentiate their products and processes. In doing so, they create *de facto* monopolies. Much of this innovation is 'knowledge-based', which creates two self-supporting competitive features. First, knowledge-based innovation involves learning-by-doing, which works to create barriers to imitators as they are only able to use the technological innovation after they have gone through a learning process. Secondly, because many types of knowledge are hard to protect and exploit, there is significant potential for applied science spillovers to others in the sector. In the first instance, the barrier to competitors helps to secure a better return to innovators while, in the second, the whole economy (regional, national and international) benefits by the externality of the innovation. Both tend to encourage restructuring by innovative enterprises.

The application of information technologies (IT), in concert with biotechnology techniques, creates incentives for industries to 'industrialize' by integrating their production chains, linking markets with genetics and coordinating the various production processes. In the past, technology was such that the only way to manage market risk was by direct vertical ownership, a process often constrained by shortages of capital and management ability. With IT now ubiquitous, the cost of acquiring the information to manage a production value chain has dropped dramatically. In the past, commodity markets typically involved arms-length trades between buyers and sellers, with price as a major deciding factor. Now, branded, differentiated products provide the base for long-term, one-to-one buyer–seller production and marketing chains. In short, the industry needs to be examined in the context of movement of product through the production chain rather than as exchange between uncoordinated firms and sectors.

As a result, trade is no longer exclusively based on traditional factor endowments; comparative advantage has become dynamic. Knowledge-based activity (e.g. research, marketing and logistics) creates significant potential for sectors or countries to develop new competitive and comparative advantages, less dependent on relative endowments of labour and capital. As sectors industrialize and innovate, the product life cycle has shortened to years rather than decades. Recognizing this, firms with innovative products or processes are driven to expand their markets by exporting and thereby capitalize on their advantage during the period in which they are the only suppliers of that product. The end result is that the flow of trade can be influenced by the actions of sectors and governments. Furthermore, although there are still potential gains from trade, the presence of imperfectly competitive enterprises removes the certainty that both parties in the trade will share the gains.

By re-introducing time, institutions and space into neo-classical economics, economic theorists have begun to model more completely the 'imperfectly competitive' markets that we see evolving in the agri-food sector. This modelling approach has been applied in four specific areas of theory: growth, institutions, trade and location. The resulting synthesized theory has significant potential to explain more fully recent developments in the agri-food sector.

One can start with the recently renewed interest in growth theory and innovation in the economy. The traditional growth model developed by Solow (1956) posits that national growth is a function of the accumulation of labour and capital, with technological change exogenous to the model. Given that labour supply is largely a function of population growth, the only stochastic variable is capital accumulation, which is a function of the marginal product of capital and the inter-temporal discount rate. The theory posits that the marginal product of physical capital declines as the ratio of capital to labour rises, so that the incentive to invest declines as an economy grows. Given that trend, at some point capital investment will converge to a constant, with the result that long-term economic growth stabilizes at the rate of growth in the labour force. Both international GDP levels and growth rates should converge due to this process. The evidence is that something is missing from this specification: growth in per capita incomes has been sustained globally and nationally for long periods above the rate of growth in labour (studies suggest that the Solow model only explains about between 20% and 50% of measured growth) and performance has varied greatly from country to country (Grossman and Helpman, 1991). Another deficiency of the Solow model is that it does not explain the role of firms in the growth process. Under perfect competition (a basic assumption in the model), firms are unable to recoup their investments in innovation because their technology is completely transferable and profits will be bid away. Without the possibility of profit, there is no incentive to innovate.

The endogenous growth model starts by re-introducing time to the analysis. Most of the new growth theorists start from Schumpeter's perspective that otherwise outwardly perfectly competitive firms pursue innovation to achieve monopoly profits during the time required for imitators to catch up. Schumpeter (1954) argued that in practice technological change is a strategic response by firms attempting to capture or create markets through product creation and differentiation. New products or new varieties of products create monopoly positions for the innovator, which allow the innovator to reap monopoly rents. But the existence of those rents creates incentives for other firms to imitate or innovate, either to match or to leapfrog their competitors. Thus monopoly rents from innovation are continuously under threat and likely to be of short-term duration. Schumpeter referred to this dynamic process as 'creative destruction'.

In this model, the focus is on innovation, which is the firm-based process of investing time and other resources in the search for new technologies and processes. Grossman and Helpman (1991) argue that innovation is undertaken for two basic reasons – to reduce costs and to develop a new product that exhibits different quality characteristics (i.e. vertical innovation) or that provides variety (i.e. horizontal innovation). Regardless of the reason, innovators will continue to innovate as long as they expect to earn a return on their efforts.

The new growth theory distinguishes innovations by two characteristics: rivalry and excludability. Rival innovations result in goods or services that can only be used by one person at one time (such as a consumer durable or personal service). Non-rival innovations involve an output (usually knowledge) that for little relative expense, or in some cases no cost, can be disseminated to and used by every producer in a country or the world, and no one's use is limited by any other's use. Excludability (sometimes referred to as separability) measures whether the innovation is protected from widespread use by legal means (e.g. patent) or whether its adoption is limited by industrial organization requirements or climate. If it is excludable, then the innovator can appropriate all the benefits from the innovation. If it is not excludable, then the innovator cannot get paid for his innovation. Table 1.1 shows examples of the different types of innovation.

The traditional case of rival innovation, with or without excludability, typifies the Solow (1956) growth model, with decreasing returns to scale and ultimately a slowing in growth. As Grossman and Helpman (1991) observe, there is limited consumer demand, so that as the number of product innovations rises, the average sales per variety will fall. Eventually profit per innovation will stabilize and innovation will converge to a stable path. Before the introduction of plant breeders' rights in 1990, almost all of the research on canola varieties was undertaken by the public institutions. Analysis by Nagy and Furtan (1977) showed the internal rate of social return to canola research in the 1980s was about 100%, which suggests that there was too little investment at that time. With the introduction of intellectual property rights for agrifood innovations (e.g. plant breeders' rights and patents) and the entry of private investment, the number of new varieties has risen sharply. Undoubtedly that should, over time, reduce the internal rate of return on canola research and at some point innovation yielding rival, excludable varieties may reach a

Table 1.1. Categories of innovation in the canola sector.

	Excluded	Not excluded
Rival	New seed varieties, e.g. varieties protected by plant breeders' rights	New seed varieties, e.g. varieties developed and marketed that are not protected by plant breeders' rights
Non-rival	Process innovations, e.g. Calgene's patented process of foreign gene expression in canola	Process innovations, e.g. use of gas spectrometer or plant genome mapping for canola

saturation point. As more than 190 varieties are now available for planting, this point may be approaching. Grossman and Helpman (1991) conclude that the stable rate of innovation ultimately is positively correlated with the taste for variety (e.g. different soil and climatic zones) and the size of the economy and the efficiency of labour, and will be negatively correlated with the intertemporal discount rate.

The more interesting case is where the innovation creates a non-rival product – either blueprints or applied science. If the firm that develops and owns the improved process acts like a pure monopolist and does not allow any other firm to use it (e.g. they don't license it), then that innovation would tend to exhibit decreasing returns to scale, as in the case of the rival innovation. Ultimately it could stifle innovation and potential growth. Some market participants expressed concern that Calgene's US patent on *Agrobacterium tumefaciens brassica* transformation and Plant Genetics Systems' patent on a hybridization system could lessen competition and lead to this result. So far, however, no firm has been able to develop a patented process that has been an effective block to other market participants.

The key factor that determines the long-term role for innovation is the non-appropriability of some of the benefits of innovation. Although economists have modelled the effect of the general or applied science innovations differently, the results converge on a common view. The new growth theory assumes that at least part of the non-rival knowledge accumulated is non-excludable. With technological change – described by Romer (1990) as an 'improvement in the instructions for mixing together raw materials' – non-excludable knowledge spills over into the economy as a whole and raises the marginal value of new innovations.[1] Hence, the positive externality associated with private investment leads to a sectoral or national production function with increasing returns to scale. In essence, the rate of growth in the economy rises with the amount of resources devoted to innovation activity (i.e. R&D, which is in turn a function of the size of the economy), the degree to which new technology is not excludable (i.e. the higher the degree of monopoly the less innovation, or, conversely, the less it is excludable, the greater are the spillovers) and a lower intertemporal discount rate (i.e. the time horizon for the investors).

Two aspects of this theory suggest that competing firms, and as a result industries, will tend to concentrate in a few locations. First, if firms innovate to earn monopoly profits, it is important to determine the possible scale of monopoly profits and to investigate how they will be used. If knowledge-based innovation is excludable solely because of legal constraints, namely patents, then the

---

[1] Theoreticians tend to assume that all innovation destroys the value of past innovations or investments. But there is also no reason to reject *a priori* the possibility that the externalities could improve the marginal productivity of existing capital and labour via more efficient production processes, especially if the innovation is in information technologies, which permits better management and new applications of existing technology. Although this would likely be a one-time upward adjustment in the marginal productivity, the adjustment would take time. Therefore, given continuous innovation, it is possible that growth would be bolstered over a long period by innovation.

period of monopoly profits will only last as long as the patent. On the other hand, if knowledge-based innovation involves extensive learning-by-doing, there would be extensive fixed costs of entering the industry. Given that knowledge-based innovations are usually transferable at low or no marginal cost (Shapiro and Varian, 1999), this creates significant economies of scale, which yields declining average costs and a major barrier to imitators. This tends to extend the period of monopoly profits. Assuming innovators are rational, they will recognize that over time their competitors will either innovate to imitate or to leapfrog the current monopolist, thereby bidding down or eliminating the monopolist's source of market power and monopoly profits. So, innovators will be driven, first, to expand production and maximize profits during the period of monopoly and, secondly, to use some of these monopoly profits to continue to innovate to keep ahead of their competitors. Having monopoly profits allows the innovator to invest a greater amount in R&D and ultimately to widen the gap between it and the nearest competitor.[2] The imperative to innovate has, in practice, tended to keep research and production units linked together in one or at most a few locations, in order to capitalize on the resulting synergies.

Secondly, although knowledge is a non-rival good among all producers worldwide, it might, at least in the short-run, be excludable between jurisdictions for a variety of reasons. In the agri-food industry, for instance, climate, soil characteristics, microbial communities and industrial structure all create natural or man-made barriers to transferring technology between jurisdictions. Some plant genetics and animals cannot survive or can produce only with wide differences in efficiency in different soil or climatic zones, certain pests or microbes limit or curtail production for other crops and livestock, while many of the new genetically altered products require a certain scale of production unit (e.g. field size) or complementary investments (e.g. mechanized seeding and harvest equipment). So it is possible, and often observed, that innovations in one country cannot be transferred elsewhere. The flip side of this is that like-types of innovation will tend to concentrate in areas where there are similar climate, soil characteristics, microbiology and industrial structure. One result is that if the final product is tradable but the innovation-based knowledge is a non-transferable intermediate factor of production, then the fact that innovation begins in one jurisdiction could forever put that site on a higher trajectory of R&D and new product development. Grossman and Helpman (1991) argue that, as a result, the high-technology share of GDP and exports will be higher than

---

[2]Grossman and Helpman (1991) argue that, because each new innovation that increases variety or quality destroys the value of previous innovations, a monopolist would not innovate indefinitely. At some point (they suggest two steps ahead of their followers) the net present value of the investment in innovation would become negative. So it is possible that the next generation of a product might not come from the leader but from a close follower. If we assume generally competitive and efficient capital markets, markets would force this result. But with only limited market discipline over uses of retained earnings, it is possible, and often observed, that monopolists continue to innovate more than two steps ahead. Either way, the monopolist would have an incentive to innovate to reduce cost if the net present value of that investment were positive.

otherwise. The authors of this study will look at whether Canada has benefited.

The distribution of these gains from innovation are seldom left up to the chance operations of the marketplace. Although excludability is defined initially as the result of the attributes of the innovation, firms can improve the odds of gaining a larger share of non-excludable benefits, depending on how they structure their operations. The evolving theory of 'institutional' economics helps to define the potential for industrial structure to adapt to the market opportunities. Coase (1937) posits that firms exist to manage risk – namely those risks and uncertainties related to price discovery, negotiation and monitoring of transactions. Risk and uncertainty creates costs. Clearly, uncertainty cannot be managed, but risk in transactions can be managed if the market transaction is replaced by some institutional arrangement. Coase hypothesizes that firms exist and operate because the cost of managing production in-house is less than the cost of transacting to buy-in. He concludes that firms will grow to the point where the cost of managing internal processes equals the cost of transacting (including the risk) with other agents.

This theoretical approach has been pursued by a number of researchers in recent years. There have been two key approaches: transaction costs and principal-agent theory. Williamson (1985) argues that contracting is not costless, for two key reasons. First, he notes that markets are best described as operating with 'bounded rationality', that is individuals act rationally but their options are limited by imperfect information or the absence of a critical actor in a market (e.g. farmers may believe they should integrate forward into processing but a facilitating mechanism may be absent).[3] Secondly, he assumes that individuals and companies act opportunistically, that is they will act in a self-interested way 'with guile' that increases their return, by renegotiating terms of agreements or by substituting lower-cost goods or services than contracted for. Their ability to succeed depends on their relative bargaining position, which is a function of the specificity of the assets each party has invested. The firm with assets that have little alternate use (e.g. hog barns) are most at risk of having their returns bid away by other actors in the production system.

The alternative approach examines the costs and benefits of principal-agent relationships. The approach assumes that firms ('principals') will contract with 'agents' to avoid market risk. Once again, there is a concern that 'opportunistic' agents will take advantage of any imbalance of power, in this case resulting from the inability to measure either their contribution to the total output (called non-separability) or their inputs to the task (called programmability). In short,

---

[3]One way of examining this problem has been to examine the question of hold-up, where if capital is specific (e.g. has little or no alternate use or value) then two economic actors may be unable to strike a bargain that secures adequate economic returns for each actor in order for each to invest to realize a potential *pareto* improving investment. The problem is that the firm with the most 'specific' capital will be at risk of its partner acting 'opportunistically' and renegotiating the arrangement – the theory suggests that the firm with the 'specific' capital will have little bargaining power *ex ante,* and will end up with simply enough return to continue to operate the asset. In this case there may need to be another actor or structure to bridge the gap.

Table 1.2. Predicting the organizational form of vertical control. (From Mahoney, 1992.)

	Low task programmability		High task programmability	
	Low asset specificity	High asset specificity	Low asset specificity	High asset specificity
Low non-separability	Spot market	Long-term contract	Spot market	Joint venture
High non-separability	Relational contract	Clan (hierarchy)	Inside contract	Hierarchy

the more measurement problems there are, the higher the cost of buying-in relative to the cost of doing-in, with the result that vertical coordination is more likely to be pursued.

Mahoney (1992) put together the two institutional economic approaches to create a synthesized transaction cost-agency model (Table 1.2). He argues that if one assumes opportunism, one can predict the organizational form of vertical integration based on the degree of asset specificity, task programmability and non-separability. Only some of the eight options are of interest for the canola case. Canola traditionally has exhibited low task programmability, low non-separability and low asset specificity, so it lends itself to spot markets. But as the production technologies have become more linked (e.g. Round-Up Ready^{TM} canola), task programmability has risen. Meanwhile, recent efforts to breed in specific market characteristics has increased non-separability. Given that asset specificity at the producer level remains low, these pressures should be leading to more contracting in the industry. In contrast, in the genetics/seed business and in the related chemical industry, where asset specificity is very high, there is real pressure for vertical integration that enforces a more traditional hierarchical structure on the industry.

The model sketched above – with industrialized production chains innovating to develop market power – involves imperfectly competitive firms. The introduction of knowledge as a critical factor of production, which creates these imperfectly competitive firms, makes it possible that comparative advantage is now endogenous and not simply predetermined by the relative endowments of labour and capital. In short, trade flows, and the resulting gains from trade, now have the potential to be driven by endogenous decisions and actions.

The neoclassical trade model needs to be reviewed in the context of the dynamics of imperfect competition to determine the resulting impact on the volume, composition and gains from trade. First, one must examine the prevailing assumption that agriculture is a 'labour-intensive' good. The modern, commercial, agri-food sector – encompassing the biotechnology industry, the input industries, farmers, processors, transport firms, logistics companies and marketing systems – now is more capital- and knowledge-intensive than basic component manufacturing. With the industrialization of agriculture, the agri-food sector now ranks in the top ten industries in terms of capital intensity, while knowledge is increasingly the defining factor in the industry (see Table 1.3). Almost all of the purchased inputs, accounting for about 23% of the total value added in the production system, are ranked as 'high' knowledge activities in a recent Industry Canada study. If one recalculates the knowledge intensity of the canola supply chain, using the relative weights of the service and supply industries in total value added, the oilseeds industry belongs in the ranks of the medium knowledge-intensity industries. Given that much of these data are from the mid 1980s, it underestimates the knowledge-intensity today of both primary agriculture and the service industries. With the introduction of new, more sophisticated machinery, farm chemicals, financing options, genetics (via seed development) and logistical control systems in grain handling and transportation systems, it is almost certain that export agriculture is vying for a place in the 'high' knowledge intensity category.

The impact of imperfect competition on trade volumes and composition depends on how far the technological externalities spread. If the spillovers are global, then relative endowments of traditional factors of production will ultimately determine trade flows. History might dictate the initial pattern of specialization as countries produce with the blueprints they inherit, but factor price equalization will drive the trading countries towards the production pattern that fits its factor-based comparative advantage. So any impact of imperfect competition arising from endogenous growth would be only transitory. Grossman and Helpman (1991) argue that technological spillovers that are limited to a specific location (e.g. due to climate or industrial structure) create the possibility that 'comparative advantage is endogenously generated' because as 'countries engage in technological competition, comparative advantage evolves over time'. If technological spillovers are geographically concentrated, initial and sequentially established conditions matter. In the extreme, if trade partners are similar in size and their endowments consist of a single primary factor, then a country that inherits even a small technological lead will come to dominate world markets for high-technology products. A productivity differential then becomes self-perpetuating. In more general circumstances, a large size, an abundance of human capital and a sizeable knowledge base contribute to a country's comparative advantage in research. In this case, there is incentive for governments to subsidize research in the knowledge good, or to protect the local market to provide an effective domestic subsidy to the home producer. If the other country does not retaliate, it is theoretically possible for subsidies or protection to assist the home producer to get the jump on the foreign competitor and thereby enable the home country to develop comparative advantage and to become sole producer of the knowledge good. This case – representative of conditions in the large-frame aircraft sector and the large-memory computer chip markets – has been much studied by strategic trade economists (e.g. Baldwin and Krugman, 1988, 1992). Their analysis shows that the actions of the US/EU and US/Japanese governments, respectively, is rational, even though they have not been overly effective because of retaliation.

Table 1.3. Knowledge intensity of the canola supply chain.

	Knowledge rating	Weight in chain (%)	R&D intensity	% Knowledge workers	Investment intensity	Patents used per Can$B sales
Seeds	High	na	na	na	na	na
Oil and lubricants	High	4.7	0.85	46.6	1.0	14.5
Chemicals	High	2.4	0.96	28.0	5.0	97.8
Power	High	2.1	1.21	29.7	17.6	14.5
Finance	Medium	7.7	0.09	25.2	3.5	0.1
Services	High	2.8	0.53	37.8	1.4	1.5
Other inputs	Medium	2.8	0.01	9.9	1.6	7.2
On-farm value	Low	64.9	0.05	10.6	10.2	11.8
Storage	Low	7.4	0.06	21.2	5.3	8.7
Transportation	Low	5.1	0.06	8.9	5.7	2.3
Total		100.0	0.15	15.7	8.2	12.0
Oilseed chain rank			31	31	10	32
Agriculture rank			40	45	8	32

Source: Weights based on Statistics Canada economic impact evaluation undertaken using the interprovincial input–output model, 1990 (1997); ranks and intensities based on data and tabulations reported in Lee and Has (1996), pp. 39–76.
na, not available.

In this book the authors hypothesize that the transgenic-based, differentiated canola sector exhibits these conditions. Given that the USA (the largest single market doing agri-food R&D) does not currently produce significant volumes of canola, that leaves Canada and the European Union, each producing about one-fifth of the total rapeseed in the world (and together producing almost all of the canola). The EU and Canada both have capacity in this area and, given relatively similar size of production and factor endowments, are possible locations for this activity to develop. This study will show that agglomeration is happening, with western Canada becoming the leader in global canola development. Although there appear to a number of possible sites for the canola industry to centre around, the effort has tended to concentrate in Saskatoon.

The theory therefore highlights three key features, which influence the development of the knowledge-based agri-food sector. First, the degree of excludability of the innovation in the canola industry both determines the rate of innovation and the distribution of its benefits. Secondly, information gaps in the production chain and resulting risks largely determine the industrial structure of the industry. Thirdly, the knowledge intensification of canola is beginning to force geographic convergence of research and production, causing greater reliance on trade to support development.

Science and Innovation

Part II of this book examines the interaction between science and the discovery process in the canola sector. In one sense, the canola story does not involve anything that had not been thought of by Mendel, the father of modern genetics and plant breeding. The science of selective breeding to enhance input or output traits goes back to his oft-cited experiments with peas. Nevertheless, there were two periods in recent times where evolutionary, and at times revolutionary, breakthroughs in the science allowed a quantum jump in the development of the rapeseed/canola industry.

In the 1950s public-sector scientists in Canada advanced the science in two important ways. Previously, breeding new varieties of canola was extremely difficult. Breeders tested a wide variety of seeds, selecting and testing those that appeared to exhibit desired traits. This was laborious as testing for oil composition, for example, took about 2 lb (1 kg) of seed and 2 weeks to undertake. Given the natural heterogeneity in canola, even when a sample tested well it was not always possible to be sure that the seed remaining in the sample would hold the same traits. The first breakthrough came in 1957 at the National Research Council (NRC) Prairie Regional Laboratory, which acquired a gas–liquid chromatography (GLC) unit and perfected the technique of assessing more quickly the oil properties of smaller and smaller samples. The NRC staff ultimately refined the technique to the point where tests took about one seed and about 15 min to complete. The NRC then assisted the breeding programmes at the Dominion Forage Lab of Agriculture Canada and the University of Manitoba to acquire and use GLC units. Keith Downey, leading a team of scientists at the Agriculture Canada centre, further refined the GLC technology to the point where they could test a half of a canola seed. His team then went on to develop a technique of cutting a single seed in half in such a way that the remaining half could germinate and produce a new plant. They proceeded to test thousands of seeds during 1962 and 1963 to find one with low erucic acid and in 1963 succeeded in finding a single seed with no measurable quantities of that undesirable trait. They grew the half left after the test into a sickly plant that ultimately yielded five whole seeds. Those seeds formed the basis for the first low erucic acid *Brassica rapa* released in 1971. The rest of the breeding programme to develop canola adopted and used those two innovations – GLC analysis and the half-seed method – forming the foundation for the modern canola industry. By 1978 both of the two varieties of canola, *B. napus* and *B. rapa*, were converted using these techniques to low erucic acid and low glucosinolates, which founded the modern industry.

Breakthroughs, beginning in 1973 and extending well into the 1990s, have further transformed the development of canola. While seed shuttling, for both breeding and multiplication purposes, and computers helped to speed up the development of new varieties and get them into the field earlier, the major breakthrough came in the area of molecular biology. Beginning with the Cohen–Boyer *in vitro* genetic engineering discovery in 1973, the modern biotechnology revolution was under way. Since then, advancements in the area of genomic mapping, isolating genes, transformation technologies, genetic markers, promoters, polymerase chain reaction and microarray technologies have all expanded the area of research and development and accelerated the search for new traits. These technologies both shortened the length of breeding for sophisticated traits and yielded new canola varieties with targeted agronomic input traits – already including herbicide tolerance and hybrid-based yield gains and soon to include insect and disease resistance – and specific output traits, such as modified industrial oils, nutraceutical properties and pharmaceutical proteins and enzymes.

In less than 40 years the science of rapeseed breeding has been converted from small plant-breeding programmes that would be almost immediately recognizable by Mendel, into a sophisticated, molecular-based, technology-driven research system involving many subdisciplines.

The shifting scientific context over the period transformed the evolution of the industry. As traditional, slower approaches were replaced by more rapid and targeted technologies, the innovation process was transformed from a relatively simple supply-push, linear research and development system into an increasingly complex, demand-pulled, dynamic and interactive research and development process, with extensive loop-backs and both programmed production and stockpiling of knowledge. In short, the shifting science created the conditions for the economic and political system to drive and manage the development process.

The State–Market Nexus

Parts III and IV of this study examine the impact of the transformed innovation system on the global rapeseed and canola industry. The purpose of this study is not to report the findings and results of scientific progress, but rather to examine how and why development proceeded as it did. Science and scientists were at the core of much of the activity but, except for a few years in the earlier period, they were acting on behalf of others with vested commercial interests. In short, these two sections apply social science tools to examine the scientific and related economic changes.

Innovation happens within and between institutions, which means it is inherently a social phenomenon. Neo-classical economists tend to suggest that there is little need to extend the analysis beyond the point of innovation, as markets will handle the production and marketing of the resulting products. In practice, however, markets often do not emerge on their own to adopt innovations. Rather, governments and industry actors, both singly and at times in partnership, actively develop markets for inputs, production, processing and consumption. The study of the role of industry and government in developing markets for innovations goes back to Alfred Marshall and his now famous discussion of the industrial development around Manchester (Marshall, 1890). More recently, this thread has been taken up by economic-growth theorists and political economists, especially those interested in institutional economics and concerned with localization of production and the related impacts on trade. Part III takes from the economics literature elements that help to explain how innovation has affected the actors in the system and the location of their efforts. The difficulty in such an analysis is that the innovation system has evolved to such an extent over time that precisely pinpointing cause and effect is next to impossible. Nevertheless, the institutional economic theory suggests a number of ways to explain how the sector has responded to innovation.

In the first instance, when innovation was managed and delivered by the public sector, the producers and rest of the supply chain were able to rely quite well on the operation of arms-length markets to marshal the inputs and manage the production, processing and marketing of the product. Even then, however, there were a number of instances of market failure. In response, the public sector supported the creation of a new set of institutions – both regulatory and participatory structures – to manage the commercialization of the innovations and development of related markets. As the innovative process became more complex, the state responded with a number of new institutions – producer check-offs, private intellectual property rights, public infrastructure and revised regulatory systems – to assist, encourage and support private initiative and investment. In short, the public sector shifted from proprietor to partner and promoter.

Meanwhile, private institutions underwent significant change. In the early years there was only limited private interest in innovation in the canola sector. The bulk of the interest was in supplying inputs or processing, distributing and marketing the output. As the innovation system became faster and more predictable, private capital began to flow into the sector, to the point that by the late 1990s more than two-thirds of the effort was financed by private capital. With the inflow of capital, it became clear that firms needed to become more involved in the entire marketing chain in order to capture enough of the value being created to compensate for the research investment. This has led over time to a massive restructuring of the sector, with the new 'life-science companies' attempting to manage their parts of the industry, ranging from the genome map to the dinner table.

The rapid and continually shifting efforts of the public, participatory and private sectors have created significant fluidity in the global industry, but there is some evidence that centripetal forces are causing aggregation of activity in those parts of the innovation and production system where economies of scale and scope exist. The main region benefiting from this agglomeration is Canada, and primarily Saskatoon, the home of the first canola variety. Nevertheless, centrifugal forces are still strong enough that it is extremely unlikely that all, or even the majority, of the activity will concentrate in one small region.

Part IV examines the impact of regulation both within countries and between countries on development in the industry. Governments around the world have adopted new rules to regulate the discovery efforts of both the public and private sectors (Chapter 13) and then expended significant effort to develop and implement new regulatory systems to manage the commercialization of those innovations (Chapters 14 and 15). Regulation of both the discovery and commercialization phases is becoming more complex. As the technologies advance more rapidly and involve new aspects (e.g. foundation science, germplasm, whole plants), government regulators are challenged to balance the need to provide incentives to innovators (through intellectual property rights or IPRs) with the desire to see those innovations spread as widely as possible. There are some concerns that the protection offered through IPRs may be greater than is socially desirable. Meanwhile, as knowledge and technology become the drivers for the global canola sector, production is consolidating in fewer countries and relying more heavily on trade. Although domestic regulations appear to be operating reasonably effectively in many countries, the international trade rules embodied in the World Trade Organization Agreement and in related mechanisms do not appear to be operating as effectively as they could.

The Impact of Research on Canola Producers and Users

The ultimate question that any study of development must ask is 'so what?'. In essence the answer is at least partly determined by *qui bono* (or who benefits). Part V examines the theoretical approaches to determining winners and undertakes some estimation of the gross benefits from canola research, and the distribution of those benefits between consumers and producers, and more specifically between farmers and others in the supply chain. We can, and do, make a few observations with confidence. First, the gross returns to canola research have been dropping with each successive year, to the point that the total social returns to canola research cannot justify the level of investment. The estimated internal rate of return is now less than the opportunity cost of this capital. Although some actors are, and will, continue to capture above-average returns on their efforts, many actors receive little or none of the benefit. Secondly, some of the direct benefit and much of the indirect benefit of the innovations for canola have been captured by consumers. Given the distribution of consumption, that means that part of the benefit has been distributed around the world, wherever the ultimate consumer lives. Thirdly, the returns to research that remain in the supply chain are not adequate to sustain the current level of private investment. Fourthly, there are some definite or indisputable losers from the innovations in canola. In particular, producers of other edible and industrial oils, such as palm and coconut producers, have lost both market share and revenues as relatively high-quality industrial rapeseed and canola oil products have pushed them from certain higher-value markets. Finally, some groups have indeterminate benefits. Farmers, for example, have invested heavily in both research (through check-offs) and in adopting the new technologies, yet the small returns mooted to be there may prove to be only transitory. Meanwhile governments, which have funded almost all of the public research and a significant share of the private effort through grants, subsidies or tax credits, have been so far unable to extract a return directly for their innovations and have some difficulty taxing the private profits from the innovations, due to the multinational nature of the industry.

General Application

In Part VI the results from the canola story are assessed against six fundamental questions that many have raised about technology-driven agriculture.

1. How has science transformed the innovation system?
2. Does knowledge-based agriculture exhibit either significant economies of scale or economies of scope that generate localized production?
3. How has recent innovation both driven and been driven by industrial restructuring in the global oilseeds industry?
4. What are the appropriate roles both for public institutions and producer-led associations in the face of an increasingly private-research-driven industry?
5. How has public and private regulation adapted to manage the changes in the industry?
6. Who wins or loses from innovation in agriculture?

The rest of this book examines the canola story to find answers to these questions and then attempts to generalize the results to the broader agri-food sector.

Innovation and Canola

2

Approaches to and Measurement of Innovation

Peter W.B. Phillips and George G. Khachatourians

Introduction

Canola is a product of innovation. From the very beginning, the development of rapeseed into a new plant variety, the products of which were suited to human and animal feeding purposes, was a science-driven process (Juska and Busch, 1994). The public sector, and more recently the private sector, have invested significant resources to change the agronomic and end-use attributes of canola to increase the value created in the industry.

This chapter examines the evolution of the innovation process in the canola industry, starting from the early years when research and development was undertaken by the public institutions, and moving into the recent period when privately funded research and commercialization is taking hold. The impetus for the research has clearly changed – initially in Canada public institutions sought new crops for western Canadian farmers; in the mid 1980s seed and agrochemical companies endeavoured to create, through plants and plant-derived products, new value for their shareholders, and now increasingly users of canola for animal or human consumption specify the attributes (e.g. fatty acid content and profile for humans or nutritative value and digestibility for animals) they seek from the seed. Furthermore, the innovation process, which has shortened from more than 15 years to 10 years or less, would appear to have evolved and benefited from the non-traditional innovation model.

Ultimately, the challenge of examining innovation is in its quantification for its contributory value to rapidly evolving user needs and significantly better return on investment. After all, innovations are the application of existing technical knowledge in a more creative manner than the previous application so as to give its originators and exploiters a competitive edge. Innovations are ideas that are generated daily in creative minds and do not subscribe to the terms of diminishing returns. It is only possible to see them at discrete points in the system, such as when they are codified either in academic literature or in patents and when they move from the laboratories into the marketplace and are produced and marketed. This chapter will examine the practical problem of measuring the stocks and flows of innovations in the canola sector.

Data reflecting various measures of innovation will be examined to determine whether canola innovation has tended to concentrate in specific geographic areas where there are similar climate, physical soil characteristics, microbiology, hydrology and industrial structure. As noted in Chapter 1, if the final product is tradeable (e.g. the canola oil or meal), but the innovation-based knowledge is a non-transferable intermediate factor of production (e.g. the canola seed may be such that it can only be grown in western Canada, either due to regulatory hurdles or due to climatic conditions), then the fact that innovation begins in one jurisdiction could forever put that site on a higher R&D and new product development trajectory. As a result, because of innovation the contribution of canola as a product of high-technology to our share of GDP and exports will be higher than otherwise.

The Characteristics of Innovation

One manifestation of innovation is the way that it yields knowledge that exhibits a number of different traits in terms of how it can be used, who can use it and how widely or narrowly it can be applied. An examination of the innovation process and the types of knowledge and their characteristics provides some insight into cause and effect parameters, such as the types of knowledge the private sector may adequately provide, against those where sustained or greater public effort may be required.

The classical innovation process has been viewed as a linear process, starting with research and leading through development, production and marketing phases (Fig. 2.1). Although this may have made some sense in earlier times when many innovations were simply the product of inventors' ingenuity, it soon became clear that the more competitive companies and industries were deploying a different strategy to develop and exploit inventions. Creating newer competitive intelligence needed a new model which turned incremental new information of markets, utilities and value on to existing inventive steps to generate intelligence, hence creating the non-linear nature of innovation and the increasingly important role in the process for market knowledge (Harvey, 1989).

Klein and Rosenberg (1986) provide an approach that identifies explicitly the role of both market and research knowledge. Their 'chain-link model of innovation' (Fig. 2.2) begins with a basically linear process moving from potential market to invention, design, adaptation and adoption, but adds feedback loops from each stage to previous stages and the potential for the innovator to

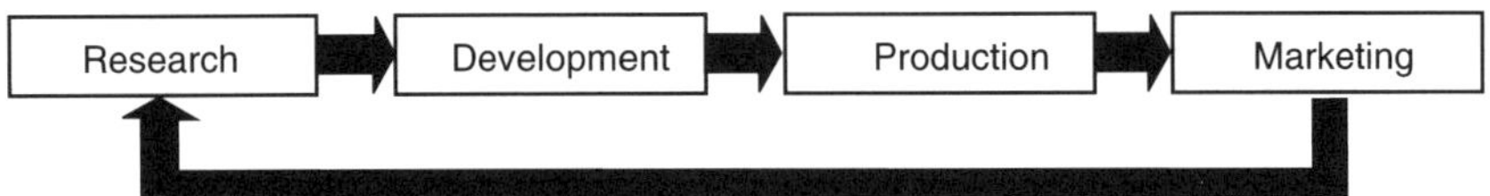

Fig. 2.1. The linear model of innovation.

seek out existing knowledge or to undertake or commission research to solve problems in the innovation process. This dynamic model raises a number of questions about the types and roles of knowledge in the process. Some of the knowledge will be available inside the institution undertaking the innovation, or could be developed within or outside the firm.

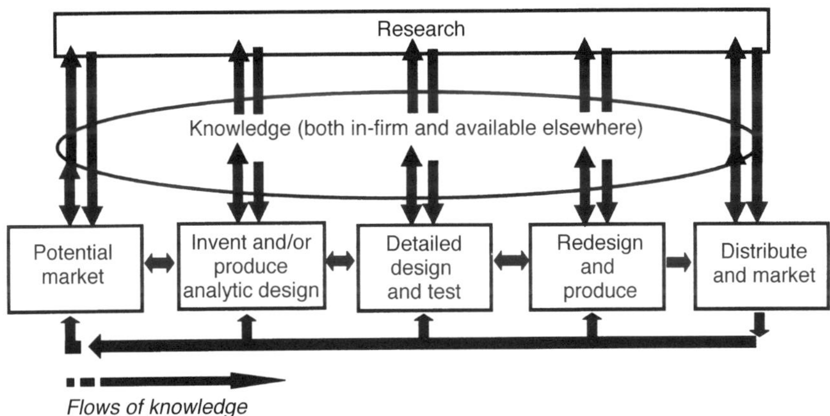

Fig. 2.2. The chain-link model of innovation.

Malecki (1997) provides a way of categorizing the types of knowledge that helps to identify which route a firm or institution might go to acquire or develop knowledge needed to innovate. He identified four distinct types of knowledge: know-why, know-what, know-how and know-who (Table 2.1). Each type of knowledge has specific features (OECD, 1996).

'Know-why' refers to scientific knowledge of the principles and laws of nature, which in the case of plant breeding relates to the scientific domains of plant physiology, genetics (theoretical and applied), molecular biology, biochemistry and newer integrative disciplines of proteomics, bio-informatics and genomics. Most of this work is undertaken in publicly funded universities and not-for-profit research institutes and is subsequently codified and published in academic or professional journals, making it fully accessible to all who would want it. This knowledge would be in the knowledge block in the chain-link model, having been created almost exclusively in the research block. In the most classical sense of scientific enquiry, very little of this knowledge would have been

Table 2.1. Classification of types of knowledge. (Adapted from Malecki, 1997.)

	Degree of codification	Produced by	Extent of disclosure
Know-why	Completely codified	Universities and public laboratories	Fully disclosed and published in scientific papers
Know-what	Completely codified	Universities, public laboratories and private companies	Fully disclosed in patents
Know-how	Not codified	Hands on in laboratories	Tacit; limited dispersion
Know-who	Not codified	Exists within firms or research communities	Tacit; limited to community

produced within firms. 'Know-what' refers to knowledge about facts and techniques: in the case of plant breeding, this includes the specific principles and steps involved in key experimental protocols of genetic crosses and selection of indicative traits after the transformation processes. This type of knowledge can often be codified and thereby acquires the properties of a commodity, being transferable through the commercial marketplace. In the case of canola, much of this knowledge is produced in private companies and public laboratories and increasingly is protected by patents and other property protection systems. The stock of know-what is in the knowledge block in the chain-link model, having been created in the research, invention, design and adoption blocks.

'Know-how' refers to the combination of intellectual, educational and physical dexterity, skills and analytical capacity to design a hypothesis-driven protocol with a set of expected outcomes, which in the canola case involves the ability of scientists to combine effectively the know-why and know-what to develop new varieties. This capacity is often learned through education and technical training and perfected by doing, which in part generates a degree of difficulty for the uninitiated and makes it more difficult to transfer to others and, hence, more difficult to codify (in some cases videotapes can codify know-how). Know-how would be represented in the research block and also in the invention, design and adaptation stages. Marketing these innovations also takes a certain skill and expertise that is not codifiable but can realistically be viewed as knowledge. Finally, 'know-who', which 'involves information about who knows what and who knows how to do what' (OECD, 1996), is becoming increasingly important in the biotechnology-based agri-food industry; as the breadth of knowledge required to transform plants competitively expands, it is necessary to collaborate to develop new products. In today's context, 'know-who' also requires industrial intelligence and tracking of private sector knowledge generators who, at times, can hold back the flow of crucial and enabling information, expertise and knowledge. In extreme cases, know-who knowledge can be critical to successful innovation; if one does not know who to work with, one may stumble into scientific pitfalls and traps that could sabotage the chance of innovative success. Know-who knowledge is seldom codified but accumulates often within an organization or, at times, in communities where there is a cluster of public and private entities that are all engaged in the same type of research and development, often exchange technologies, biological materials and resources, and pursue staff training or cross-training opportunities. This type of knowledge would be represented by the arrows in the chain-link model, as building relationships that lead to trusting networks of know-who. A major challenge in trying to examine innovation is finding some way to monitor and measure the stocks and flows of these different types of knowledge.

Measuring Innovation in the Canola Sector

No definitive set of measures for knowledge has yet been developed. Nevertheless, significant work has been undertaken in a number of areas using proxies for knowledge and transmission of knowledge. Taking the four types of knowledge, and the resulting products, one can construct a package of empirical measures that approximate the flow of innovations into the marketplace.

First, starting with know-why knowledge, it is clear that while it is quite difficult to identify the inputs to the research effort, one can look at 'bibliometric' estimates to measure the flow of knowledge from the initiators/originators, generally the universities, research institutes and private firms. There is general acceptance of the view that publications such as academic journals are the primary vehicle for communication of personal and institutional findings that become the vehicle for evaluation and recognition (Moed *et al.*, 1985). Hence, in general in the past, and to some extent even in current practices, most if not all of the effort put into a research area will be presented for publication. The common catch phrase, 'publish or perish' captures the essence of the past practice, while the more progressive modality is 'patent and then publish', especially for a large number of research universities. There have been a number of efforts (by the National Science Board, 1988; Industry Commission, 1995; Katz *et al.*, 1995) to develop and use literature-based indicators to evaluate science effort. The ISI-based evaluation system for connecting the scientific effort of anyone's publication and a journal's placement in the world of publications is becoming a more quantitative indicator, which is presently used for analysis of progress and evolution of science and innovative steps.

In the canola area, Juska and Busch (1994), sociologists from Michigan State University, developed a database of scientific and technical journal articles related to rapeseed or canola, published between 1970 and 1992 and cited in AGRICOLA and CABDATA, the two major bibliographic databases for agriculture literature. They also searched by hand the Bibliography of Agriculture for references in the 1943–1970 period. They found 12,456 discrete references. This data, while useful, could only be manipulated in limited ways; it could not, for instance, identify links between articles based on citations or collaborations.

Furthermore, the databases include working papers and technical reports, which leads to the possibility of over-counting output as many of these less formal papers ultimately become journal articles.

For the purposes of this study, Juska and Busch's general 'bibliometric' approach is adapted to a more refined database. Initially a manual search of the Institute for Scientific Investigations Scientific Citations Index for 1965–1997 was undertaken. The manual search identified 3646 articles over the period, with 648 in the 1965–1980 period. The ISI was then contracted to undertake an electronic search of their databanks, which then covered the period from 1981 to July 1996, with a few entries in the following months. They were instructed to search their database, which included approximately 8000 journals in the sciences and social sciences, for seven key words/phrases: *Brassica campestris*, *Brassica napus*, *Brassica rapa*, canola, canola meal, rapeseed and oilseed(s). The special tabulation identified 4908 individual articles in 650 journals meeting the criteria (hereafter called the canola papers) produced by approximately 6900 authors in approximately 1500 organizations in 79 countries (see Table 2.2 for the types of papers).

Secondly, know-what knowledge is most commonly examined using patent information. Trajtenberg (1990) argues that 'patents have long exerted a compelling attraction on economists dealing with technical change ... The reason is clear: patents are the one observable manifestation of inventive activity having a well-grounded claim for universality'. As of 1990, there were approximately 4 million patents issued in the US (about 70,000 new patents were granted annually) and over 25 million patents worldwide. Trajtenberg points out that because patents vary enormously in their technological and economic importance, it is not possible merely to count them and use that as a measure of the magnitude of the value of knowledge created. Griliches *et al.* (cited in Trajtenberg, 1990) calculated that simple patent counts explain less than 1% of the variance in value of companies. Nevertheless, Trajtenberg concludes that

Table 2.2. The canola papers database.

Type	Papers	Citations	Average citations per paper
Journal article	3,800	24,238	6.38
Note	364	1,106	3.04
Proceedings paper	163	876	5.37
Review	50	516	10.32
Meeting abstract	448	139	0.31
Letter	18	36	2.00
Editorial	48	32	0.07
Book review	7	2	0.29
Correction/discussion	10	1	0.10
Total database	4,908	26,946	5.49

Source: ISI Citations Database, special tabulation based on keywords.

in the context of specific, clearly demarcated innovation (in his case CT scanners), patents 'play an important role in studying the very emergence of new markets, which seems to be the period when most of the innovative activity takes place'. He likens patents to working papers in economics. Papers and patents are produced roughly in proportion with effort: a larger number of papers/patents indicates a larger research effort. 'Patent counts can thus be regarded as a more "refined" measure of innovative activity than R&D, in the sense that they incorporate at last (*sic*) part of the difference in effort, and filter out the influence of luck in the first round of the innovative process.'

For the purposes of the canola study, two patent systems were searched. First, the US patent system databank, available on-line, was searched for canola-related patents. The server provides access to the bibliographic data and text of all claims of numerous patents issued by the United States Patent and Trademark Office from 1974 to present, plus some patents issued during 1971 to 1973. The search uncovered 186 patents that had specific reference to canola or rapeseed. Secondly, the Canadian Intellectual Property Office (CIPO) database of Canadian patent bibliographic data was searched. That database contains issued patents and applications for patents that have been made available to the public from 1921 to the present. More than 630 patents for canola-related work were provided by the CIPO as of January 2000.

Know-how and know-who types of knowledge, as discussed above, are often inseparable and are tricky to track at the best of times. Nevertheless, this type of knowledge can be mapped by looking at a number of different sources. The regulatory systems in Canada and elsewhere provide one means of identifying who is converting the know-why and know-what knowledge into actual products. The regulatory systems for genetically modified organisms during the detailed design, testing and redesign periods provide an insight into who is doing what and where. As well, this particular information suggests intentions of producer, manufacturer or technology innovators and their willingness to support the financing of the next steps. These data are available in Canada through the Canadian Food Inspection Agency (CFIA) authorizations for field trials for 'plants with novel traits' and internationally through the OECD website on field trials. Moving along through the innovation system, the resulting products can be observed through the varietal registration system in Canada under the Seeds Act and in Canada and elsewhere through the registration of new canola varieties for plant breeders' protection, as provided under the UPOV Agreement. These data must at times be supplemented by industry data to identify public varieties that are not protected by breeders' rights.

In addition to investigating the regulatory records to determine who is working with whom, this study has also used the ISI canola papers to map capacities and linkages. The advantage of using the ISI database over the AGRICOLA or CABDATA systems is that the ISI database provides the capacity to look both forwards and backwards from the target articles to determine where the key knowledge inputs come from and where the resulting knowledge is being used. The database identifies 17,995 papers from 1294 journals, produced by approximately 28,800 authors in 3816 organizations in 107 countries, which were cited a total of 28,946 times by the 4908 papers that relate to canola research. Although the average paper is cited only 1.6 times, approximately 300 papers were cited between 10 and 96 times. At the other end of the system, the 4908 canola papers were cited 26,946 times, for an average citation rate of 5.49 and a median citation rate of 2. As a further point of reference, it is worth noting that the average citation rate for the 690,000 publications within the biological and natural sciences literature during 1992–1996 was 4.1.

One-third of the papers were never cited in any other paper and as such represent either relationships that are quite distant to the mainstream of canola research or represent end points or discontinuities in particular research lines. The database can also be sorted and searched by author, institution, subject and country of the researcher, and then cross-tabulated for collaborations, allowing one to examine both the stocks and flow of knowledge. In this way, one can investigate the know-who linkages that underpin the innovation system.

Finally, the ultimate measure of innovative success is market adoption. The challenge is that marketing information is getting more difficult to find. Aggregate data for canola acreage and yields are available nationally and through the Food and Agriculture Organization (FAO) but production information on specific varieties is difficult to obtain. Nagy and Furtan (1978) provide variety market shares for Canada for 1950–1963. In 1963, a consortium of the three Prairie Pools began to produce and release a survey of their seed business that provided estimated market shares for each available variety in the three prairie provinces. When the seed industry began to go private, the quality of the Prairie Pool survey became suspect (Alberta Wheat Pool, 1998, telephone conversation) because many of the new varieties were only marketed by selected seed merchants, who often were unwilling to share market information with their competitors. So in 1991 the Alberta Wheat Pool withdrew from the prairie pools survey, and in 1992 the last survey, covering only Manitoba and Saskatchewan, was produced. After that point, the only publicly available organized source of market shares comes from the Manitoba Crop Insurance Corporation, which records acres planted to different varieties in 1992–1999. For the purposes of this study, these data have been used to develop a proxy for western provincial varietal market shares by extrapolating Manitoba planting decisions to the other two provinces, weighted by the historical shares of *B. rapa* and *B. napus* and adjusted for varieties that are only available in Manitoba. The resulting statistics, while not perfect, provide a base for testing market adoption rates for new varieties.

Next Steps

The following chapters use the chain-link innovation model and the constructed data sources to examine the structure and impact of the innovation system with respect to canola, looking for areas of stability or change. Rothwell

Table 2.3. Rothwell's system integration and networking paradigm applied to canola. (Modified from Gibbons, 1995.)

	Canola period	Characteristics
First generation	1944–1971	Technology push; simple linear, sequential innovation; markets simply 'receptacles' for resulting products
Second generation	1971–1985	'Need pull' starts to replace technology push; simple linear, sequential innovation continues; greater emphasis on marketing
Third generation	1985–1990	Linear innovation model but interactions involve a number of feedback loops, creating both push and pull combinations; marketing increases in importance
Fourth generation	1990–1995	Introduction of parallel product and process development with multi-skill, cross-functional development teams; integrated model; collaboration increases; developing links with customers
Fifth generation	1995 onward	Multiplication of linkages has effectively destroyed linear model; strong links with leading-edge customers; strategic integration with suppliers; horizontal link via alliances and collaborations; increased focus on quality and other non-price factors

(cited in Gibbons, 1995) puts forward a paradigm for innovative development that defines five generations of sophistication. When applied to the case study of canola, it is possible to see those five 'generations' (Table 2.3).

The first stage, spanning 1944–1971, involved a simple, linear, technology-pushed innovation system (characterized by the model in Fig. 2.1), with markets simply receptacles for the resulting products. The second generation began in 1971 and lasted until 1985. The key change from the earlier period was that 'need pull' entered the system as a loop-back from the market to the research level, so that the technology push, linear system was ultimately being driven by market needs, especially the desire for a rapeseed with low erucic acid and low glucosinolates. At the same time there was enough of a momentum in research results and investment that the publicly supported institutions (e.g. National Research Council (NRC) and Agriculture and Agri-food Canada (AAFC)) and universities, funded through the same bodies, had to make a strategic decision to continue or change their research programmes. The big change happened after 1985, with the granting of generally regarded as safe (GRAS) status in the USA. The linear innovation model became more complex, with more feedback loops and the beginning of a chain-link model evolving. More importantly, newer tools of molecular genetics and a new generation of technically trained scientists had replaced the more conventional plant breeders and agri-food researchers. The fourth generation began about 1990, with the introduction of plant breeders' rights in Canada and the rapid acceleration of private investment. The innovation system began to evolve into an intricate web of parallel product and process developments with multi-skilled, cross-functional development teams. Collaboration and integration accelerated, with the early signs of stronger links between developers and customers. The fifth and final generation in Rothwell's paradigm began in the mid 1990s, with the multiplication of linkages effectively displacing the linear innovation system. Every research company is now seeking strong links with leading-edge customers, strategic integration with suppliers, horizontal linkages via alliances and collaborations. The result is that the system is evolving into one with an increased focus on quality and other non-price factors.

An Introduction to the History of Canola and the Scientific Basis for Innovation

3

George G. Khachatourians, Arthur K. Sumner and Peter W.B. Phillips

Introduction

It is only recently that canola or, as it was previously known, rapeseed has been in the top three oilseeds. Canola represents a major source of edible oil and industrial oil production. Prior to the 1970s the high erucic acid and glucosinolate content were an important impediment to the use of rapeseed oils and meal, but newer varieties, selected for the absence of these traits, made possible a significant change in the production and use of rapeseed for human consumption and other uses.

Brassica oilseeds typically produce, on a dry weight basis, 40–42% oil. The residual cake, or meal, contains 38–42% protein and is used in livestock, poultry and fish diets. The efficacy of oil extraction based on current technology is 98%. *Brassica* oilseeds require a particular temperature range, rich soils and a moist climate available on five continents – China, India, Canada, France, Germany, Poland, Pakistan, Sweden and the USA dominate production. In spite of fluctuations in production due to changes in environmental conditions, demand, price, trade and other governmental determinants, rapeseed production has continually increased.

With the evolution of various trade associations and national and international trade policies, starting from the 1960s and up to date, rapeseed/canola demand remains high and therefore the incentives for investment in R&D have remained at a premium. Because of researchers' ability to introduce either microbial, plant or animal genes into this plant, a new era in the production of high-value, low-volume commodities is reshaping the modern history of the rapeseed. Indeed from its scattered and humble origins, canola is occupying an enviable position and has become the Cinderella plant for many jurisdictions.

This chapter draws on a number of sources, including the Canadian Food Inspection Agency regulatory directives on the biology of *B. napus* L. and *B. rapa* L., to explain in layperson's terms the key features of the plant's biology, agronomy and breeding techniques.

Introduction to the Biology of *Brassicaceae*

It is useful for this study to know something of the biology of rapeseed/canola and to have a basic understanding of the different technologies used for breeding and processing the resulting crops. The name 'canola' was given to represent well-defined characteristics of particular plants within the genus *Brassica* belonging to the *Cruciferae* (*Brassicaceae*, mustard) family. The word *Cruciferae* is derived from the Latin word *crux* (*cruc-*), meaning cross, in reference to the arrangement of the flowers, which is that of four diagonally opposite petals. The seeds of some members of this plant are the source of the oil. Most of the varieties belong to either *Brassica napus* L. or *B. rapa* L. The designation 'L.' represents the first letter of the last name of the famous Swedish botanist, Carolus Linnaeus, who identified the first rape plants. He was the first to enunciate the principles for defining a uniform rule for naming plants and animals. He used a binomial system, named the rape plants and reported their occurrence in Sweden in 1741.

There are two species of plants that qualify as canola-quality rapeseed – *B. napus* L. and *B. rapa* L. – and a wide number of technologies. This chapter provides a tour of the state of the science related to those two particular plants. This review of the biology of the two plants is drawn from the Canadian Food Inspection Agency's regulatory directives for the two plants that are used in the environmental assessments for plants with novel traits.

First, *B. napus* L. is an ancient crop plant, belonging to the *Cruciferae* (*Brassicaceae*) family, also known as the mustard family. *B. napus* has dark bluish-green foliage, glaucous, smooth or with a few scattered hairs near the margins, and partially clasping. The stems are well branched, although the degree of branching depends on variety and environmental conditions; branches originate in the axils of the highest leaves on the stem, and each terminates in an inflorescence. The inflorescence is an elongated raceme, the flowers are yellow, clustered at the top but not higher than the terminal buds, and open upwards from the base. There are two types, the oil-yielding oleiferous rape, often referred to in Canada as Argentine rape, of which canola is a type having specific quality characteristics, and the tuber-bearing swede or rutabaga. The oleiferous type can also be subdivided into spring and winter forms.

Fertilization of ovules usually results from self-pollination, although outcrossing rates of 20–30% have been reported (Rakow and Woods, 1987). The pollen, which is heavy and sticky, is moved from plant to plant primarily through insect transmission, mostly by bees. Cross-pollination of neighbours can also result from physical contact. Successive generations of *B. napus* arise from seed from previous generations.

The origins of *B. napus* (an amphidiploid with haploid chromosome number, $n = 19$) are obscure, but were initially proposed to involve natural interspecific hybridization between the two diploid species *B. oleracea* ($n = 9$) and *B. rapa* ($n = 10$) (Table 3.1). Recent evidence suggests that *B. napus* has multiple origins and that most cultivated forms of *B. napus* were derived from a cross of a closely related ancestral species of *B. rapa* and *B. oleracea*.

The second species is *Brassica rapa* L. *B. rapa* (previously *B. campestris*) was first described ˘ species by Linnaeus, with *B. rapa* being the turnip form and *B. campest* ˘wild, weedy form. These were determined in 1983 to be the same species and were combined under the name *B. rapa*. *Brassica rapa* L. also belongs to the *Cruciferae*. It has green foliage, leaves glabrous or slightly hispid when young, and the upper leaves partially clasping the stem. The stems are well branched, although the degree of branching depends on biotype/variety and environmental conditions. Branches originate in the axils of the highest leaves on the stem, and each terminates in an inflorescence. Lower leaves are sparingly toothed or pinnatifid and petioled, while upper leaves are sessile, subentire, oblong lanceolate and often constricted above the base. The inflorescence is an elongated raceme, the flowers are pale yellow, densely clustered at the top with open flowers borne at or above the level of terminal buds, and open upwards from the base of the raceme.

There are three well-defined groups of *B. rapa* (A genome, $n = 10$), based on their morphological characteristics: (i) the oleiferous or oil-type rape, often referred to in Canada as Polish rape or summer turnip rape, of which canola is a specific form having low erucic acid in its oil and low glucosinolate content in its meal protein; (ii) the leafy-type *B. rapa*, including pak-choi, celery mustard, chinese cabbage and tendergreen; and (iii) the rapiferous-type *B. rapa*, comprising turnips, which are important as vegetable sources and forages for sheep

Table 3.1. *Brassica* species origin and chromosomal characteristics.

Brassica species	Origins	Chromosome number ($n =$)
B. aba L.[a]		12
B. carinata Broun		17
B. napus L.[b]		19
B. nigra		8
B. juncea L. Czern and Cross	Middle East	18
B. oleracea		9
B. rapa L.[c]	Central Europe or Central Asia and India	10

[a] Also known as *B. hirta* Moench.
[b] A cross of *B. rapa* × *B. oleracea*.
[c] Formerly *B. campestris*.

and cattle in many parts of the world. *B. rapa* can be subdivided into Indian and western European/North American forms.

B. rapa, with the exception of the Indian form, is an obligate outcrosser due to the presence of self-incompatible genes. As a result, little or no seed is set when *B. rapa* is self-pollinated. As with *B. napus*, relatively high pollen levels can result from synchronously blooming *B. rapa* fields, and the concentration of pollen, as indicated by outcrossing studies, decreases rapidly with increasing distance from the source of pollen (Stringam and Downey, 1978; Raney and Falk, 1998). Under field conditions, *B. rapa* pollen is transferred from plant to plant primarily through physical contact between neighbouring plants. It can be also transferred over longer distance by wind and insects. To minimize contamination of *B. rapa* foundation seed plots, the Canadian Seed Growers Association has set a distance of 400 m between seed plots and other sources of *B. rapa* pollen. Pollinating insects, in particular honey bees and bumble-bees, are believed to play a major role in the transfer of pollen over long distances.

Breeding of *B. rapa* is commonly achieved using simple recurrent selection procedures to prevent inbreeding. *B. rapa* shows considerable heterosis for yield, with up to 40% yield increases, and there is consequently a strong interest in developing hybrid or synthetic cultivars (Falk *et al.*, 1998). Synthetic cultivars are the result of random matings between selected parental lines, which give rise to both hybrids and progeny from matings within the same parental lines (Buzza, 1995; Falk *et al.*, 1998). The first *B. rapa* synthetics were registered in Canada in May 1994 under variety names CASH, Hysyn 100, Hysyn 110 and Maverick.

Today 'double-low' commercial varieties of both *B. napus* and *B. rapa* dominate the oilseed *Brassica* production area in developed countries. These 'double-low' varieties are characterized as having a very low (<1%) content of erucic acid in the fatty-acid profile of the seed-storage lipid and a very low content of glucosinolates (<18 μmoles per gram seed at 8.5% moisture) in their seed and meal. The term 'canola quality' is normally applied to seed, oil and meal from such varieties.

Early History of Rapeseed Cultivation: Asia, Europe and the Americas

The earliest records of rapeseed growth in Asia belong to ancient Sanskrit writings of 2000–1500 BC. Varieties of *B. campestris* represent the oldest on record for cultivation. As to where the seed source for this early agriculture came from, or if it was a discovery of local farmers, is anyone's guess. Japanese literature, on the other hand, shows the introduction of rape via Korea and China *c.* 1000 BC. Both of these records, being somewhat sketchy, lack information on specific locale, agronomic and other utility indicators.

Unlike in Asia, parts of Europe and northern Africa have better historical descriptions of the earliest cultivation and use of turnip rape (*B. campestris* L.) and rape (*B. napus* L.). The presumable rationale for interest in rapeseed in parts of European countries outside the Mediterranean region was the inability to cultivate either olive or poppy oils (Appelquist, 1972). There are records of both seeds and the milling of rape in older central European settlement records and about 35 seeds of *B. campestris*-type rape from the Bronze Age were found near Zurich, providing further proof (Appelquist, 1972).

In the 13th century the cultivation of rapeseed and turnip rape north of the Alps was for the provision of lamp oil. Thereafter, up until the 1700s, some records of farm locations, field size and cropping dates for rape have been found for Belgium and Holland (Appelquist, 1972). At the end of the 17th century, Germany and Spain also cultivated rape. With the appearance of mineral oil as a substitute oil for rape oil in lamps in the 19th century, the cultivation of rapeseed in Europe declined, except in Poland, Russia, Denmark and Sweden, where it increased. Increases and decreases in rape production occurred from the early 19th century. The Danish acreage of rapeseed amounted to 15,500 ha (about 35,000 acres) in 1866, which was probably the highest level attained during that time. In 1866, Sweden produced about 3000 tons. However, the price reduction due to the importation of other vegetable oils for margarine and soap production from 1866 onwards caused production to decline rapidly; at the end of the century only about 2 tons were produced in Sweden. After the First World War and until 1923 rape and mustard were cultivated again in Sweden but production of rape ceased afterward, although brown mustard (*B. juncea*) and white mustard continued to be grown (Appelquist, 1972).

The history of rape in the Americas was different. None of the *Brassica* vegetables are native to the American continent. Some production of rape in North America (e.g. Canada) during the 1920s to the late 1930s is reported to have occurred, while in Argentina, average production of 26,000 metric tons of rapeseed between 1935 and 1939 has been reported.

World Production Between 1940 and 1965

To meet the demand for edible and cooking oil during the years 1934–1938, many European countries (e.g. Sweden, Poland, Bulgaria, France, Germany, Romania, the USSR and Yugoslavia) led the production of rapeseed of various species. European production was about 7.7% of the total tonnage of 3.9 million tons worldwide. The majority of the production, namely 74.5%, occurred in Asia, while the Americas contributed 1.3%. With the onset of the Second World War, oil demand increased and the situation changed. Research efforts on rapeseed increased significantly and the area of cultivation, and hence harvested acreage of oil crops including rapeseed, increased. Immediately after the war, production either kept increasing rapidly or remained static in most countries.

After the Second World War and continuing into the mid 1960s there was continued interest in planting and utilization of the rape and mustard seeds to meet the oil market demand. France, during 1963–1969, increased production not only because of internal consumption as a table oil, but also for export to

Algeria and The Netherlands. Germany, during the same period and up to the 1970s, was a big importer of oil and meal. Because of agreements reached between the margarine industry and the Danish government, the regulation of rapeseed prices and the purchase of the resultant meal was a matter of law. A similar situation in Sweden in 1933 resulted in a law regulating the price and tax structure on margarine and restrictions on cultivation of oilseed crops. As a result of the Rapeseed Agreement in 1956, rapeseed oil had to be used in the manufacture of margarine in Sweden or be exported, which resulted in the internal purchase of 105,000 tons of oilseed for the 120,000 tons of margarine produced. In terms of a conversion factor, 40% of the fat requirement of margarine was coming from the oilseeds (Ohlson, 1972). Poland, with its unique arable-land base and capacity for production of rape and turnip rape, has been a dominant oilseed producer. Its capacity after the Second World War increased, with much larger oil extraction and milling facilities. In part this increase paralleled the domestic high demand for vegetable oils during 1957 to 1966.

Production in Canada

The earliest record of production of rapeseed in Canada began at Shellbrook, Saskatchewan, by Mr Fred Solvoniuk, an immigrant farmer from Poland. He obtained an envelope of seed from a friend or relative in Poland, from where he had emigrated in 1927. Mr Solvoniuk planted this seed in his farm garden and found it well adapted. He continued growing a small plot for a few years. Later it was established that this rapeseed was the *B. campestris* (later *B. rapa* L.) species.

In Canada, commercial growth of rapeseed for production of oil for lubrication of marine engines, and a prioritized research agenda of the federal government, provided much of the boost. The crisis of the Second World War initiated commercial production of the crop in Canada because of the blockade of European and Asian sources of supply of rapeseed oil. The head of the Forage Crop Division of Canada Department of Agriculture, Ottawa, Dr Stevenson, was given the mandate for Canadian production in 1942. The harvest in 1942 of 2600 lb (52 bushels) of seed of the *B. napus* species was required for seed for planting in 1943 to relieve the serious shortage of rapeseed oil. Additionally, Canada purchased a total of 41,000 pounds of *B. napus* of Argentine origin from US seed companies. A selection of the *B. napus* rapeseed that came from Argentina at that time was adopted as the so-called Argentine type rapeseed. Both the Argentine and the earlier Polish types were found to be well adapted to the prairies of Canada and grew fast with good yields.

The interest amongst farmers was increasing, especially among those looking for additional crops to diversify away from wheat and barley. Concomitantly, there was even further enhanced production after establishment of oilseed crushing and extraction facilities between 1956 and 1968. With these positive trends there was enhanced research into the quality of oilseed of *B. napus* (e.g.

production of Oro which contained no erucic acid and only traces of eicosanoic acid). This variety, processed by Canbra, generated good salad oils.

Within Canada, the primary production areas are the prairie provinces of Manitoba, Saskatchewan, Alberta and the Peace River area of both Alberta and British Columbia, although there is also some production in Ontario and Quebec. Production in Canada began at Shellbrook, Saskatchewan, in 1936 and spread during the Second World War to Manitoba. Production began in Alberta in 1955. *B. rapa* was the dominant species in western Canada in the early 1970s (Downey *et al.*, 1974). Similar acreages of *B. napus* and *B. rapa* were grown in the prairie provinces in the late 1980s but in the 1990s the proportion of the production area sown to *B. rapa* diminished to about 15–20%. The main *B. rapa* production regions in Canada are the province of Alberta, the Peace River region of British Columbia and northern Saskatchewan. The Manitoba acreage is primarily *B. napus*. *B. rapa* withstands spring frosts better than *B. napus* and generally requires fewer frost-free days to complete its life cycle. *B. rapa* varieties therefore tend to be grown north of the areas in which *B. napus* is grown. Both spring and winter annual forms occur but only spring forms are grown. The winter forms are more productive than the spring forms but are not sufficiently winter hardy to be grown in western Canada (Downey and Röbbelen, 1989). Rapeseed production partially relieved the economic stress generated by the overproduction of cereals. In the span of about 25 years rapeseed progressed to become Canada's fourth most extensively sown annual crop.

With aggressive breeding programmes that changed the rapeseed product qualities, there was also a need to change its name to communicate the progressive image. Although many types of rapeseed oil were being marketed as late as the 1970s, the Canadian product (with low erucic acid oil and low glucosinolates meal) had the opportunity to become differentiated. Already there was one brand-named rapeseed oil, 'colza oil', marketed by France, which stood for the French iteration of the combined derivative from *kool* (cabbage in Dutch), and *zaad* (seed in Dutch). The Rapeseed Association of Canada (see Chapter 6 for details) delegated the task of finding a new and appropriate name to a search committee, which recommended the new oil be called 'canola'. Although the word itself meant nothing, to the consumer it sounded like an abbreviation of Canadian, can-, and a suffix (-ola) which could have stood for -ol, or a chemical compound containing a hydroxyl group, or equally for -ole, -oleo (French) for oil. By definition, canola was to be understood to have been the oil derived from *B. napus* or *B. rapa*, containing low erucic acid oil (less than 2% in terms of 1996 standards) and low glucosinolates in the meal (it should contain less than 30 μmol of glucosinolates g^{-1} of meal at a moisture content of 8.5%).

The word 'canola' was a trademark registered in 1978, and over time the industry came to view it as Canada's 'Cinderella' crop, especially when Canada became the world's leading exporter. Precedent-setting factors helped with the initiation of commercial production and in the crop's attainment of such a significant place in the Canadian economy. Growing farmers responded to market demand, and enhanced industrial capacity for processing and of utilizing the products stimulated R&D efforts. Universities and public laboratories, and increasingly private companies, focused on the collective goal of developing new varieties and improved oil extraction technology to fully capitalize on this new crop.

The Agronomics of Canola Production in Canada

A discussion of agronomics of canola is relevant because the earliest breeding exercises were attempts to fit temperate climatic adaptation into the canola seed. *B. napus* requires more frost-free days than *B. rapa* to mature. Whereas *B. napus* varieties may require on average 105 days from seeding to harvest, *B. rapa* varieties require on average only 88 days. Consequently, *B. napus* varieties tend to be grown south of the areas in which *B. rapa* is grown: the central parts of Alberta and Saskatchewan, and the southern part of Manitoba.

Both species are cool-season crops and are not as drought-tolerant as cereals. They are widely adapted, and perform well in a range of soil conditions, providing that moisture and fertility levels are adequate. Air and soil temperatures influence canola plant growth and productivity. The optimum temperature for maximal growth and development is just over $20°C$, and it is best grown between $12°C$ and $30°C$. After emergence, seedlings prefer relatively cool temperatures up to flowering; high temperatures at flowering will hasten the plant's development, reducing the time from flowering to maturity. Most cultivars grown in Canada are of the annual type, the species showing poor survival at temperatures lower than $-6°C$, although there is some production of autumn-sown winter-hardy types in the warmest part of southern Ontario.

In Canada, due to an increased awareness of soil conservation issues, minimal or no-till canola production is practised nowadays. This practice allows for most of the crop residue and stubble to be left on the soil surface to trap snow, reduce snowmelt runoff, stop erosion and increase soil water storage. However, with no till, rigorous and systematic weed control programmes are needed. Pests, weather, weeds, insects or fungi can limit canola production. Some of the closely related cruciferous weeds (wild mustard, stinkweed, shepherd's purse, ball mustard, flaxweed, worm seed mustard, hare's-ear mustard and common peppergrass) are often problematic as they outcompete canola during early growth stages. Weeds must be controlled early to avoid yield loss due to competition. Insect pests including diamondback moth larvae, flea beetles, bertha armyworm and, depending on the absence of cereal crops, grasshoppers can attack and injure cruciferous plants, including canola. Several fungal diseases of canola can cause disease and damage, especially if repetitious cultivation has occurred. Canola is recommended not to be grown on the same field more often than once every 4 years, to prevent the build-up of diseases, insects and weeds. In selecting a planting site, volunteer growth of buckwheat and chemical residues from herbicides from previous cropping year(s) are important factors to consider.

Rapeseed is commonly swathed when 30–50% of the seeds on the plant have begun to turn from green to brown or yellow. The seed is then combined when nearly all the seeds have turned colour. *B. rapa* is more resistant to shattering than *B. napus*, making *B. rapa* easier to straight combine. The use of desiccants on *B. napus* allows a reduction of shattering, thus allowing direct combining. Most recently, shatter-resistant varieties of *B. napus* have been bred.

Traditional Rapeseed Breeding in Canada

Traditional techniques for plant breeding reside in identification of traits for which natural variants could be observed and selected. Some of the traits for which variations can be found are fatty-acid composition, height, maturation period post-seeding, oil and protein content of the seed, seed colour, seed shattering, seed size, tolerance for late sowing, winter hardiness, yield and others. Many traits can also be manipulated and placed in a desirable background through crossing and selection. Finally, use of mutations and selection are powerful instruments for production of varieties.

Saskatoon must be acknowledged as the key starting point for Canadian rapeseed breeding efforts. Since 1944, when Dr W.J. White began a programme at the Canadian Agriculture Research Station, there has been a continuous programme on rapeseed. In 1998, the Research Center was named the Center of Excellence in Canola, by the government of Canada. The earliest organized team effort to this end belonged to researchers Dr W.J. White, Dr H.R. Sallans, Mr G.D. Sinclair, Drs B.M. Craig, L.R. Wetter and C.G. Youngs at the Prairie Regional Laboratory (PRL) of the National Research Council, in Saskatoon. Characterization of rapeseed oil in terms of iodine and acid value, refractive index and oil colour was accomplished. By the mid 1950s, oils for edible use were developed and passed through the Food and Drug Directorate of the Canadian Department of National Health and Welfare.

One of the earliest improvements in canola was the reduction of erucic acid content. High amounts of erucic acid were reported to have a possible damaging effect on the heart muscle of animals. Erucic acid, whose systematic name is docosa-13-enoic acid, is an unsaturated fatty acid with a shorthand designation of 22 : 1 (i.e. 22 carbon atoms in the fatty-acid chain, with 1 double bond). Seeds of the rapeseed plant contain high amounts of erucic acid and small amounts of eicos-11-enoic acid. Erucic acid is converted to oleic acid (18 : 1) by the digestive process in animals. Drs Craig and Downey used radioactive isotopes and separation techniques to provide information on the biosynthetic pathways used by the plant to synthesize rapeseed fatty acids. Conventional plant breeding generated the first low erucic acid rape in Winnipeg and Saskatoon, in 1960 and 1963, respectively. Further improvements led to the registration of the rapeseed lines Oro (1968), Midas (1973) and Tower (1974) from Argentinian seeds, and Span (1971) from Polish seeds.

In 1956, researchers at PRL learned of a new analytical method known as gas–liquid chromatography, which could rapidly identify the components, and amounts of the components, in a mixture such as fatty acids in rapeseed. It was first necessary to hydrolyse the oil into its fatty acids, which were then separated into the GLC column. The PRL researchers refined the method and its sensitivity so that an analysis could be carried out on a single seed. A further refinement by Drs Keith Downey and Bryan Harvey of Agriculture Canada, Saskatoon, allowed the analysis of one-half a seed, and if that seed had the desired composition (such as low erucic acid), the spared half could be grown and bred into a full plant. This discovery sped up the plant breeding time. An additional innovation was to reduce the innovation cycle time by growing the crop in the winter in California, Australia or New Zealand (more recently in Chile).

These new techniques and information were extremely timely and valuable to plant breeders. Additional breeding programmes to that in Saskatoon were initiated at the universities of Alberta and Manitoba. Dr Keith Downey and Dr Baldur Steffanson, University of Manitoba, undertook breeding of low erucic acid, but nutritionally superior material, compared to most other plant or vegetable oils. This discovery led to the wide acceptance of rapeseed oil in foods such as margarine, salad dressings and cooking oils in the food system. The differentiation of high and low erucic acid oils allowed food and other industrial markets to expand independently. High erucic rapeseed oil, which was used as a lubricant in sheet metal milling, as a non-smoking machinery oil and/or as a special grease, was a commodity of its own. Today 'double-low' commercial varieties of both *B. napus* and *B. rapa* dominate the oilseed *Brassica* production area in developed countries. These 'double-low' varieties are characterized as having a very low (<1%) content of erucic acid in the fatty-acid profile of the seed storage lipid, and a very low content of glucosinolates (<18 μmol g^{-1} seed at 8.5% moisture) in their seed and meal. The term 'canola quality' is normally applied to seed, oil and meal from such varieties. Table 3.2 shows the varieties of rapeseed developed during this period.

Biotechnology of Canola Breeding

Biotechnology involves a group of technologies, which aim, in part, at the manipulation of cellular components (McHughen, 2000, provides a good overview of the technologies). From the plant's perspective, usually this means improved genetic traits or identification markers, and from a consumer's context it means improvements in the food and feed value. Thus, for the purposes of canola improvements, most of these manipulative techniques are those of molecular genetics. From the perspectives of canola oil, there could be in addition many other tools and techniques of biotechnology associated with process technology and microbial biotechnology that would impact here (Hills and Murphy, 1991; Carr, 1995; Hui and Khachatourians, 1995).

To begin with, the production of anther and microspore cultures were

Table 3.2. Known commercialized rapeseed varieties released worldwide, 1950–1985.

Variety	Origin	Improved trait	Year and country registered or released
B. napus			
Golden	Argentine	Oil content and lodging resistance	1954 Canada
Bronowski	na	Low erucic acid and glucosinolate content	1955 Poland
Liho	Argentine	Little or no erucic content	1960 Canada
Nugget	Golden	Oil content and lower iodine value	1961 Canada
Tanka	Golden	Yield and seed size	1963 Canada
Diamant	na	Tolerant to late sowing	1964 W. Germany
Target	Tanka	Height, higher yield and oil content	1966 Canada
Oro	Lihox Nugget	The first Canadian low erucic acid variety	1968 Canada
Panter	na	Tolerant to late sowing	1968 Sweden
Marcus	na	Resistant to *Phoma* spp.	1969 France
Norde	na	Winter hardy	1969 Sweden
Turret	Oro	Yield and oil content	1970 Canada
Zepher	Orox Target	Improved oil and protein	1971 Canada
Midas	Target	Low erucic acid	1973 Canada
Tower	Turret	Low erucic acid and first Canadian low glucosinolate in the oil and meal	1974 Canada
Regent	na		1977 Canada
R-500	na	HEAR variety; high glucosinolates	1980 Canada
Andor	na		1981 Canada
Qinyou 2	na	Cytoplasmic male sterile	1981 China
Reston	na	HEAR variety; low glucosinolates	1982 Canada
Westar	na		1982 Canada
Triton	na		1984 Canada
Global	na		1985 Canada
Pivot	na		1985 Canada
Tribute	na		1985 Canada
Pollux	na	Tolerant to late sowing	na E. Germany
Slapska	na	Tolerant to late sowing	na Czechoslovakia
B. campestris			
Arlo	–	The first variety licensed in Canada	1958 Sweden
Echo	Polish	Yield	1964 Canada
Polar	Polish	Oil and protein content of the seed	1969 Canada

Table 3.2. *Continued.*

Variety	Origin	Improved trait	Year and country registered or released
Span	Polish × Arlo	First low erucic acid Polish rape variety	1971 Canada
Torch	Span	Yield	1973 Canada
Candle			1977 Canada
Tobin			1980 Canada
–	–	High palmitic and stearic acids	1985 Sweden

na, not available.

perfected for the production of pure, double-haploid plants (Swanson *et al.*, 1987). To accommodate genetic exchange, without resort   classical mating, protoplast fusion techniques were employed. In protop'   'on, two plant cells with selectable genetic markers are fused, through the use of chemicals known as fusants, to generate cytoplasmic (mitochondrial) or chromosomal traits. This approach was used for the production of triazine resistance and cytoplasmic male sterile plants (Pelletier *et al.*, 1983). Somaclonal variation was used for selection of resistance to the fungus *Leptosphaeria maculans* (Sacristan, 1982). *In vitro* genetic engineering has been much of the basis for selection of canola resistant to glyphosate, sulphonurea and other herbicides. None the less, irrespective of the biotechnological pathway(s) required, there is an inevitable need for conventional backcrossing before the new variety or recombinant canola reaches the market. Figure 3.1 shows the pathways and methodologies used for development of new varieties of canola.

Table 3.3 (see also Table 8.4, p. 140) identifies the array of significant new traits developed in canola. These breeding programmes can be grouped into four areas: (i) seed yield; (ii) seed quality (i.e. oil type and composition), meal quality (glucosinolate content) and seed fibre and colour; (iii) plant resistance to pests (i.e. resistance to microbial phytopathogens and resistance to insects); and (iv) agronomic traits (i.e. winter hardiness, herbicide resistance, height, lodging, maturity time, shatter resistance and others).

The methodology for breeding programmes varies, and the primary choice is the particular *Brassica* sp. For example, both *B. napus* and *B. juncea* are self-fertile and can pollinate in the absence of insects. To transfer an inherited but poorly adapted donor trait into a well-adapted and stable line, backcrosses are performed. In a backcross, the traits of a donor parent are placed into a cultivar that is in use by repeated crossing between the trait donor and the target line.

B. rapa shows considerable heterosis for yield, with up to 40% yield increases, and there is consequently a strong interest in developing hybrid or synthetic cultivars (Falk *et al.*, 1998). Synthetic cultivars are seeds produced from mixtures of two or more cultivars and hybrids (e.g. Syn 1, Syn 2 seeds and

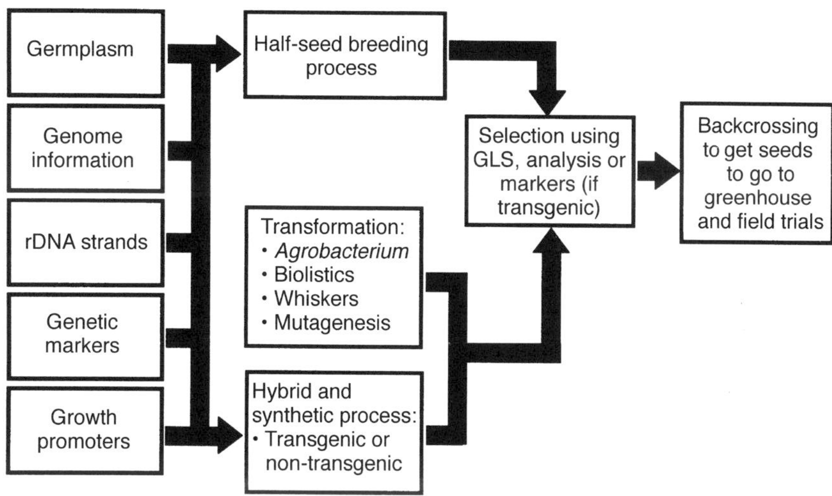

Fig. 3.1. The traditional and biotechnological canola development processes.

so on; Buzza, 1995). Synthetics in cultivars of *B. napus* and *B. juncea* are the result of random matings between selected parental lines, which give rise to both hybrids and progeny from matings within the same parental lines (Buzza, 1995; Falk *et al.*, 1998). The first *B. rapa* synthetics registered in Canada in May 1994 were CASH, Hysyn 100, Hysyn 110 and Maverick. Other methods of breeding include hybrid breeding, e.g. the cytoplasmic male sterility (CMS) system which requires three genetic lines: the A (male- or female-sterile parental) line, the B (or maintainer) line and the R (or restorer) line. The A and B lines have the same nuclear but different cytoplasmic make-up, in that there is an incompatibility between the incoming (guest, foreign) mitochondria and the host species' nuclei. As a result, the A cell has a cytoplasm which confers male sterility, which the B cytoplasm allows for normal pollen development. The cytoplasm of the R line allows for checking on the presence of a restorer gene. Through this breeding, the lines 'Polima', developed in China, and 'Ogura', developed in France, are the only CMS lines of canola.

A surprising innovative development has been taking shape, centred on the oleosins. Oleosins are a novel class of oilseed proteins, which are specifically associated with the oil bodies where triacylglycerols are stored (Hills and Murphy, 1991). They have been purified from several species of rape and soy. Oleosins act as emulsifiers by inserting their hydrophobic domains into the triacylglycerol, with the hydrophilic domains protruding into the aqueous environment. They probably prevent coalescence of oil droplets during desiccation of the seed and may also contain a lipase-binding site to allow lipase access during the early growth of the plant after germination, as non-viability of some stored seeds has been shown to be due to the coalescence of oil during storage.

Table 3.3. Known transgenic *Brassica napus.*

Herbicide tolerance
  Basta, phosphinotricin, glyphosate (*N*-phosphonomethyl) glycine,
  acetohydroxyacid synthase, oxynil herbicides

Pathogen and insect tolerance
  Blackleg, pea defence gene, damping off, PR peroxidase of *Stylosanthes humilis*,
  sclerotinia, viruses (BWYV, CaMV, TYMV, TMV), *Bt* toxin gene for diamondback
  moth larvae and coleopteran insects, cysteine proteinase inhibitor

Protein production
  Increased amino acids (lysine, cysteine and methionine), Brazil-nut storage
  protein albumin, oil body protein, soybean oleosin, oleosin hirudin, seed protein
  with phytase, root proliferation, peroxidase production, enkephalins, low
  glucosinolates

Seed oil composition
  Laurate accumulation, altered saturated fatty acids, medium-chain hydrolase and
  fatty-acid modification (increased palmatate, erucic acid content,
  polyunsaturated, eicosanoic acid content, 8: and 10:0 fatty-acid levels, stearate
  levels, saturated fatty acids)

Miscellaneous
  Seed chlorophyll content, cadmium resistance, pollen expression, modified
  cytokinin levels, reduced gravitropism

PR, pathogenesis-related; BWYV, beet western yellow virus; CaMV, cauliflower
mosaic virus; TYMV, turnip yellow mosaic virus; TMV, tobacco mosaic virus;
Bt, Bacillus thuringiensis.

Since the oleosins act so efficiently in preventing oil coalescence, even under the
large pressure present in the dry seeds, they have found applications in the food
and pharmaceutical industries (Hills and Murphy, 1991; Plant *et al.*, 1994;
Parmenter *et al.*, 1995; Van Rooijen and Moloney, 1995; Chaudhary *et al.*,
1998; see also Chapter 8, p. 146).

Conclusions

In this chapter, we have summarized the history and evolution of scientific
knowledge leading to innovations in rapeseed/canola. With the continued
expansion of the market for oilseed products, there continues to be increasing
recognition of the need to develop new varieties. This need will be realized by

combining methods of conventional plant breeding with biotechnology (i.e. the new technologies for molecular genetic manipulation). For the biotechnologist this science-based development is challenging but relatively easy. These plant species have been around for a long time and a reasonably diverse gene pool is at hand. Further, the major model plant system, *Arabidopsis thaliana*, can provide a gene system for much of the new plant developmental and molecular genetic research. Fortuitously, *A. thaliana* is closely related to the rapeseed species. While the scientific foundation is clear, the real challenge is to manage the erratic drivers of the market, economics and politics.

Nevertheless, there are grounds for optimism that rapeseed/canola will complete the transformation into a knowledge-based industry and that with advances in biotechnology and orthodox tools of genetics and plant breeding it will provide a clear example and potential template for a new generation of knowledge-based innovation in the agri-food world.

Innovation in the Canola Sector

Peter W.B. Phillips

Introduction

In this chapter the objective is to answer two key questions. First, how has the innovation system for the canola/rapeseed industry changed over the years? To do so, we measure the length and amplitude of the stages in the innovation cycle to determine the impetus for research and the impact of innovations in the research process on the length of the product life cycle. These data are then used to examine linkages in the canola research effort. Secondly, where has the research been done? The chapter contains an examination of the evolution of the research across different geographical locations, in order to determine how stable the competitive market position of existing producers is. These data will be examined from a geographic perspective to determine who is doing the work and where is it being done, in order to consider whether innovation in the canola sector, as suggested by industrial location theory, is becoming localized in one or more centres. This will then be used in following chapters to examine how industrial and public action have influenced the comparative advantage of different locations in the production of canola.

Innovation in Canola: Analysis of the Innovation Process

Using the chain-link model outlined in Chapter 2 and Malecki's typology of knowledge, we can examine the scale and flow of research effort focused on canola since 1980.

Basic, know-why research

As indicated by the chain-link model, basic, know-why knowledge directed at canola is created by drawing upon both canola-related research from previous years and on more fundamental research that is not specifically targeted to canola. The 4908 articles published between 1981 and 1996, included in the ISI canola databank, together cited a total of 17,995 papers in 1294 journals. The journal producing the most papers and citations was *Plant Physiology* (5.2% and 6.7%, respectively). The top 15 journals accounted for almost 27% of the papers that were used by canola researchers, and accounted for 39% of all the citations made by the canola papers (Table 4.1). Twelve of the top 15 journals are predominantly basic science journals, with specific research related to genetics, organic chemistry and molecular biology. Taking the top 11 journals, all basic science journals, the market share has risen dramatically over the period, from about 18% in 1981–1985, to 23% in 1986–1990 and to 26% in 1991–1996.

A comparison of the basic research in the cited papers with the flow of canola-related research as measured by the canola papers shows that there appears to be a 3–5 year lag in the implementation of basic research into pure agricultural research (Fig. 4.1).

Canola- or rapeseed-specific know-why work has ebbed and flowed over the past 30 years. The actual research effort in many cases will have been

Table 4.1. Basic research cited by canola researchers by source journal: percentage of total papers and citations in source research in the top 15 journals. (*Source*: ISI, 1997.)

Journal	Papers	Citations
1. *Plant Physiology*	5.2	6.7
2. *Journal of American Oil Chemists Society*	2.7	4.1
3. *Theoretical and Applied Genetics*	2.3	5.4
4. *Plant Molecular Biology*	2.0	2.9
5. *Planta*	1.8	2.3
6. *Physiologia Plantarium*	1.7	2.1
7. *Proceedings of the National Academy of Sciences, USA*	1.5	1.9
8. *Nucleic Acids Research*	1.4	1.8
9. *Molecular and General Genetics*	1.3	2.2
10. *Journal of Biological Chemistry*	1.3	1.7
11. *Phytochemistry*	1.2	1.4
12. *Journal of Dairy Science*	1.1	1.6
13. *Plant Cell*	1.1	1.6
14. *Journal of Food Science*	1.1	1.4
15. *Journal of the Science of Food and Agriculture*	1.0	2.0
Other journals (1279 titles)	73.4	61.1
Total	17,995	26,946

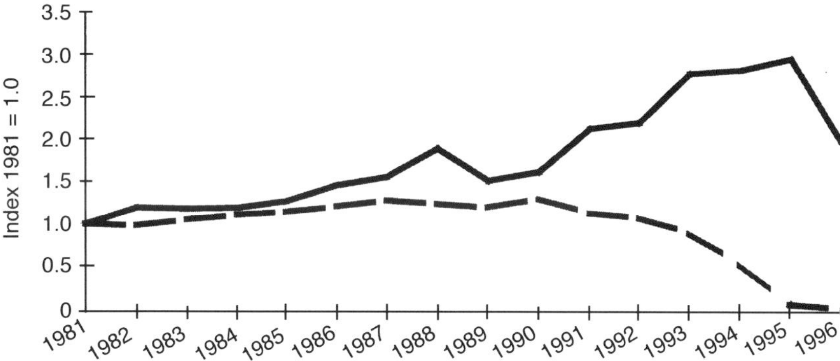

Fig. 4.1. Relationship between source articles (dotted line) and canola research (solid line).

completed about 2 years before the ISI recorded date, due to four built-in lags: the research takes time to write up, the peer review process for many scientific journals adds approximately 1 year to the lag, press time adds up to 8 months and, finally, the ISI citations system lags actual publication by up to 6 months, which could put some journal articles published in the second half of a year into the next year. Keeping in mind these limitations, the data show a number of informative trends. Total canola-related research has risen sharply over the past 30 years. With new molecular, biochemistry and genetic developments flowing from basic research in the late 1970s, the pure agricultural research output accelerated. The publications rate shows that the number of canola papers peaked in 1995, which suggests that research may have peaked sometime around 1993. This would appear to be at least partly driven by a decline in the transfer of know-why knowledge from the basic sciences into the canola research effort. Beginning in 1990, there was a drop in the number of new basic science articles cited. Over the following 3 years, the number of citations dropped by more than 25%, which suggests that this slowdown may be fundamental rather than transitory.

Although canola-related research rose sharply in absolute terms after 1980, relative to the underlying industry the rise is not as large. Comparing canola research to global canola production, the relative annual research effort has oscillated around a mean of about 14 articles per million tonnes (Mt) of canola/rapeseed produced globally; the range has been from a low of about 12 articles per Mt to a high of about 19 articles per Mt. Pure applied research peaked in 1983, 1988 and again in 1993 in relative terms.

The focus of research shifts periodically between different subjects. Therefore the aggregate data mask the individual innovation cycles that relate to narrower research areas. One basic distinction that can be examined is the relative research into the two basic types of rapeseed, which has varied depending on the flow of new processes and technologies from the basic research community. *Brassica napus* (Argentine rapeseed), which is self-pollinating and therefore easier to create pure lines of, has been the prime focus of much of the research to date, while *Brassica rapa* (Polish rapeseed), which is self-incompatible and therefore needs cross-pollination, has only become more attractive as a target with the recent development of commercially viable transgenic and hybrid technologies. Looking at the data, one can begin to see the cycles that are embedded into the larger research agenda. The peak-to-trough cycle in effort on the two varieties appears to average about 4–5 years since 1970, ranging from as short as 2 years to as long as 7 years. Although the absolute effort has fluctuated more in recent years, the peak-to-trough activity has narrowed in relative terms, to about 25% maximum from about 40% in the early years.

Two other readily identifiable research subjects that can be tracked in the data are aspects of oil characteristics and feed use. Both of these issue-specific research streams have exhibited short-, medium- and long-term trends. Oil research has been by far the most active, with an average of about 40% of the identifiable research targeted to this issue. Once low levels of erucic acid became the product standard in the mid 1980s, work on oil properties tailed off until the combination of new biotechnology methods and renewed market interest in speciality oils triggered greater oil-based research, which has rebounded to more than 40% of total research. As long as glucosinolates were an issue, research into feed represented as much as 30% of the work. Once again, as low glucosinolates became the product standard, research effort tailed off, to less than 15% in 1996.

Looking beyond the simple amount of research, the ISI canola database can be used to determine whether there are any economies of scale in the canola research business. It is reasonable to assume that the number of articles simply reflects the inputs invested in canola research. In contrast, the number of citations of those articles is a measure of the volume of output produced with those inputs. Looking at the canola research business, there appear to be three distinct relationships between the average citation rate for an institution's output and the number of articles produced by that institution (Table 4.2). For those institutions that produced between 1 and 11 articles over the 15-year period, there was a strong positive correlation, suggesting that there are scale economies of pure agricultural research. Institutions that are not regularly doing research into canola on average have higher costs (or a lower output) than firms pursing an extended research programme. For institutions producing between 11 and 22 articles (i.e. institutions with a small but sustained programme), there is a somewhat less, but still strong, correlation between quality of output and the number of articles, suggesting that economies of scale continue in this realm. After about 23 articles per institution, there is only a very weak negative correlation between number of articles and quality of output, suggesting that there are no discernible economies of scale beyond the basic size.

Looking further along the innovation chain, the canola research in any one period becomes the stock of knowledge to be used in subsequent periods. This know-why research on canola becomes new ideas or approaches that are used

Table 4.2. Correlation coefficients for average citations versus average number of articles per institution (1981–1996).

1–11 articles	+0.903
12–22 articles	+0.837
23–388 articles	−0.059

Source: Author's calculations based on ISI, 1997 data.

by two main groups. First, the research published in journal articles is absorbed into the larger body of academic knowledge. The ISI database notes that as of July 1997, the 4908 articles published related to canola have been cited 13,575 times, for an average citation rate of 2.77 citations per article. The second market, the applied research community, is more difficult to quantify as the ideas can only infrequently be linked directly to the applied research. Nevertheless, when one looks at the major public institutes that do both pure agricultural research and applied R&D (e.g. NRC, AAFC), one can see a linkage between pure and applied research (see Chapter 7).

Another role of basic research is as a training exercise for researchers. A significant share of the know-why research published in journals comes from doctoral research or from postdoctoral research in collaboration with academic faculty. The knowledge that is developed in those two research efforts ultimately flows with the researcher when the student graduates, leaves the university environment and goes to another university, to a public research agency or to the private sector as a researcher. UMI's Dissertation Abstracts database of doctoral dissertations and Master's theses was searched for the keywords canola, rapeseed, *Brassica napus*, *Brassica rapa* and *Brassica campestris*. The database includes citations for materials ranging from the first US dissertation, accepted in 1861, to those accepted as recently as last semester. The database represents the work of authors from over 1000 North American graduate schools and European universities. The number of dissertations on canola in key research communities around the world gives some indication of the potential flow of researchers from academic institutions to the broader research community. The search for the entire period identified 581 theses, of which 384 were doctoral dissertations. Before 1977 there was only the occasional thesis annually, but in the 1970s the rate of production rose to about three per year, almost all of them in Canada (Table 4.3). In 1980, the rate of production rose to about 12 per year, with more than half of the research being done in other countries. After 1990, the average rate of production rose to 53 per year, with about two-thirds of the papers being produced in Canada. The main producers of graduate theses are the Universities of Manitoba, Alberta, Guelph and Saskatchewan. The US contributed about 25% of the papers and the EU about 13%, mostly coming from Sweden, Denmark and the UK.

Table 4.3. Dissertations by period and region.

	Canada	USA	EU	Australia	China
1925–1959	0	3	0	0	0
1960–1969	3	2	0	0	0
1970–1979	27	6	0	0	0
1980–1989	53	38	24	1	1
1990–1998	259	108	54	1	0

Source: Keyword search of the UMI Dissertations Abstracts on 17 September 1998 for the terms canola, rapeseed, *Brassica napus*, *Brassica rapa* and *Brassica campestris*; http://wwwlib.umi.com/dissertations/search

Integrating know-what, know-how and know-who into new products

Applied research and development is the stage where the know-why information is combined with know-how and know-who knowledge to create new know-what data. In short, institutions and people take the basic concepts and convert them into potentially marketable goods or services. The funds for know-why research may come from producers, industry or the public sector, but the research usually does not have direct commercial application. The next step involves taking the technologies and genetic mapping from the pure agricultural research stage and developing new proprietary technologies or breeding specific varieties for commercial use. This involves a mix of different types of actors, including universities (e.g. Manitoba, Alberta, Guelph), publicly funded research agencies (e.g. NRC and AAFC) and a mixture of seed, grain and agrochemical companies (e.g. Limagrain, AgrEvo, Monsanto and Saskatchewan Wheat Pool).

As one would expect in a sector that has significant private-sector investment and highly competitive firms with products that have a short economic life, it is often difficult to gather definitive data on the investments made by firms for creating the capacity to develop either new technologies or new varieties. Nevertheless, there are two main sources of data that measure at least part of the effort in this stage of the innovation system.

Companies, universities and government laboratories invest a significant amount of time and money in the development of new processes and products involving canola. Given that both of these types of effort yield intellectual property with some value, there is a strong incentive, at least at this stage in the innovation chain, for developers to attempt to protect and recover from the marketplace some of the value of their innovations. Thus, a review of patent databases can give us some indication of the net outcomes of the foundation research undertaken by many of these firms. Given that patent applications come at the end of a research effort and are a time-consuming process, the patent data should be viewed as a lagged indicator of effort (see Trajtenberg,

1990). Although companies will apply for a patent within 1 year of completing the invention, the average applied research programme is estimated to last 3–5 years and the lag between patent application and patent grant averages about 27 months in the US and 45 months in Canada (Gold, 2000). As a result, the patent information (based on the year of patent grant) reflects the results of applied research that was being undertaken approximately 3–5 years earlier.

A review of the patents issued in Canada, the US, Europe, the EU and through the World Intellectual Property Office (WIPO) shows that more than 1400 patents were issued in those countries over the period 1978 to early 2000. The on-line patent servers for each of those systems were searched for the keywords: *B. napus, B. rapa, B. campestris,* canola and rapeseed. The data from all the systems show that the patenting activity (i.e. the commercially directed applied research) accelerated in the late 1980s, after canola acquired generally regarded as safe (GRAS) status in the US. Given an estimated 4-year lag inherent in patent data, these data (Table 4.4) suggest that applied research into canola-related products began to accelerate in about 1985. Canada has issued a total of 634 patents for canola-related innovations, the single largest number of patents in any country. Using Canada as the proxy for the industry, one can see the acceleration. Before 1990, only an average of ten patents per year were issued, which indicates only minimal applied research before 1985. The number of patents then rose to an average of 60 year^{-1} in 1990–1994 and dipped to an average of 40 year^{-1} in the 1995–1999 period, which suggests that applied research accelerated after about 1985 and then peaked in the early 1990s.

As one would expect if technological innovations are harder to exploit than product innovations, public-sector agencies have disproportionately invested in developing new processes (64% of their patents are for process innovations) while almost 60% of the patents issued to commercial companies were for new products. Private-sector innovation appears to have peaked in or around 1989 (based on patents peaking in 1992–1994). A review of the specific patent claims suggests that most of the product patents approved were for industrial or food uses of canola rather than for new varieties. The intellectual property

Table 4.4. Canola patents by type and patent holders, Canada.

	Technology		Product		Total	Total	Total
	Public	Private	Public	Private	public	private	
Total patents	66	220	37	311	103	531	634
% Total patents	10%	35%	6%	49%	16%	84%	100%
% Public patents	64%	–	36%	–	100%	–	–
% Private patents	–	41%	–	59%	–	100%	–

Sources: Author's calculations using CIPO patent search for canola, rapeseed, *Brassica napus* and *Brassica rapa*, December 1999–January 2000 (http://patents1.ic.gc.ca/srch_sim-e.html).

resulting from breeding of new varieties can be protected through plant patents in the US, but patents are not issued for multicellular whole organisms in Canada. Given that Canada is a far more significant producer of varieties than the US, varietal development is better measured through the varietal testing and registration system.

Once a new variety has been developed in the laboratory there is an extensive process of field trials to select the specific line that has the desired yield, oil content, and tolerances to temperature, moisture stress, insects and disease. In addition, the adaptation phase involves the replication of the foundation seed into registered and certified seed for commercial planting. That process takes a minimum of two growing seasons and for many varieties (especially those developed through traditional breeding methods) it can take four or more seasons. The process of shuttling seed between growing areas to accelerate this stage, developed by CIMMYT wheat breeders in Mexico in the 1960s and perfected by Pioneer Hi-Bred, has been adopted by some of the companies in the canola sector, with significant shuttling occurring between Canada and California, the southern US states, Chile and Australia.

Field trial data, unlike the data for the previous three innovation stages, are not lagged. The trials are recorded in the year in which they take place. The field trials for traditionally bred varieties are not recorded in the same way as those for varieties developed using biotechnology methods. Looking only at the data for field trials for novel varieties, we can see that adaptation began for those crops in 1988 and peaked in 1994. The bulk of the early activity was directed to modifying *B. napus* varieties but once regulatory approval was achieved for herbicide-tolerant varieties in 1995, *B. napus* activity diminished; since then activity on *B. rapa* has risen significantly, at least partly filling the gap caused by the lower level of research into *B. napus*. This lagged relationship between *B. napus* and *B. rapa* conforms with the relationship in pure agricultural research in universities and institutes for the two crops, with *B. napus* work accelerating past *B. rapa* in the 1980s and finally peaking in 1994, which, given the publication lag, really means that the know-why type research peaked in 1992. The results of that research hit the fields with about a 2-year lag.

Moving along the product development system, one can measure both flow of activity and relative capacities by examining the varietal registrations records required by the Seeds Act and administered by the Canadian Food Inspection Agency. An average variety requires at least 2 years of field trials to meet the Canadian regulatory requirements of registering a variety. The varietal data in Table 4.5, compiled from Canadian Food Inspection Agency (CFIA) records (supplemented with data from the Canola Council of Canada and other sources), show that until about 1985–1989 there was less than one new variety registered per year. By the early 1990s, the rate of registration had risen to 12 per year and in 1996 peaked at 32 new varieties. The time to breed and register new varieties ranges from about 6 to 10 years (depending on whether transgenic or traditional breeding methods are used), which suggests that the last peak breeding effort in the canola industry began in the 1989–1991 period.

Table 4.5. Field trials for transgenic canola in Canada.

Total	1988	1989	1990	1991	1992	1993	1994	1996	1997
Total field trials	23	53	96	172	241	362	578	387	438
Trials by private company									
B. napus	23	49	90	161	168	265	394	223	253
B. rapa	0	0	0	0	0	6	82	148	166
Trials by public agency									
B. napus	0	4	6	11	73	91	100	13	15
B. rapa	0	0	0	0	0	0	2	3	4

Source: CFIA (1999) special tabulations.

The varietal registration data for both *B. napus* and *B. rapa* show that the efforts and capacity to breed new varieties has expanded significantly, with more than 17 active breeding institutions in the past decade, compared with only one public breeding programme in the early years (Table 4.6). As with most data, this source is not perfect. As the industry evolves, companies merge or split, causing a disjointed picture of the research capacities. Although there appear to be 17 agencies actively breeding *B. napus* throughout the 1990s, the actors changed significantly between the early part of the decade and the last few years, with ten new entrants and nine entities either merging with others or ceasing canola research altogether. The data in the late 1990s are also somewhat misleading as many of the smaller breeding programmes (e.g. Plant Genetics Systems (PGS)) continue to register varieties under their own name even though they are now wholly owned subsidiaries of other companies. The research capacity around *B. rapa* has always been less dispersed and proportionately more dependent on government investment. In the 1990s, three private companies, both individually and in collaboration, now dominate the seed business for *B. rapa*.

The flow data are useful but the best picture of effort is based on independent public data, supplemented by survey data from industry. This study did just that, asking both public and private agencies to identify the number of scientists and other research staff involved in canola development and to describe their in-house versus contracted research efforts. The answers to these questions confirmed, as is discussed in Chapter 8, that the research capacity dispersed after about 1985 as many private companies entered the business but, contrary to the varietal registration data, has since consolidated into a few relatively large and independent operations and a number of smaller, niche breeders and developers.

Table 4.6. New varieties developed by institution and by period. (*Source:* CFIA (1999) variety registration records.)

	1950–1959	1960–1969	1970–1979	1980–1984	1985–1989	1990–1994	1995–1998
B. napus							
Total varieties from public institutions	1	4	5	4	8	8	10
Total varieties from private institutions	0	0	0	0	12	39	76
Total varieties	1	4	5	4	20	47	86
Number of active institutions	1	2	2	3	11	17	17
B. rapa							
Total varieties by public institutions	1	2	5	1	1	4	2
Total varieties by private institutions	0	0	0	0	3	7	16
Total varieties	1	2	5	1	4	11	18
Number of active institutions	1	2	1	1	3	7	4

Adoption

The final stage in every successful innovation chain is the adoption of the product in the marketplace. Varietal demand and usage has become more difficult to determine as that information is now viewed as proprietary and has value as the market goes private and segregates. Nevertheless, Canada has been fortunate in having a varietal survey, produced by the Prairie Pools for canola, that covers the 1960–1991 period for the three prairie provinces. In addition, Manitoba Crop Insurance Corporation (MCIC) has gathered extensive varietal records for Manitoba for use in insurance premium setting and has made those records available through the Internet. By taking the Prairie Pools survey as the base and using the MCIC records for Manitoba to extrapolate for the prairies, we obtained an estimate of varietal adoption and diffusion in recent years.

Looking simply at the rate of adoption of new seeds, we observe that seeding of new varieties usually starts slowly but accelerates rapidly, at times significantly changing the market shares of seed developers. Few private varieties existed as recently as 1985. As the number of new private varieties grew, the market share of private seed companies rose dramatically. In the 12 years after 1985, the private seed companies developed and introduced 96 new varieties, compared with 29 new varieties from the public breeders (Tables 4.6 and 4.7). As a result, the public breeders' share of the canola seed market fell from almost 100% in the 1985–1989 period to only about 26% in 1997. Private

Table 4.7. Canola market share: % of acreage seeded to varieties developed by selected institutions (Polish varieties in brackets; zero unless given).

	1960–1969	1970–1979	1980–1984	1985–1989	1990–1994	1995	1996	1997
All public institutions	100.0	99.8	99.3	98.3	49.3	26.5	26.1	10.6
	(75.6)	(59.6)	(42.0)	(42.4)	(27.3)	(18.1)	(11.5)	(na)
All private companies	0	0.2	0	0.4	43.2	56.6	60.6	63.1
		(0.2)			(10.3)	(17.5)	(23.9)	(na)
Unallocated	0	0	0.7	1.3	7.5	16.9	13.3	26.3

Sources: Nagy and Furtan, 1978; Prairie Pools, 1963–92; Manitoba Crop Insurance Corporation (1998) varietal data weighted by historical shares for 1993–96.
na, not available.

breeders captured the lion's share of the market as their varieties have embodied new commercially attractive innovations.

Production records tell us quite a bit about the absolute length of the innovation chain. Table 4.8 shows that as the number of varieties has risen due to innovation, both the maximum and minimum market shares have diminished. In short, increased variety reduces both the theoretical and practical market share for any new product. As Grossman and Helpman (1991) point out, at some point the market for new varieties that are only marginally different from existing varieties will become satiated, which should at some time cause a slowing in the development of new varieties. It is not clear from the data that this point is reached, but clearly it is closer than it was. Related to this is the fact that each new variety has the potential to destroy the value of earlier varieties that have poorer agronomic traits. As the rate of introduction of new varieties rises, the average age, and by implication the expected average economic life of the variety, declines. Putting these two trends together, one can see that the adoption phase of each individual innovation has shortened significantly, with the expected economic life now about 3 years, down from an average of as much as 14 years in earlier decades.

Furthermore, the peak market share for each variety is reached much faster for the new varieties than previously (Table 4.9). In the 1954–1984 period, the peak market share, which was 20% for *B. napus* varieties and 43% for *B. rapa* varieties, was reached either in the third or fourth season after introduction. As innovation accelerated, the average peak market share dropped sharply, until in the 1990s the average peak was now less than 5% for new varieties. The drop in the peak market share occurred more rapidly for *B. napus* as the product was subject to earlier and more intensive research effort. At the same time, the average lag between introduction of the variety and its peak market share narrowed to less than 2 years, from almost four seasons in the earlier period.

In short, canola has truly become an innovation-led product, with basic

Table 4.8. The evolution of the canola seed industry.

	Number of active varieties	Weighted average age	Average market share
1960	4	13.9	25.0
1965	6	9.5	16.7
1970	7	6.1	14.3
1975	8	3.4	12.5
1980	8	4.6	12.5
1985	9	2.8	11.1
1990	16	5.2	6.3
1995	48	3.9	2.1
1996	50	3.6	2.0

Source: Author's calculations using canola variety database; the average age is the number of years since introduction for each variety weighted by the market share for that variety in that year.

Table 4.9. Adoption rates for new varieties.

	B. napus		B. rapa	
	Lag between introduction and peak market share (years)	Average maximum market share (%)	Lag between introduction and peak market share (years)	Average maximum market share (%)
1954–1984	3.5	19.9	3.9	43.2
1985–1992	2.5	5.4	3.5	13.1
1993–1995	1.5	1.2	1.4	3.1

Source: Author's calculations on the canola variety database.

and applied research and development setting the pace for commercial development, which has correspondingly radically shortened the economic and commercial life of both past and each successive invention.

The herbicide-tolerant canola example

The general ebb and flow of research, development, adaptation and adoption examined above in aggregate for the canola sector involves a large number of nested innovation cycles that loop back and support or cross-cut the overall trend. The challenge in such an analysis is to identify these embedded developments. It is possible to identify all four types of innovations as categorized by Paul Romer (1990): rival and non-rival, excludable and non-excluded innovations.

Two key non-rival, non-excluded innovations provided the platform for the development of canola. The NRC Prairie Regional Laboratory in Saskatoon provided the means for identifying specific plant characteristics when, in 1957, they acquired and perfected the use of a gas–liquid chromatography (GLC) unit. This allowed the analysis of as little as a seed's worth of oil in only 15 min; previously scientists needed at least 2 lb (1 kg) of seed and 2 weeks to do the analysis (Kneen, 1992). The Agriculture Canada laboratory in Saskatoon provided the rest of the necessary technology – they developed and perfected the method of dividing individual seeds into equal halves and growing new seeds from a half seed. Thus, breeders could test half a seed using GLC and then grow the rest of the seed if the oil characteristics were desired. This innovation allowed more precise development of 'rival' varieties with targeted characteristics. Nevertheless, in spite of the improvement in breeding techniques, breeding still took from 6 to 12 years to perfect a variety and get it into widespread use. That, plus absence of any effective way to exclude the use of these new varieties (given that all new varieties were open pollinated and plant breeders' rights were not available in Canada before 1990), was undoubtedly the reason for there being only a modest rise in the rate of introduction of new varieties and the virtual absence of any private applied research.

The introduction of excludability for both non-rival innovations (patents for transgenic processes, beginning in 1980) and for new varieties (plant breeders' rights began in Canada in 1990) set the stage for the expansion of private applied research in the late 1980s. During the 1980s a number of companies adopted biotechnology processes (i.e. genome maps, cell fusion, genetic recombination and polymerase chain reaction) and significantly reduced the average time to undertake the applied R&D to develop new canola varieties, in some cases reducing the lab work to 1 year from 5 years. These non-rival innovations, although technically excludable through patents, were only partially excluded as no single patent unequivocally controls access to the research approaches; the patents instead simply provide a base for negotiating access to the technology. Each of the first new varieties developed using this new technology did not realize the full benefits of this reduced development time as they required more time in the adaptation phase, doing field trials to confirm the vigour of the plants and to conform with the more stringent regulatory requirements established for plants with 'novel' attributes (see Chapter 13). Nevertheless, companies were willing to make these investments as they could see the unprecedented opportunity of this new technology to reduce the time and cost of subsequent innovations and the potential to introduce new genetic materials to differentiate the seeds (e.g. herbicide-tolerant genes and novel oil characteristics). As it turns out, the development period for the first varieties using new technologies was not shortened at all. For subsequent varieties from an approved genetic package, the time to introduce derivative varieties has been almost halved, to an average of about 3 years from the previous 6-year minimum, which makes commercial investment much more attractive. The resulting varieties, however, will soon begin to exhibit the characteristics of rival, excludable innovations. In short, the growth resulting from that innovative effort will begin to slow.

The development of herbicide-tolerant canola represents one readily available example of the innovation process in the canola sector. The basic research into transgenic processes was undertaken in the 1970s and involved the gene-splicing technology work at the University of California in 1973 and the first successful gene transfer in 1976. Pure and applied agricultural research into herbicide tolerance began in Canada in the mid 1970s at Guelph and led to applied research by Agriculture Canada to develop a triazine-tolerant canola variety. This variety was field tested in 1982–1983 and was introduced to the market in 1984. Although it was not commercially successful because it threatened to create triazine tolerance in other plants, this research was a forerunner to work that began in the private sector in the late 1980s (interestingly, the technology is now being used in Australia for triazine-resistant canola). Pure applied research into herbicide tolerance (HT) was being undertaken throughout the 1980s and the number of publications about HT development peaked in 1987 (which, with the lags of academic publishing, suggested the laboratory work peaked in 1985). Commercial applied R&D began in some places as early as 1985 but was well under way in the late 1980s, peaking about 1989–1990.

The adaptation phase, as would be expected for a new process, began while the applied research was still under way. Some companies began trials as early as 1988 but activity accelerated in 1992–1994 as the new varieties were field tested for vigour and to gather the data necessary for the regulatory process. By 1994 the varieties were chosen, the domestic regulatory approvals received and limited unconfined releases were under way. The adoption of these crops was slower than one might anticipate because the industry adopted a voluntary identity-preserved, close-looped production (IPP) system to segregate genetically modified crops from traditional varieties in order to maintain access to the EU and Japan, where the new varieties had not been approved for importation or consumption. With the removal of the IPP rules in 1997 (following Japanese approval and in anticipation of rapid EU approval for genetically modified (GM) canola), production of the new HT varieties took off. In the intervening few years, new HT varieties have been developed and the number of varieties has increased rapidly.

Innovation in the Canola Sector: Analysis of the Location of Knowledge

The second question related to knowledge-based growth is where the value is being added. As one would expect with such a large research effort and such a wide variety of specific research issues, the core research capacity related to canola since at least the 1970s has not been limited to one or even a couple of institutions. The following geographic and institutional review of the stages of innovation suggests that the locus for the knowledge-based growth is highly

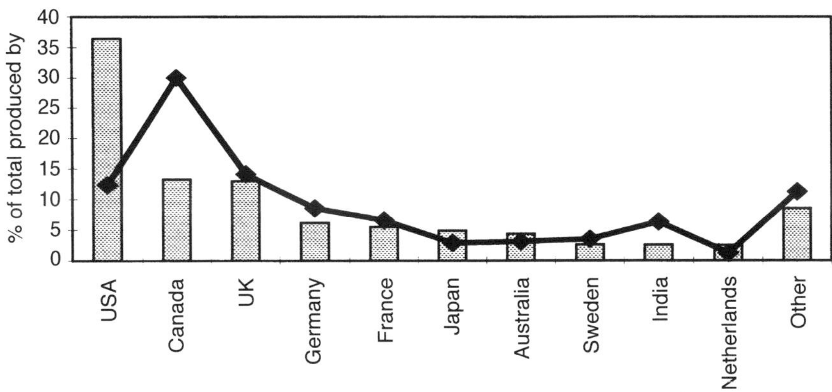

Fig. 4.2. Relative production of source articles (bars) and canola research (lines), by location.

mobile and can, and does, migrate across international boundaries, depending on various circumstances.

Basic science

Although Canada is the single largest country producing pure agricultural research into canola, the basic research that underlies that effort comes disproportionately from the US (35%) and other countries. Canada is the only significant canola researcher that 'imports' significant amounts of basic research relative to its canola research. This could be viewed as either a strength, in that Canadian scientists are open to, and do use, ideas from elsewhere, or as a weakness, in that without a full appreciation and hands-on participation in basic sciences, researchers might lag behind others in the use of that work.

Looking downstream at the use of the canola papers (Fig. 4.2), the single largest market for this research is the US, which accounted for more than 25% of the total citations. Canadian academics accounted for about 17% of the citations while the UK, Germany and France accounted for 13%, 9% and 8%, respectively. In aggregate, the five top markets accounted for about 73% of the citations of the canola papers.

Canola-specific know-why research

The dominant position of Canada in canola-specific research is declining. Before 1980, Agriculture Canada, NRC and the University of Manitoba did the bulk of the work in Canada. Even in the early 1980s, all Canadian locations combined contributed between one-third and 40% of the pure agricultural research on

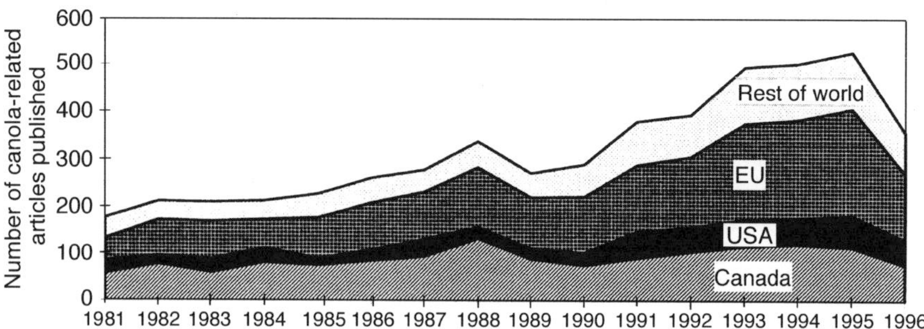

Fig. 4.3. Pure research into canola by country.

canola being done in the world. More recently, however, the global research volume has expanded greatly, while Canadian research efforts have been relatively steady. The result is that canola research has disseminated into a wide variety of locations and institutions (Fig. 4.3).

Canada's relative decline has been matched by an increase in activity in the US, the UK, France and a wide variety of other countries. In 1981 only about 22 countries were doing any research into canola. By the mid 1990s more than 40 countries were doing research into canola on a continuing base. Over the intervening period more than 70 countries did some work on canola. Each country has specialized to some extent, with France, Germany and the UK working almost exclusively on *B. napus* while India does the bulk of its work on *B. rapa*. Canada has devoted more than 80% of its effort on *B. napus* while the US does almost an equal amount of work on each (Table 4.10).

Along with the diffusion of activity by country, the research has been dispersed among a much larger group of research institutions. In 1981 only about 110 organizations were doing canola-related work. By 1995 the number had risen to about 350. Over the intervening period more than 1500 institutions undertook or collaborated on pure agricultural research on canola. Thirteen institutions produced about 36% of the total papers on canola over the period.

Table 4.10. Canola related know-why research by genus. (*Source:* ISI, 1997.)

Country	No. of papers on *B. rapa*	No. of papers on *B. napus*	% *B. napus*
Australia	33	62	65.3
Canada	71	340	82.7
France	9	97	91.5
Germany	8	183	95.8
India	103	67	39.4
Sweden	28	84	75.0
UK	27	305	91.9
USA	96	113	54.1

Five Canadian universities (Guelph, Saskatchewan, Manitoba, Alberta and Calgary) and two Canadian public institutions (AAFC and NRC/PBI), combined with six international institutions (INRA in France, AFRC in Germany, the Swedish University of Agricultural Sciences, UC Davis, USDA and the John Innes Research Centre in England), together produced more than 1750 papers over the period. While the role of those institutions in aggregate has been relatively steady, the role of any individual institution has undergone significant shifts. AAFC, and the four traditional canola research universities in Canada have lost market share, while the NRC, the University of Calgary and the international institutions have gained market share.

Ten of the 13 centres have average citation rates higher than the overall average. Combined, the 13 centres have an average citation rate of 6.53, which is almost one-third higher than the 4.91 citation rate for the rest of the papers. The averages, however, mask significant differences between the institutions. All of the top foreign institutions, except the USDA, have had citation rates that exceed the average for all papers, and in many cases are much higher. The volatility in their output share and their citations rates suggest that they are closely tied to specific innovation streams rather than to the core industry. In contrast, the Canadian universities, except Calgary, have lost both market share and have had a relative decline in citations over much of the recent past. The University of Saskatchewan, for example, was a leading research centre for issues related to the quality and digestibility of canola meal, but since that issue has waned in importance, the institution's output and citation rate has dropped. In contrast, the John Innes Centre in the UK and UC Davis have targeted new breeding technologies and oil modification processes, which has caused a sharp jump in their productivity in terms both of output and citations.

As the work becomes more specialized, collaborations are playing a greater part in the evolution of research capacity globally. Research by the ISI (Science Watch) shows that international collaboration tends to raise the quality of the work (as measured by citations). The top nine canola research countries undertook between 5% and 22% of their research in collaboration with others in other countries (Table 4.11). A correlation coefficient calculation on that data shows that the citation rate is strongly positively correlated with international collaborations. This poses a challenge for Canada, which, among the top producers, collaborated on only about 8% of its work, which is less than half of the average of all the other countries; only India collaborated less.

Applied Research

Although the international data on applied research is sketchy, one can conclude that Canada is the main country involved in developing new varieties of canola but that a significant share of the applied research to develop the processes used to develop those varieties has been done in other countries, and that much of the applications-based research (e.g. uses for new oils) is happen-

Table 4.11. International collaborations and the citation rates. (*Source:* ISI, 1997.)

	% Papers involving international collaborations	Average national cite rate	Total number of articles	International collaborations
Australia	22.1	6.5	154	34
Canada	8.4	5.9	1435	121
France	18.1	5.1	320	58
Germany	21.3	8.3	417	89
India	4.9	1.6	307	15
Japan	17.4	6.1	144	25
Sweden	19.1	9.0	173	33
UK	12.3	7.6	690	85
USA	17.5	8.3	605	106

ing elsewhere. This suggests that Canada may have found a niche in this knowledge-based industry – the know-why, know-how and know-who of varietal breeding and primary production – but that the bulk of the activities up- and downstream of that stage in the production system are now, and may continue to be, done elsewhere.

Looking first at the processes, we can observe through the US patent database that since 1985 there have been 311 patents issued for canola-related innovations, the vast majority of them to US-based research teams and companies (Table 4.12). At this level there is no evidence of any ability of research programmes to predictably or consistently deliver innovations. Rather, the evidence suggests that innovations in processes are largely like a lottery – you need to invest in research to have a chance of success, but there is no guarantee. Canada accounted for only 33 patents, most of them to research by a couple of multinational firms operating in Canada and to the federal research programmes in AAFC and the NRC. The EU comes third, with about 28 patents. As discussed further in Chapter 11, virtually all of the patented technologies used by the canola sector are the proprietary products of non-resident companies.

Moving downstream in the product development system, we can observe where adaptation and field trials are happening. Adaptation is closely linked to

Table 4.12. Patent count by geographical area. (*Source:* IBM Intellectual Property Network, Patent Database search, June 2000.)

	USA	Canada	EU	Japan	Other	Total
1980–1984	13	5	2	2	0	22
1985–1989	5	6	0	5	0	16
1990–1994	34	2	1	1	0	38
1995–1999	181	20	25	7	1	235
Total	233	33	28	15	1	311

the applied research stage, in that the results of the earlier stage need to be verified in the field. Thus, these two stages tend to go together. Looking at international field trial data, we can observe that Canada had an early and dominant lead over any competing country at this stage in the innovation process. The OECD field trial database for transgenic canola shows that although 13 countries have field tested at least one transgenic variety between 1988 and 1997, Canada accounted for 57% of all the field trials undertaken and did almost four times the number of trials of the nearest country (USA). Nevertheless, the almost overwhelming dominance of Canada has begun to diminish. By 1995 the five countries reporting field trials from the EU combined did almost as many trials as Canada and by 1997 the US was doing three-quarters the number of trials as Canada (Table 4.13).

Data on investments in varietal development is even more difficult to come by. Preliminary evidence suggests that the vast majority of the applied research to develop varieties is happening in Canada. Canada has registered 180 varieties of rapeseed/canola since the 1940s, more than two-thirds of which have been developed domestically (Table 4.14). Germany (East and West) ranks second, with 150 varieties, about 57% of them domestically developed. China ranks next with about 122 varieties developed between 1947 and 1991, of which approximately 20% incorporated foreign germplasm but only four were direct imports. Most other countries are relatively self-sufficient, with the few varieties they use coming from smaller domestic breeding programmes. Only the US, with 62 varieties introduced since 1985, has been a major importer of foreign-bred varieties (mostly from Canada).

That is all the data that is available publicly. A survey of all the companies in the industry was also undertaken. The results, which likely have missed some of the smaller entities in other countries but capture all of the main public programmes and large private efforts, show that almost half of the research on canola undertaken globally in 1997–98 is conducted in Canada, with about 31% of the research being undertaken in Saskatoon (Table 4.15). A key feature of these numbers is the proportion of the relative effort that is conducted by public-versus private-sector entities. Globally, just less than three-quarters of the effort is undertaken by private companies, with Australia and the US having a large share of their research in the private sector. In contrast, most of the research in India is in the public sector. The private sector share of the effort in Canada is at the low end of the range for the larger countries. Saskatchewan, however, is clearly the focal point for public research efforts, being the home of the *Brassica* programmes for both the NRC and AAFC. Private companies have not been as drawn to Saskatoon as location theory might suggest. Chapter 8 will discuss the motivations of private actors further.

Table 4.13. International field trials for transgenic canola.

	1988	1989	1990	1991	1992	1993	1994	1995	1996	1997	1988–1997
Australia					1					–	1
Canada	7	27	39	29	23	53	63	57	76	95	469
European Union	1	4	13	13	30	20	46	56	24	3	210
Belgium			*5*	*5*	*13*	*5*	*4*	*4*	*2*	*1*	*39*
Denmark					*1*		*1*			*–*	*2*
Finland									*1*	*–*	*1*
France	*1*	*3*	*6*	*0*	*12*	*6*	*16*	*22*	*8*	*–*	*74*
Germany						*1*	*9*	*6*	*4*	*–*	*20*
Netherlands			*1*	*1*	*0*	*1*	*4*	*0*	*1*	*1*	*9*
Spain						*1*				*1*	*2*
Sweden		*1*	*1*	*1*	*1*	*1*	*2*	*2*		*–*	*9*
UK				*6*	*3*	*5*	*10*	*22*	*8*	*–*	*54*
Japan							1	3	1	5	10
USA				5	7	8	16	11	18	71	136
Total	8	31	52	47	61	81	126	127	119	174	826

Source: OECD Database of Field Trials http://www.olis.oecd.org/biotrack.nsf

Table 4.14. Dispersion of rapeseed and canola varieties by country.

	Period	Total number of varieties	% Domestically bred
Australia	1998	19	95
Canada	1944–28 June 1998	180	67
China	1947–1991	122	97
Germany	1944–1998	150	57
India	1992–1996	14	100
Sweden	1944–1998	92	76
UK	To 1998	22	na
USA	1985–1998	62	21

Sources: Canada CFIA Seeds Act Registrations; USA website; China, India website; Australia, website; UK Plant Variety Protection office.
na, not available.

Table 4.15. Estimated distribution of canola-related employment by country and type of institution, 1998.

	PY devoted to canola-related research	As % of global total	Private share (%)
Canada	398	51	60
of which Saskatchewan	247	31	46
EU	242	31	77
USA	84	11	76
Australia	24	3	75
Rest of world	38	4	24
Total world	786	100	74

Source: Data collected by P. Phillips through survey of industry during 1997–1998.
PY, person-years.

Adoption

The rate of adoption of new varieties is governed by a number of factors that tend to relate closely to the individual sites. The regulatory system clearly determines how quickly farmers can and will adopt new varieties. Although the Canadian regulatory system is generally viewed as somewhat slower than the US system (although this lag may have diminished with the experience gained with early transgenic varieties), it is clearly faster than either in Europe or Australia, where, taking transgenic varieties as an example, no transgenic canola varieties were approved as of June 1999. The first transgenic varieties were approved for unconfined release in 1994 in both the US and Canada and began commercial production in 1995. Beyond that, farm programmes partly determine the pace of adoption of new varieties. In Canada, new varieties of canola are automatically eligible for crop insurance and other stabilization assistance, without limits on acreage seeded or location. In the EU, where the Blair House Accord and the WTO agreement limit canola acreage, incremental planting is treated less favourably than existing acreage, which limits its adoption in some countries and regions. Furthermore, canola is not yet suited agronomically to most parts of the US and Australia, which limits adoption there.

Farmers respond to varieties partly based on the regulations, partly based on the markets and partly based on their prior knowledge and experience in adopting new varieties. The openness of the regulatory system and extensive experience in Canada with canola and with new canola varieties makes farmers more receptive here than in any of the competing markets (see Chapter 7). Ultimately, the market opportunities and prices will determine uptake of new varieties. So far in the 1990s there have not been any constraints flowing from these. The result has been the rapid adoption of new varieties. Once limits on HT canola production were removed in 1996, farmers rapidly adopted these new varieties. An estimated 50% of the acreage in 1998 was planted to HT varieties and some estimate that up to 75% of the canola acreage in 1999 went to HT canola.

Conclusions and Future Trends

One can conclude from the foregoing that canola exhibits many of the features predicted by theories of innovation and growth and that Canada has carved a niche in the knowledge development part of the industry but is not the only significant actor. Universities and public research laboratories elsewhere, and multinational companies operating both in Canada and abroad, are now significant developers of pure agricultural science, technology and varieties.

The obvious next question is what does the past tell us about the future? So far in the canola sector the vast majority of innovations has been targeted on agronomic improvements or general product attributes, such as oil content. Now the focus is beginning to shift to targeted agronomic improvements and specific product attributes that have the potential to expand the market for canola-based products. On the agronomic side, HT varieties are now in the fields and account for a major share of the market. Given the success of HT varieties, seed and chemical companies have gone to work on other agronomic factors. Between 1994 and 1999, there were 156 field trials for transgenic varieties that involved introducing stress, insect, virus and fungal resistance into canola (CFIA, 1999). The next, and possibly ultimate, phase of development has also begun. End users, such as Procter & Gamble, Nabisco, Frito Lay, Lubrizoil and Ciba Geigy, among others, are showing more interest in canola oil for its nutritive value as well as for its value as a vector for other transgenic features. In the 1994–1999 period in Canada there were 290 field trials of transgenic varieties of canola that have been manipulated to modify the oil composition, change the nutritional balance of the seed or to produce nutra- or pharmaceutical products (CFIA, 1999).

As far back as the 1960s, a number of industrial users demanded varieties with novel characteristics, such as high erucic acid or low linolenic oils. The change has been that biotechnology has now created the opportunity of adding significantly more novel traits, such as engineered fat chains, industrial oil genes (e.g. laurate), nutraceutical characteristics or pharmaceutical proteins and enzymes. The possibilities appear almost endless. Canola, for example, can be modified to produce a wide variety of products: low-value, commodity, end-of-the-scale, proteins for improved nutritional value of the seeds; intermediate-value, bulk proteins such as industrial and food enzymes; and high-value proteins, mainly of interest to the pharmaceutical industry.

The challenge of mapping and measuring the evolution of this new market is that much of the pure agricultural research is now being paid for by contracting end users, so that academic citations analysis may miss any references to this work until well after the work is being applied in varietal development. This will force analysts to look to different indicators of knowledge-based innovation, particularly patents and field trials, both of which are, at best, coincident indicators of development.

The Actors

The Evolving Industry

5

Peter W.B. Phillips

Introduction

Successful, sustainable innovation requires that the capacity to innovate be matched with an ability to take the results of any innovative process and position them in the marketplace in such a way as to capture a return that both compensates for the investment in the innovation and yields a positive return to the risk taker. Those two different abilities do not always go together. There has been extensive research in other countries and in other product areas to determine the elements of such a sustainable system.

The political economy literature has labelled this the 'national systems of innovation'. Metcalfe (1995) defines an NSI as comprising 'that set of distinct institutions which jointly and individually contribute to the development and diffusion of new technology and which provides the framework within which governments form and implement policies to influence the innovation process. As such it is a system of interconnected institutions to create, store and transfer the knowledge, skills and artifacts which define new technologies.' Mowery and Oxley (1995) point out that any definition must include more than the research actors, but also must include public programmes intended to support technology adoption and diffusion, and the array of laws and regulations that define intellectual property rights and manage the discovery, production and marketing systems.

The OECD (1992) points out that the first step in defining such a system empirically must be to locate the boundaries, the component institutions and the way in which they are linked. Many institutions are involved, including private firms working individually or collectively, universities and other educational bodies, professional societies, government laboratories, private consultancies, industrial research associations and other collaborative ventures, and related and supporting industries.

The next challenge is to identify how the institutions are linked. Metcalfe (1995) argues that 'in practice, connectivity is achieved via a variety of mechanisms. Mobility of scientists and technologists in the labour market and collaboration agreements to develop technology are important formal mechanisms linking firms. Links between firms and universities are often instituted through grants and contracts for research, especially in the transfer sciences', such as plant breeding and computer science. Those and others will be explored in this section of the study.

The combination of the actors and the connective structures makes up the innovation system and effectively determines the 'absorptive capacity' of an economy to exploit either domestically developed or imported technologies (Mowery and Oxley, 1995). This capacity includes a broad array of skills, reflecting the need to deal with both the explicit and the tacit components of the new technologies.

The Actors

As noted by the OECD (1992), there are many actors involved in a national system of innovation. For the purposes of this study, they have been divided into larger categories: the public sector, the private sector firms, the collaborative associations, and the related and supporting institutions, including farmers, servicing and supporting industries and the people who cross-over the institutions, creating the fine web of tacit linkages that make the system work.

Firms, motivated by profit, are the key to a sustainable innovative sector. Their search for market advantage creates an inexorable drive to invest and search for new processes, new products or new varieties. As they pursue new opportunities, they operate as innovative, fast-growth enterprises; if the innovative effort matures, the firms will almost inexorably go through a variety of stages ranging from competitive firms to oligopolies to mature, often declining, firms (Table 5.1). The more innovative the firms, the more tied or attracted they are to skilled labour pools or potential collaborators or competitors. As the firms mature, their attachment to their research location can weaken, often to the extent that relocation is feasible. For that reason, regions desiring to build sustainable industrial capacity are increasingly driven to support and nurture sustained innovation as a fundamental part of any targeted industry.

Metcalfe (1995) notes that Malerba's study of Italy in 1991 identified two discrete, independent systems of innovation. One, typified by the computer software industry, is based on flexible networks of small and medium-sized firms, often co-located in distinct industrial districts (such as Silicon Valley). These firms exhibited both significant volatility and rapid growth. The other type of system, which perhaps better reflects both the system in canola so far and likely all agri-food research, is based on the universities, public research laboratories,

Table 5.1. A neo-Schumpeterian model of industrial development. (*Source:* Lundvall, 1992.)

Factor	Innovative stage	Competitive stage	Oligopolistic stage	Declining stage
Localization pattern	Close to skilled labour pools or founder's residence	Firms attracted to least-cost site	Relocations retarded for early firms but encouraged at later stages	Relocation from old plants to modern plants in new regions
Importance of proximity	Agglomeration economies high	Proximity declining	Firms operate in larger input and output markets	Low
Growth	High	High	Low	Negative
Technological development	Innovation is key	Products are standardized; economies of scale	Product differentiation and process development	Limited; profit maximization dominates

and large firms performing and commercializing R&D. Metcalfe (1995) further argues that, regardless of which model prevails, no institution can be, or is, self-contained in its technological activities. All firms, large or small, have to rely on knowledge from other sources. Systems that support a firm's ability to access, absorb and use external knowledge can be critical to the growth of firms, sectors and regions. This is especially so in the early stages of the development of a technology or whenever a technology has a rapidly changing knowledge base. As shown in the previous section, this is clearly the case in canola development to date.

Although Metcalfe (1995) hedges many of his conclusions about innovation systems, he states firmly that while firms are the primary actors in the generation of technology 'their activities are supported by the accumulation of knowledge and skills in a complex milieu of other research and training institutions'. At the core of this network are universities, which create fundamental knowledge and invest in 'transfer sciences' (e.g. computer science, plant breeding), each one tied to identifiable technological activities while drawing on insights from a range of fundamental disciplines. In addition, as discussed in Chapters 4 and 7, universities educate and train the workers in these industries. Universities therefore create new knowledge, act as repositories of the stock of knowledge and disseminate that knowledge both directly through commercialization efforts and indirectly through graduates (Metcalfe, 1995).

Chapter 7 contains an examination of the critical role played by public research programmes and agencies in the development of canola and in the development and use of transfer sciences such as plant breeding, genomics and other know-how technologies underpinning the innovative process in the canola sector. Public research laboratories, however, are not the only public actors in the canola research effort. A number of other development and regulatory agencies also have managed development of the industry through the regulatory process, using legal and policy mechanisms to moderate incentives, set standards and manage the processes of discovery, production and marketing. All of those are critical elements in a successful innovation system. A discussion and examination of those roles is left for Part IV of this volume.

Other actors, including private consultancies, professional associations, industrial research collaborations and collective action groups all can, and in the canola case clearly do, play significant roles as bridging institutions between both industry and academe and between industry and markets.

The Patterns of Interaction

The second issue addressed in the following four chapters is how these actors operate and interact with others. Zilberman *et al.* (1997) undertook a conceptual analysis of agricultural biotechnology, proposing a five-stage linear development process (discovery to marketing) that is consistent with the chain-link innovation model if the knowledge links are removed (i.e. if the model is looked at in only two dimensions). They then looked at the California biotechnology industry in search of different patterns of the division of responsibility between entities.

Taking that structure and applying it to the canola sector since the 1950s, one can trace an evolution of the leadership from public laboratories to private corporations (Table 5.2). Pattern one, representing the period between 1950 and 1985, was characterized by public agency leadership, through AAFC and NRC. As discussed earlier, in this period the innovation system probably was reasonably represented by the linear model, as there was little collaboration beyond the core public agencies and a few universities. Pattern two began to emerge in the 1970s as the not-for-profit Canola Council of Canada worked to expand the number of research actors. Universities (Manitoba, Guelph and Alberta) began during this period to develop the capacity to develop new canola varieties. This pattern still mirrored pattern one, with a relatively linear development path and limited collaborations. In both cases, after the varieties were registered, the seed was marketed to farmers through spot sales, and farm output was purchased and processed or marketed by downstream companies that had little or no interest in the research system.

After 1985, three new patterns emerged. Patterns three and four replaced patterns one and two, with corporations increasingly setting the objectives of the research programmes and then conclusively taking over responsibility for the registration, multiplication and sale of the seeds to farmers. Downstream, however, the system remained unchanged, with farm output continuing to be

Table 5.2. Leadership responsibility for various stages of product development. (Adapted from Zilberman *et al.*,1997.)

Pattern	Period	Discovery	Development	Registration	Production	Marketing
1	1950–1985	Public laboratories/CCC	Public laboratories	Public laboratories	Farmers	Corporations
2	1970–1985	University/ CCC	University	University	Farmers	Corporations
3	1985–date	Public laboratory	Public laboratory	Corporation	Farmers	Corporations
4	1985–date	University	University	Corporation	Farmers	Corporation
5	1985–date	Corporations	Corporations	Corporation	Corporations	Corporation

CCC, Canola Council of Canada.

marketed by companies for the most part unrelated to the innovation system. Pattern 5 represents a truly new departure from past experience. Private corporations are increasingly assuming leadership and responsibility for all of the stages of product development. This pattern reflects the dominant role of the agrochemical and global seed companies in the breeding business in recent years. Based on proprietary market assessments, these large firms undertake the discovery, development and registration steps, either on their own or through tightly controlled collaborations where they establish the research objectives, pay for the research and assume ownership of the resulting products. In many, but not all cases, firms have taken control right into the production and marketing stages, at times supplanting or absorbing traditional seed merchants and marketers. These companies have developed or acquired seed marketing arms, offering seed to farmers often only on a closed-loop contract system (especially when the seed has some differentiated product attribute, such as modified oil properties), with the corporation buying back the farm product and either marketing or processing it itself.

Trust – the Intangible Factor

As Joseph Stiglitz, Chief Economist of the World Bank, has noted (1999), 'it has long been recognized that a market system cannot operate solely on the basis of narrow self interest. The information problems in market interactions offer many chances for opportunistic behavior. Without some minimal amount of social trust and civil norms, social interaction would be reduced to a minimum of tentative and distrustful commodity trades.' Arrow (1988), Putnam (1993), Fukuyama (1995) and a wide variety of others have examined the role of norms, social institutions, social capital and trust in creating and sustaining markets. Trust is not simply utilitarian economics. It is based on common history, family ties and political institutions – in short, on a sense of community – which are based on many motivations, of which economics is only one. This study will, in a variety of places, examine the history, institutions and communities that influenced the development of canola. Although there is no econometric test to ascertain the degree of causality, we are convinced that trust is one of the most critical elements in the creation of knowledge-based centres and industries (Smyth and Phillips, 2000). This factor is examined in Chapters 7–10.

Outline for Part III

This section uses the 'national systems of innovation' structure, augmented by the 'new' institutional economics (NIE) to deal with the economics of institutions and institutional change. Unlike traditional theory, NIE pays attention to the determinants and the evolution of different institutions and contracts over time. This approach focuses primarily on the costs of alternative types of transactions. In a great many instances in the marketplace, a simple exchange of goods and services at an agreed-upon price is a low-cost transaction that provides the correct incentives for the buyer and sellers. But transactions are seldom without cost. When the marketplace fails, then a market failure is said to exist and institutions evolve to overcome market failures. Particular institutions tend to be best suited to govern particular types of transactions. In Chapter 6, the theoretical dimensions of this approach will be discussed further and applied to the canola sector in Canada, with a particular focus on the role of collective institutions in furthering industry development. The evolving role of the public sector is examined in Chapter 7. Chapter 8 is devoted to the rise to dominance of the private sector. Chapter 9 contains analysis of the influence of other institutions – such as local suppliers, the labour market – and the idiosyncratic role of individual leaders on the location of activity.

Industrial Development and Collective Action

6

Richard S. Gray, Stavroula T. Malla and
Peter W.B. Phillips

Introduction

The best place to start an investigation of the industrial structure of the canola industry is with the ever-shifting relationships between the not-for-profit associations and the public and private sectors. Although the private and public sectors have contributed the greatest amount of effort over the history of the sector and control most of the results of that effort, the extent and array of collective action pursued by firms, producers and other actors have, to a great extent, defined the evolution of the industry.

Collective action tends to arise whenever the markets or the public sector have been unable to act alone. Industrial development is one explicit case where various private or government investments and activities can be, and regularly are, held up. The development of an industry inevitably involves the provision and delivery of a large number of factors that are critical for the success of the sector. Storey (1998) in his examination of the development of the soft-fruit industry in Atlantic Canada, argues that when starting from scratch, a new agri-food sector requires: research and development to create the product; extension work with producers to enable them to adapt and adopt the new crop; marketing infrastructure to distribute product-specific inputs (e.g. seeds) and to assemble and move the resulting product into the marketplace; information on markets, including supply, demand and price discovery mechanisms; market development; and standardizing and grading of the product. Many of these are often missing in new industries.

A number of economists have argued that many of these features are not forthcoming because of a variety of market failures. Incomplete markets, public-good attributes, common pool resources and technical externalities all inhibit development. In addition, if investors believe they have a weak post-investment bargaining position with respect to users of their transaction-specific investments, and as a result will not be compensated adequately for their investment, then they are likely to withhold investment. These problems that cause investors to hold up investments are more often than not the result of incomplete institutions.

The development of an industry inevitably involves the provision and delivery of a large number of factors that are subject to market failures. Taken together, the up-front costs of introducing a new crop are large and it is highly uncertain that any individual investor acting independently could recoup any early investments in industry development. In the case of canola, some of the elements were most effectively provided by the public sector, as the new crop was able to piggyback on capacity already developed and being managed by public agencies. For example, adding canola to the ongoing crop reporting system was far less expensive than developing an entirely new system. By the same logic, Agriculture and Agri-food Canada and the Canadian Grains Commission were well placed to extend their regulatory capacity to canola (e.g. Seeds Act). Other parts of the system represented little asset specificity and therefore presented few concerns about opportunistic behaviour. Private firms were able to take on those tasks, at times totally independently and at other times with either implicit or explicit agreements with others. This included the provision of the logistics and marketing system for providing inputs and moving outputs to markets, and the provision, by the Winnipeg Commodity Exchange, of canola trading contracts and options.

A number of critical elements, however, were not forthcoming from individual industry participants due to market failures. First, little research and development was done by industry before there was effective legal protection for the innovations (ultimately granted via Plant Breeders' Rights in 1990). Second, market development, which benefited everyone, would not have been done by any one participant alone as there was no way to exclude others from the markets that were developed. Third, extension to encourage greater farmer adoption of canola varieties, equally, was in the collective public and industry interest but unlikely to be provided by any private firms as there was no way to earn a share of the benefits of an expanded market. Fuglie *et al.* (1996) presented a series of studies that showed that the annual returns to extension ranged from 20% to 110%, which suggests that too little extension is undertaken, even in fully developed product markets. These gaps were compounded by the absence of intellectual property rights and the underdeveloped state of the market.

In this chapter the history of the development of the canola industry is examined to highlight the critical role of collective institutions in its development, both to illustrate the role that not-for-profit associations have played in furthering industrial development and to assess how these new institutions may need to adjust, given the significantly changing role of the private sector in discovering, producing and marketing new canola varieties.

The Theory

The new institutional economics (NIE) framework is employed in this and subsequent chapters to interpret the existing structure of institutional and contractual arrangements used to fund canola research. The creation of wealth is heavily influenced by institutions. NIE, a field of economics that deals with the economics of institutions and institutional change, pays close attention to the determinants and to the evolution of different institutions and contracts over time. According to Jacquemin (1987), 'hierarchies, federations of firms, and markets compete with each other to provide coordination, allocation and monitoring. It is only when one organizational form promises for specific activities a higher net return than alternative institutional arrangements that it will survive in the long run.' Transaction costs often provide a clear signal of institutional efficiency. Williamson (1979) argues 'if transaction costs are negligible the organization of economic activity is irrelevant, since any advantages one mode of organization appears to hold over another will simply be eliminated by costless contracting'. However, as Coase (1937) pointed out, the price mechanism is seldom free of costs; different firm structures exist to manage and reduce the highly variable and uncertain costs of contracting.

In the transaction-cost economics, the control variable of analysis is the transaction. According to Williamson (1981), a transaction is said to occur when a good or service is transferred across a technologically separable interface. For instance, transactions occur when one stage of processing or assembly activity terminates and another begins. This usually happens between institutions, but in some of the more sophisticated companies this can happen within single institutions.

Transactions are not without cost. Dahlman (1979) identifies three cost elements: search costs (the costs of locating individuals whose reciprocal relative subjective valuations make them candidates for a transaction); negotiation costs (the costs of reaching agreement on the terms of exchange); and enforcement costs (the costs of monitoring the performance of parties to the transaction). Hence, transactions incur both *ex ante* negotiating costs and *ex post* compliance costs. According to Williamson (1985), it is the difference between transaction costs rather than their absolute magnitude that really matters and, most critically, the optimal institution will minimize the sum of both production and transaction costs rather than simply transaction costs – trade-offs can be, and are, recognized.

The principal dimensions in which transactions differ, according to Williamson (1979), are their uncertainty, the frequency with which they recur and their asset specificity. Uncertainty arises either from lack of communication or from strategic behaviour. In the former, a decision maker usually cannot anticipate concurrent decisions made by others and is thereby limited by 'bounded rationality'. In the latter, uncertainty may arise from strategic non-disclosure, disguise or distortion of information, all forms of opportunistic behaviour. The frequency of a transaction also bears on its cost: occasional transactions generally do not warrant specialized mechanisms while recurring transactions are often governed by either short- or long-term contracts. Finally, assets which have a lower value in alternative transactions create higher risks of opportunistic behaviour.

Asset specificity is particularly important under conditions of bounded rationality, opportunism and uncertainty (Williamson, 1985). There are four types of asset specificity: site (e.g. successive stations are located close to each other in order to economize on inventory and transportation expense), physical (e.g. specialized dies to produce a car part), human (e.g. specific skills that employees gain from learning-by-doing) and dedicated (e.g. discrete investment in generalized production for the prospect of selling to a specific customer).

This all comes to bear in industrial development because arm's-length markets are unable to strike contracts that minimize the specific risks and uncertainties inherent to the investments. In a competitive marketplace made up of many informed buyers and sellers, market exchange is an institution that effectively governs the production and consumption of goods and services. The prices generated in a market create Adam Smith's 'invisible hand' to match the marginal cost of providing a good to the marginal value of that good to society. In a great many instances in the marketplace, a simple exchange of goods and services at an agreed-upon price is a low-cost transaction that provides the correct incentives for the buyer and sellers. When the marketplace operates in a manner such that the marginal social benefit is not equal to the marginal social cost of the transaction, then a market failure is said to exist.

Those market failures from standard economic theory most relevant to this study are associated with public goods, common pool resources and technical externalities. Markets fail to provide adequate public goods because no one can be excluded from their consumption and, hence, there are no feasible means for a firm to charge the users for the provision of the goods. Common pool resources also suffer from the lack of exclusion, in that the resource is 'subtractable' or rival and overuse can result in the depredation of the resource. Both positive and negative technical externalities, such as knowledge or pollution, also represent market failures because they are unpriced in the market. The key factor in each of the market failures is the lack of marginal cost pricing, often due to the inability of producers to exclude consumers from using their good without paying the price.

One market failure that has recently attracted attention in the investment literature is referred to as the hold-up problem (Williamson, 1983; Grossman and Hart, 1986; Tirole, 1988; Choate and Maser, 1992; Koss *et al.*, 1997). Milgrom and Roberts (1992) define it as 'the general business problem in which each party to a contract worries about being forced to accept disadvantageous terms later, after it has sunk an investment or worries that its investment may be devalued by the actions of others'. With asset-specific (specialized) investments, the value of the asset in its intended use is far greater than its value in the next-best use. In order for the initial specific investment to be undertaken the real rents to each party (returns in excess of the investment) must not be negative. However, when a investor can see that its post-investment bargaining power is weak and that it is less likely for the party to cover its initial investment, it will be unwilling to incur the initial investment cost. Hence, if the cost of the investment is large and the post-investment return is risky, the initial investment may not be undertaken by that party. Thus, market failure occurs as the specific transaction is Pareto superior to all alternative transactions.

Institutions are essential to solving market failures. They encompass a set of rules, both formal (e.g. statues) and informal (e.g. norms), which constrain the behavioural relationship among individuals or groups (North, 1991). Institutions are not simply nominal rules, they are defined by those effective rules which can be enforced (Eggertsson, 1994). Institutions can be established, enforced and policed either by an external authority or they can be voluntarily accepted, but the key is that they are predictable, stable and applicable in repeated situations. Ultimately, institutions define decision makers' utility choice sets and responses to incentives. Property rights are a profound example of an institution which creates economic incentives (e.g. private companies to invest in R&D) (Eggertsson, 1994).

Particular institutions tend to be best suited to govern particular types of transactions. Picciotto (1995) classifies institutions into three general types: hierarchy or the government sector; the private sector; and the participatory sector (Fig. 6.1). First, the hierarchy sector or government, whose stockholders are all citizens of a country, seeks to gain and maintain power, so it pursues policies (usually in the best interest of society, or at least groups in society) that contribute to re-election. Second, the private sector owns property and seeks to maximize their profit on that investment. Third, the participation sector involves those who voluntarily join to obtain the benefits of collective action (Olson, 1965). Participants in collective ventures either seek to put forward their views and ideas or to pursue more material goals that cannot be realistically obtained through individual action. In a research context, this need for 'voice' involves coordination among multiple researchers and between researchers and the production and marketing chain. Each sector represents different individuals, involves different incentives and is effective in producing goods or attributes with specific characteristics. The government sector produces public goods (e.g. justice, defence, public health) usually characterized by low excludability,[1] low rivalry[2] and low voice,[3] that are involuntarily consumed by all citizens equally (Table 6.1). On the other hand, the private sector provides market goods (e.g. farming) which exhibit high excludability, high rivalry and low voice, and are consumed voluntarily by individuals. In contrast, the participation sector specializes in common pool goods (e.g. market development services) with low excludability, from low to high rivalry and high voice (e.g. coordination).

[1]Excludability is a circumstance where individual consumers can be excluded without incurring substantial cost.

[2]Non-rival, or low subtractable, goods are ones where the consumption by one person does not diminish the ability of other persons to benefit from the good.

[3]Voice is the ability of members in a sector to have their opinion heard by those who make decisions.

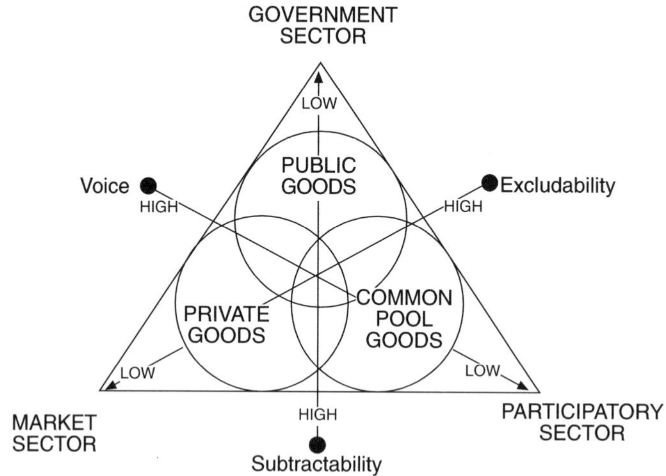

Fig. 6.1. The nature of goods and services (from Picciotto, 1995).

Table 6.1. Taxonomy of attributes for goods produced by different institutions.

	Excludability	Rivalry	Voice
Government-provided public goods	Low	Low	Low
Privately produced market goods	High	High	Low
Association-produced collective goods	Low	Low to high	High

At the practical level, institutional economics offers a number of routes for actors in an industry to overcome market failures arising from potential hold-up. Williamson (1983) suggests that common ownership (e.g. vertical integration) can address asset specificity. Klein and Crawford (1978) have observed that 'integration by common or joint ownership is more likely the higher the appropriable specialized quasi rents of the assets involved'. Joskow (1987) suggests that, for many types of asset specificity, long-term explicit contracts can reduce the potential for *ex post* hold-up. The problem, however, is that it is often very costly to identify all the contingencies of those investments. Williamson (1983) argues that an alternative solution for transactions (potentially) subject to hold-up is for the potentially opportunistic party to make a credible *ex ante* (upfront) commitment or investment to the exchange. This commitment usually takes the form of partial redistribution of specific investment costs to the investor in the specific asset, to compensate for anticipated incomplete *ex post* capture of the benefits of that investment. This has occurred explicitly in the canola sector, where the government, the Canola Council and the growers' associations have collected resources and made up-front payments to the research community to undertake the effort, knowing that the market would not provide adequate returns for private investors.

In short, there are significant lessons for development practices that can come from institutional economics (Nabli and Nugent, 1989; North, 1989). Development, as a long-term incremental process involving investments in specific assets with uncertain paybacks, depends critically on predictable, effective institutions to mobilize resources, to develop objectives and to produce benefits. It is crucial that the right institution be in place for the stage of development.

The rest of this chapter is devoted to an examination of the ebb and flow of public, private and not-for-profit activity in canola development, in order to identify and define how the roles for the public and the participation sectors have shifted when confronted by a larger, more proactive role for private companies.

The Institutional Story of Canola

The institutional story of canola spans four relatively distinct periods, where different market imperfections led to different institutions. The glaring need for market development at the beginning of the story, combined with the absence of intellectual property rights, led to a largely publicly led development effort. In the early years almost all of the resources invested and all of the resulting outputs (i.e. varieties) were from the public sector (Table 6.2).

As more institutions were developed to overcome the market failures (i.e. not-for-profit collective associations, contracts, hybrids and finally intellectual property rights), the role of the public and not-for-profit sectors shifted, giving way to the private sector. Tables 6.2 and 6.3 show that by the 1990s the private sector dominated the industry, contributing approximately 60% of the total investment and controlling 86% of the resulting varieties, and more than 80% of the new technologies. Partly because the sector is now dominated by private actors, there has emerged a number of new potential hold-up problems that may rejuvenate and refocus the public and not-for-profit sectors in the coming century.

Table 6.2. Investment in canola research, by funding source, over selected periods.

Investment by	1944–1966	1967–1973	1974–1989	1990–1998
Public sector	85	80	58	39
Private sector	15	17	40	59
Associations	0	2.5	2.2	2.0
Average global industry investment per year (1998 Can$)	$3 million	$16 million	$38 million	$98 million

Source: Authors' calculations from the canola industry survey, 1997–1998.

Table 6.3. Attribution of new canola varieties to public and private sector over selected time periods. (*Source* CFIA, 1998a.)

New varieties by	1944–1966	1967–1973	1974–1989	1990–1998
Public sector	6	7	15	23
Private sector	0	0	15	139
Total	6	7	30	162
of which hybrids	0	0	1	14
% Distribution				
Public sector	100	100	50	14
Private sector		0	0	50
86				
of which hybrids	0	0	3	9

First Stage: 1944–1967

In Canada, canola is probably the most recent and pronounced example of how R&D can result in the rapid ascent of a crop to a multi-billion dollar status from humble origins. The initial impetus for development came from the public sector. As discussed in Chapter 8, the public sector sought both a Canadian-sourced edible oil and a new crop to diversify western Canadian agriculture.

During this early period rapeseed was not established as an economic product and almost all its attributes were unknown (McLeod, 1974). There were no significant quantifiable returns to research to be captured by a research effort, which made it impossible to justify private research investment. Furthermore, with no legal protection or effective corporate strategies for managing intellectual property rights, the private sector left the business of research and development in the agri-food sector to others. The almost complete absence of any formal institutions left the public sector with virtually total control over the development of the sector (Table 6.4).

The result was 20 years of public research into rapeseed, involving Agriculture Canada, the NRC and a number of universities, with only a small amount of effort by a few selected Canadian companies. Research funding in the early years came almost exclusively from the public sector, and the research activity and the resulting product resided in the public domain. New technologies and new varieties were released for use without any restrictions. The only significant role played by private companies was in the area of oil processing, with a low level of participation by Edible Oils Ltd, Saskatchewan Wheat Pool and Canada Packers. Nevertheless, information, including germplasm and technologies, freely moved between the parties involved in the search. All information developed for the crop was shared throughout the industry. Researchers in various disciplines and agencies examined the problem and worked collectively. The first years of rapeseed research were characterized by general research into the agronomics of rapeseed and its properties as an edible oil and animal feed. The research was fundamentally public in character – there were no significant

Table 6.4. The role of the three sectors, 1944–1966.

	Research investments	Share of research results	
	% of total inputs	% new technologies	% new varieties
Government and universities	85	95	100
Associations	0	0	0
Private companies	15	5	0

Source: Data collected by P. Phillips through survey of industry during 1997–1998.

or predetermined capturable returns. The small return from this early research (in terms of yield and improved oil and meal content) was dispersed between growers and all other social groups. The public, as consumers, was the main beneficiary; producers only captured a small portion of the returns, which raised fundamental concerns about the industry's capacity to become self-sustaining. In short, in that period very basic research was driven by government and the industry was dominated by government priorities and action.

The first stage ended auspiciously, with the development of standards for rapeseed oil in 1965 through the Edible Oil Institute. Samples of oil from the four western crushers were examined in six refiners' laboratories. Specifications – for free fatty acids, moisture and impurities, flash point, refined bleached colour, green colour in crude oil, refining loss and phosphatide content – were approved and published by the Canadian Government Specification Board (McLeod, 1974).

Second stage: 1967–1973

The late 1960s brought a fundamental change in the development of the sector. The slow but steady development by the public sector of rapeseed as an oil crop had reached a threshold, where more investment in both product development and in market structures (e.g. extension, foreign market development) was required, but no single institution (public or private) had the means or incentive to undertake the work alone. The industry faced a true hold-up problem, with the benefits of any individual's investments likely being shared with a wide variety of free-riders.

Necessity became the mother of invention. The absence of adequate individual incentives or an effective institution to develop the industry further spurred a collective response, culminating in the creation of the Rapeseed Association of Canada (RAC). Established in 1967 to serve as an umbrella organization for the groups that had a stake in the Canadian canola industry, the RAC played an increasingly significant role in development of the industry over the next two decades. The impetus for the establishment of the Association came from all parts of industry as they realized the potential of the crop and the need for some central body to work for the development and betterment of the rapeseed industry. In short, the industry came to recognize that a number of investments were needed to secure the development of the sector that no one entity had the wherewithal or incentive to provide. The campaign to launch the organization began in 1966, and by 1967 the new organization was granted a federal charter to deal with three key hold-up problems – market development, product research and extension.

The Rapeseed Association, as a non-profit organization, for a period of time was effective in mobilizing a relatively small amount of financial resources from industry and government, which came to leverage and direct a large volume of activity in support of industry development. At the beginning, 70% of the Association's budget came from crushers and exporters through a voluntary Can$0.50 per tonne levy on rapeseed exports and seed crushed domestically. Within a short period of time government contributions rose to almost half of the resources. In conjunction with the Association, the federal department of Industry, Trade and Commerce set up the Can$1.25 million Rapeseed Utilization Assistance Program, which was funded by the federal department but administered by the research committee of the Association (McLeod, 1974). Together, the levy and government funds comprised the major source of funding for the Association until the mid 1980s, when private sector interest in breeding expanded, increasingly through collaborations with the Council, and growers introduced their own check-offs to fund research and programmes.

In the early years, the Council devoted the majority of its resources and effort to research into new varieties that were low in erucic acid and glucosinolates. By 1968 Downey (AAFC) and Stephanson (University of Manitoba) had bred Oro, the first low erucic acid *B. napus* variety, and in 1971 Span, the first *B. rapa* variety, was released. The timing was fortuitous as in 1970 at a conference in St Adele, Quebec, a team of European scientists presented the results of a health study that showed that high-erucic rapeseed oil consumption by young animals caused a short-term fat build-up around the heart and kidneys, which appeared to cause long-term muscle lesions of the heart. The future of the rapeseed industry was in serious doubt. Both Japanese and European buyers were uncertain about continuing to use Canadian rapeseed.

The new association was instrumental in resolving the problem. Their major focus was to provide financing to Agriculture Canada and the public universities to fund research into new varieties of rapeseed with low erucic acid and low glucosinolate. During these years the Council increasingly set the research direction for the whole industry and coordinated the research activities in the various public laboratories. The Council did not do any of its own research; rather it allocated funding to support and leverage research by others. In spite of the significant flow of funds into the public research effort, the government sector continued to dominate the research effort, mainly because much of the work was done by public sector scientists in public laboratories (Table 6.5).

Perhaps more importantly for the long-term future of the industry, follow-

Table 6.5. The role of the three sectors, 1967–1973.

	Research investments	Share of research results	
	% of total inputs	% new technologies	% new varieties
Government and universities	80	95	100
Associations	2.5	0	0
Private companies	17	5	0

Source: Data collected by P. Phillips through survey of industry during 1997–1998.

ing the St Adele conference the Association took the lead in market development, crop production and public relations, in an effort to more firmly position rapeseed in the market. Given the problems with high erucic rapeseed oil identified at the 1970 conference, it was vital for the Canadian industry to adopt as quickly as possible the recently available low-erucic varieties. Due to the extensive efforts of Agriculture Canada and the Association, the changeover to low-erucic varieties was 86% complete by 1973 and 95% complete by 1974 (NRC, 1992). The Association noted that the European studies that had threatened rapeseed in 1970 had a silver lining – they confirmed that low-erucic rapeseed oil did not have any health problems – which it was able to use as evidence to encourage farmers to rapidly adopt the low-erucic varieties and as part of the active campaign to market the product to major oil users. Given that the Association did not engage in actual market transactions or handling of the product, and did not take a position on the marketing system, it was able to act as a credible voice in the market. Without the efforts of the Association, it is highly unlikely that any of the firms or actors in the sector would have been able to put together the necessary package of programmes both to push rapeseed research forward at that critical juncture and to lay the groundwork for expanding production and export markets. The market development problems were simply too large.

A small but significant development in this period was the effort by growers to form separate provincial associations to mobilize producers to have a say in the development of the sector. These associations in Saskatchewan (1969), Manitoba (late 1960s), Alberta (1970s) and Ontario (1988) started small, with limited funds coming from producer membership fees. Their primary purpose when they started was to focus on extension, agronomy and policy development, in order to accelerate development of the sector.

During the period, rapeseed made significant strides in becoming the oil of choice in Canada. In 1967 rapeseed accounted for approximately 25% of the total edible vegetable oil consumed in Canada. Soybean oil accounted for 38%. By 1973 the situation had reversed, with rapeseed accounting for 41% of total consumption and soybean oil accounting for only 30% (McLeod, 1974).

Third Stage: 1974–1990

During the 1974–1990 period the traditional commodity-based canola industry matured. The institutions established in earlier years provided an adequate level of research, extension and market development, so that by 1988 approximately 4.2 Mt of seed were produced on 9.1 million acres in western Canada, and canola had earned regulatory and market acceptance as a premium oil in all the major export markets. Clearly, the institutions fit the needs of the industry for that period, effectively overcoming the market failures apparent in the earlier periods.

A major watershed in the industry came in 1974, when Agriculture Canada registered Tower, the first *B. napus* variety with both low erucic acid and low glucosinolates. By 1978 the research effort had produced Candle, the first double-zero *B. rapa* variety. With the final piece of the breeding puzzle now in place, the roles of both the public research community and the Rapeseed Association began to shift. Although public research continued to work to lower the levels of erucic acid and glucosinolates, the push was on to improve the yields and extend the effective planting range for canola. From the Rapeseed Association's perspective, with double-zero rapeseed now available, the biggest challenge was to increase both the production and market for the new product. Further investment in research and infrastructure required a greater flow of product, which was in everyone's best interests but impossible for anyone individually to pursue. Meanwhile, the private sector began to invest in new breeding technologies. With successive breakthroughs in US universities related to the technologies of manipulating genes, the impediments to commercial investment diminished. Calgene's breakthrough patent on the *Agrobacterium* transformation technology for *Brassica* ultimately led the way to intensive investment and research by private companies. By the early 1980s private companies had positioned themselves to dominate varietal development in the following period (see Chapter 8). The first few private varieties were released in this period (Table 6.6).

In 1978, the Rapeseed Association took what may have been the most astute and fundamental step in developing the market for the new product, registering the name 'canola' as the trademark for rapeseed varieties with low erucic acid (5% or less) and low glucosinolate content (3 mg or less). With continuing research through the following 8 years, the levels of erucic acid and glucosinolates continued to drop, so that in 1986 the canola trademark was amended to restrict the designation to rapeseed varieties with less than 2% erucic acid and less than 30 μmol of glucosinolate (Dupont *et al.*, 1989).

The Rapeseed Association of Canada formally completed its shift to the new product when it changed its name in 1980 to the Canola Council of Canada, acknowledging the development and acceptance of canola varieties. The purpose of the Canola Council was set out in its by-laws:

> For the advancement of the canola industry in all its aspects, including the producers, processors, handlers, manufacturers, exporters, dealers, and other parties interested in canola and canola products, and more particularly: to explore poten-

Table 6.6. The role of the three sectors in research, 1974–1990.

	Research investments	Share of research results	
	% of total inputs	% new technologies	% new varieties
Government and universities	58	20	50
Associations	2.2	0	0
Private companies	40	80	50

Source: Authors' calculations from the canola industry suvey, 1997–1998.

tial markets and to conduct promotional and servicing activities of any kind conductive to the expansion of markets throughout the world for the canola industry of Canada; to improve, through education, research studies, and promotional programs the production of canola; to collect information, to contract research and to disseminate information with respect to producing, handling, marketing, processing, utilizing and promoting canola and canola products; and to achieve and maintain good public relations and to assist and co-operate with other persons or organizations in the furtherance of the objectives of the Council.

(Canola Council of Canada, 1995)

During this period the Council's research funds came increasingly from private sources (Table 6.7). Following the negotiation in 1978 of a new UPOV (International Union for the Protection of New Varieties of Plants) agreement, Canada began to talk domestically about implementing plant breeders' rights (PBR), a new form of intellectual property rights for the agri-food sector. Although the Canadian PBR Act was only passed in 1990, the indication of intentions was enough to attract a number of companies to initiate or relocate canola research programmes to Canada. Corporate sponsorship of Council research began during this period. Nevertheless, in this period, research direction and control remained dominated by the Canola Council and research remained highly coordinated between the Council, the public laboratories and industry.

More importantly, throughout this period the Council worked with researchers and marketers to position canola as a premium human oil. The Council funded extensive research into the health benefits of canola, with a successful outcome. By 1984 a number of health studies showed that consumption of canola oil, which was low in saturated fats, provided significant health benefits compared to consumption of palm, coconut and corn oils (Gray and Malla, 1998). These results, plus longitudinal food safety studies, contributed to the evidence that the United States Food and Drug Administration (US FDA) used to grant canola 'generally regarded as safe' (GRAS) status in 1986.

Meanwhile, the provincial growers' associations intensified their extension efforts to increase the rate of adoption of the new crop and to steadily improve the quality of the product. In Saskatchewan, for example, the provincial Canola Growers Association began the 'Grow with Canola' programme, which provided

Table 6.7. Sources and uses of funds by the Rapeseed Association/Canola Council of Canada, various years. (*Source:* Adolphe, 1998.)

	1972	1982	1998
Sources of funds			
Crushers and exporters levy	46	62	37
Government	50	33	11
Membership dues	4	2	0.1
Sales and service revenues	–	3	26
Corporate sponsors	–	2	11
Grower organizations	–	–	15
Uses			
Research	83	40	40
Public relations	8	20	9
Market development	4	5	4
Crop production	2	10	36
Corporate affairs and administration	3	26	11

an extensive set of agronomic services, including basic varietal, agronomic and fertility information and demonstration test plots that were harvested with standard farm equipment. Many participants in the sector credit such programmes with the rapid expansion of canola in the prairies. Without such a rapid take-up, export market growth would have been severely limited and further investment curtailed.

The institutions of the 1970s worked well for the 1980s but, by increasing market size and attracting the attention of large private actors, sowed the seeds for their own descent in the 1990s.

Fourth Stage: 1990–1999

The current stage of development is dominated by private actors (Table 6.8). The ascendancy of the private sector was assured in 1990, with the adoption of plant breeders' rights in Canada. This came on top of the US Patent and Trademarks Office decision in 1985 to grant patents for whole plants, and the introduction of canola hybrid technologies and the first hybrid variety in 1989. Together, these efforts strengthened private control over intellectual property in the breeding and seed business, removing one of the main hold-up problems that necessitated the massive public research effort of the past generation and the creation of the Canola Council. Furthermore, as discussed in Chapter 8, private investments in new technologies and seeds created market returns that were often difficult to extract given the long supply chain with widely divergent levels of market power. As a result, many of the firms investing in canola research sought to integrate or coordinate the supply chain that related to their investments. This has resulted in significantly greater use of contracts, often involving a web of relationships that link research units with seed companies,

Table 6.8. The role of the three sectors, 1990–1998.

	Research investments	Share of research results	
	% of total inputs	% new technologies	% new varieties
Government and universities	39	10	14
Associations	2	0	0
Private companies	59	90	86

Source: Data collected by P. Phillips through survey of industry during 1997–1998.

other input providers, a select group of farmers and specific processors or marketers. These industrial networks, by solving many of the market failures, have begun to break down the public nature of the industry, both undermining the historical roles for the public and not-for-profit sectors and creating potential new roles.

By 1990, the playing field for development had changed dramatically. The private growth in breeding and the seed business was now actively encouraged by federal and provincial governments (see Chapter 7). Although producers in the three prairie provinces and Ontario introduced check-offs to raise funds for more extensive farmer-directed programming and research, their efforts came too late to offset the move to a privately driven industry. The public-sector research agencies refocused their efforts to complement rather than to compete with private efforts.

The combination of new proprietary technologies, patented genes and hybrid technologies greatly increased private interest and investment in canola. Many of the seeds developed during the 1990s had attributes that created the potential for hold-up. Herbicide-tolerant canolas required the use of a specific herbicide in order to be useful, while canolas with particular oil characteristics needed specialized processing and marketing chains in order to be viable. The most dramatic change in the private sector was the introduction of large agro-chemical companies into the plant genetics industry. AgrEvo (now part of AgriEva), Dow, Monsanto and Zeneca (now part of Syngenia), for example, have entered canola breeding on a significant scale. The very large capital base and international network of these companies have introduced a whole new level of capacity in canola genetics. These multinationals have vertically integrated much of the plant breeding and herbicide production intra-company in an effort to address the potential hold-up problem and to capture the economic value of these new technologies. Those firms which did not vertically integrate made credible *ex ante* investments to get asset-specific research and development (e.g. Proctor & Gamble invested in Calgene to develop laurate canola and AgrEvo paid AAFC to develop Liberty-Link™ canola). In addition to the vertically integrated production of genetics, herbicides and seeds, these companies rely on contracts with producers to maintain control over their property once it enters the market and have privatized much of the extension effort as an adjunct to marketing. The most important contract used to maintain this supply chain is the production contract with producers. A production contract usually specifies that the farmer is to use registered seed and herbicides bought from designated dealers, pay a technology use fee (for one company) and, for some, deliver all production derived from that seed either to a licensed elevator or end user. Most production contracts also include an Act-Of-God clause, which relinquishes producers of contractual obligations in the case of crop failures. In 1998, an estimated 50% of the canola acreage was planted to herbicide-tolerant varieties and about half of that acreage was managed under partial production contracts. For much of the rest of the production, restricted access to herbicides provided an equivalent level of protection to the companies. These contracts ensure the owners of intellectual property that the natural propensity to practise moral hazard, with respect to holding back the final product for future seed use, is minimized. The risk is not, however, eliminated.

Despite growth in the use of property rights and contracts to protect the owner of the genetic material, there is significant potential for breeders to lose control of genetic material once it sold to producers. A single pound of canola seed can produce approximately 300 pounds of seed a year later and 90,000 pounds in 2 years. If a high price has to be paid for new registered genetic material, this creates a very strong incentive for producers to retain some of the product for their own seed in the subsequent year or to sell some of this product to their neighbours in what is referred to as the 'brown-bag' market. The small volume required to cheat makes this black market impossible to eliminate. This is not a problem for those herbicide-tolerant varieties that require the annual purchase of a specialized, patented chemical (e.g. Liberty™, Pursuit™), because the price of the herbicide rather than the price of the seed is used to capture the rents for the breeders. It is also not a problem for designer-oil canolas because the product must be sold to a particular processor in order to have value. The real problem of unenforceable property rights is in the case of herbicide-tolerant varieties that use generic herbicides (e.g. Round-Up™, which is off the patent in Canada), and in the case of open-pollinated canola varieties with improved agronomic properties. In these cases, the breeders must capture much of their rents in the first year or two of release, rather than over the life of the product. Incidentally, the brown-bag market creates a huge potential problem for segregation of non-transgenic varieties. Producers who have grown brown-bag transgenic seed are not going to declare it upon sale, given that they may face prosecution if they do so.

The diversity of the technologies makes industry-wide institutional solutions increasingly difficult. The development of hybrid varieties, for example, has given some private firms a greater ability to capture value for their genetic material. These hybrid technologies, which make second generation seed less viable, effectively eliminate the incentive for producers to retain production for future seed use. The first hybrid variety was introduced in 1989. Although often protected with plant breeders' rights and production contracts, these varieties do not require the enforcement of contracts to maintain control over the use of the genetics. Because hybrid varieties produce a very poor second generation with very mixed genetic properties, there is no ability for a grower to retain its agronomic benefits.

Virtually all of the seed sold in 1997 was subject to some form of contract or property rights. Almost every breeder applied for plant breeders' rights for varieties developed after 1990, even though many of the applications were abandoned if the market share for the variety did not justify the effort. In 1997, only about 75% of the acreage seeded could be assigned to specific varieties. Of the acreage that could be assigned, almost all of it was covered by one or more intellectual property rights system. There was an estimated 70% of total acreage covered by PBRs, 35% managed through production input contracts for herbicide-tolerant varieties, up to 30% planted to hybrid or synthetic seeds and less than 5% under identify preserved production contract.

Research, which was clearly the first and dominant priority of the Council in earlier periods, became less of an imperative with the rise of private interest. The Board of the Council previously was very receptive to new ideas and proposals from researchers, so much so that any budget constraints were passed along to marketing and administration, rather than research. This is no longer the case. The large increase in private investment in research and the refocusing of public research towards partnerships with private firms has reduced the importance of Council funding for research.

The size, role and use of the levy became a focus for debate. As private property rights for research increased private investment, the imperative for collective research came under attack. Exporters and crushers, believing that the amount that they contributed for canola research via the levy to Council was too high, were increasingly divided in their opinions on the size and use of the levies they had to pay. Exporters and crushers both argued that the traditional levy makes them less competitive both in international markets and in domestic markets against imported oils, as their competitors do not have any levies to pay. They argued that the levy puts Canadian exports at a disadvantage as the Canadian canola offer prices need to be higher to compensate for the levy, while domestic canola oil prices are also forced higher due to the levy. These concerns ultimately resulted in a reduction of the levy: the crushers reduced their levy to Can$0.30 per tonne in 1994 from Can$0.50 per tonne and the exporters followed their lead, cutting their levies by the same amount in 1997 (Gruener, 1999). This reduced the cash flow of the Council, slowing investments in research through the early parts of the 1990s.

Meanwhile, farmers raised questions about their role in the system. Farmers felt that their participation in the allocation of research dollars was very important. Although the levy was paid initially by crushers and exporters, producers believed that ultimately they ended up paying for it – they argued that crushers and exporters simply passed along the cost through lower prices to producers. As a result, each group considered the levy as their own source of funding for the Council and tried to influence the direction of research for the benefit of their members. Farmers, for example, argued for much more research at the farm level while exporters and crushers sought a broader research and development programme. The end result of this debate was that as the exporters' and crushers' levy was lowered and farmers introduced producer check-offs at the delivery point, which have worked to shift the power structure within the Council. The producers, through their provincial canola commissions, now bring their own money to the table to help to influence the direction of the research.

Beginning in 1983, farmers gained some new leverage in the research community with the creation of the Western Grains Research Foundation. The WGRF, endowed with Can$9 million of funds remaining in the Prairie Farm Emergency Fund in 1983, has approximately Can$1 million of annual revenues which it allocates to research on grains and oilseeds. Since 1983 the farmer-directed organization has invested approximately Can$2.5 million in 26 research canola projects involving public researchers in Agriculture Canada and at the Universities of Manitoba, Alberta and Saskatchewan (Table 6.9). Although the WGRF does not have any formal links with the Canola Council or the provincial commissions, it shares its canola-related research applications with CANODEV, the research arm of the Saskatchewan Canola Development Commission, to ensure that no overlaps or duplications are funded.

After 1988 the provincial associations began to implement producer check-offs, each using different provincial enabling legislation to collect the funds. The first to establish a check-off was the Ontario Canola Growers Association, which in 1988 assessed a Can$3.00 per tonne charge against deliveries, with Can$1 per tonne going to a financial protection fund and Can$2 per tonne going to the operation of the OCGA (small donations came from the Canola Council in the early years). Given the small production base (less than 50,000 t annually),

Table 6.9. Producer-financed research and industry development.

	Western Grains Research Foundation	Alberta CPC	Sask. CDC	Manitoba CDC	Ontario CDC
Year funding began	1983	1989	1991	1996	1988
Endowment	Can$9 M	–	–	–	–
Canola producers' check-off	0	$0.50/t	$0.50/t	$0.50/t	$3.00/t
Total resources allocated to date for canola	Can$2.5 M	$5.3 M	$6.8 M (?)	$530 K (?)	$1.1 M (?)[a]
% to research	100	44	40	45	5
% to extension, market development, admin.	–	56	60	55	95[a]

[a]includes Can$1 per tonne for financial protection plan ($0.50 per tonne after 1996).

CDC, Canola Development Commission; CPC, Canola Producers' Commission.

there were little actual funds available for activities. Until 1996 the Association concentrated primarily on developing and testing production methods to improve the yield and quality of Ontario canola and to lower the unit cost of production. In 1989 the organization began to run small crop production centres as demonstration sites to encourage farmers to grow canola. In 1996, the Association negotiated a change in its allocation of the check-off, reducing the share going to the financial protection fund and diverting Can$0.50 per tonne to research. So far this has contributed only a small amount of money, which has been used to develop a research programme at the University of Guelph.

The prairie provinces then developed their own programmes. Alberta was next, with its producers' commission introducing a Can$0.50 per tonne check-off in 1989, 44% of which has been used for research. The rest of the funds have been used for extension and market development. Saskatchewan followed suit shortly afterward. The provincial growers' association created an arm's-length Canola Development Commission in 1990, which then introduced a Can$0.50 per tonne producer check-off in 1991. Since then approximately 40% of the check-off funds have been used for research; the rest have been used for extension and market development. In 1995 the Commission created CAN-ODEV, a wholly owned, private-sector subsidiary to undertake the research effort and take advantage of the research tax credits available for private companies. Finally, Manitoba in 1996 introduced a Can$0.50 per tonne check-off, 45% of which is used to undertake research. All three provincial efforts work in close cooperation with the Canola Council. Producer contributions to the Council research programme in 1998 equalled about 15% of the total Council budget, which is the equivalent of 38% of the Council's research programme.

One research problem that these new institutions, markets or contracts have not resolved is the difficulty of coordinating research to develop new germplasm or application of new technologies. The Council filled part of that role, as producers pooled funds with governments and the private sector (which contributed about 11% of the Council budget in 1998) to conduct pre-commercial and non-competitive research through the Council programme. With the increasing complexity of the technologies and tools, the Council programme was not enough. The public sector, through Agriculture and Agri-Food Canada, the NRC and a few universities, has also acted as a coordinator. By providing platform technologies and collaborative research opportunities, the public laboratories have become a focal point for both discovery research into canola-specific technologies and the base for acquiring germplasm, plant breeding transfer technologies and the general know-how that makes a successful research programme.

To sum up the research story during this period, the introduction of effective private property rights and the increase in contracting caused a dramatic increase in private funding, forcing major changes in the roles for the other actors. As governments shifted the bulk of their effort towards collaborations with, and matching grants for, private research, the Canola Council and the producer commissions were forced to restructure their finances in order to capture some of that matching funding for producer interests. Although the research effort through the Council decreased only modestly in absolute terms over the period, it dropped sharply in relative terms. There was an unambiguous shift in emphasis from coordinated industry growth to increasingly private research. Coordination became more difficult as information and knowledge became confidential. Although all three sectors continued to undertake research, the private sector increasingly came to dominate.

Meanwhile, the Council expanded its programming to include an extensive crop-production extension programme, in 1998 involving sites in the Peace River country, three in Alberta, three in Saskatchewan and two in Manitoba. In addition, the Council collaborated with the Minnesota Canola Council on sites in the USA. This activity was largely coordinated with the continuing extension programmes in Alberta, Saskatchewan and Manitoba but in competition with an increasing effort by private companies to use extension as part of their marketing programmes. With the development of private seeds that, especially in the case of herbicide-tolerant and novel oil varieties, had more specific agronomic requirements, private companies saw both a need and value to provide greater information to farmers. Most of the larger private seed developers report that since 1993 they have undertaken demonstration seed trials in competition with the cooperative system. The private companies assert that their private trials, which tend to be head-to-head competitions in larger plots (10–20 ha), are more likely to influence farmer's seed purchases than the smaller-plot trials run by the Canola Council. Some companies have declined to participate in the Council programme while others participate but supplement that activity with their own private trials. Most of the companies market their seeds using the results of their own trials. In contrast, public breeders suggest that the private trials are less rigorous and more open to manipulation by selecting either differential quality seeds or by setting up competitions with foregone results. Increasingly, given the proprietary nature of the seed and input packages on the market, it is becoming difficult for farmers to get unbiased agronomic advice. Once a producer has decided to purchase a seed with these novel traits, he is often forced to take the related agronomic advice, simply because extension agents from the Canola Council, the provincial canola growers associations and provincial agriculture departments often do not have access to the full and timely information required to give appropriate advice. Hence, the increasing proprietary nature of the seed business and the increase in *de facto* contracting between producers and seed companies raise some doubt about the need for such extensive third-party extension programmes.

Conclusion

The canola industry is the most pronounced and recent example of agri-food industry development through extensive research, market development and extension. This examination illustrates the significant potential for hold-up problems – including asymmetry of information, opportunistic behaviour due to investments in highly specific assets, idiosyncratic transactions causing high negotiation and enforcement costs and ineffectual assignment of property rights – and the vital role that different institutions have played in overcoming the hurdles.

The NIE framework has been applied to examine the structure of the institutional and contracting challenges facing the emerging industry and to evaluate the evolving role for those institutions. There would appear to have been some potential redundancy in the development of the new institutions, with overlapping and, at times, competing producer organizations directing funds to research and extension work, but on the whole, the new institutions did resolve the market failures as they arose.

Beginning with extremely weak institutions in the 1940s to 1960s, the full responsibility fell to the public sector. Once a critical mass in the sector was reached in the mid 1960s, other collective and private institutions began to form. First, the not-for-profit associations were created, followed closely by producer groups, each seeking to push forward research into edible oils and to expand both production and markets for that product. By the late 1980s, however, the sheer scale of the sector and the promise of private plant breeders' rights and effective hybrids, combined with greater contracting, brought forward private investment, which by dwarfing public and collective resources have forced those institutions to seek new roles in the industry.

The future of the associations and other third parties warrants some further discussion. The three biggest hold-up problems – research, market development and extension – appear to have been at least partly resolved, which raises a number of possible changes in the role of the associations.

The extension of private property rights to seed varieties and the extension of patents into the biotechnology industry have encouraged a sharp rise in private research into canola development. As discussed in Chapter 8, the problem now may not be too little investment – it may be inefficient and wasteful competition in research. Furthermore, the results set out in Chapters 15 and 16 show that with a sharp rise in investments, the net return to research has fallen, possibly to the point that some of the investment may not be justified. Clearly, there may be a strategic role for collective or producer-financed research. One option suggested by many is that collective or producer-financed research could be directed to maintain public germplasm that will effectively reduce the oligopolistic power of the increasingly concentrating multinational seed companies. Additionally, producer funds might also be directed to sustaining public domain technologies or ready access to proprietary biotechnologies, thereby maintaining freedom to operate for new market entrants. Each of these elements have high voice requirements.

The most problematic area is market development. There are diverging trends that will influence the future effectiveness of the role of associations. On the one hand, the commodity canola market has the potential to be segmented into a wide array of differentiated product markets for novel canola oils, feeds and industrial ingredients. Association-delivered commodity-market development programmes likely will be ineffective or even possibly commercially harmful to the proprietary interests of the producers of these products. There is every reason to believe that only the owner of the proprietary product will be able to position that product to earn the optimal return on the research investment. On the flip side, the risk of market failure rises when differentiated products exist. For example, in 1998 Canadian canola was shut out of the EU market because the Canadian marketing system did not segregate genetically modified seeds from the traditionally produced varieties. This marketing problem may become more complex in future years as a greater number of non-edible canola-based oilseed varieties enter the market. Given the potential for market failure, the Canola Council and related provincial grower associations could play a significant role in developing and managing a quality assurance programme within the marketing system. This would necessarily start with the seed developers and span the supply chain right up to the consumer markets, but would most certainly require extensive work with farmers to implement the appropriate system. The Council, working with the Canadian Grains Commission and federal and provincial agricultural departments in the 1970s, successfully transformed the Canadian industry from rapeseed to canola. It could have a similar role in managing the transformation from being a commodity edible oil to a differentiated, high-value product.

Finally, as noted above, there is a likelihood that the Council and related provincial growers associations are earning a declining return for their investments in extension. At some point, a restructuring may be in order. The extension programme might more usefully be shifted towards extension programmes in support of quality-assured marketing.

This analysis continues in the next chapters, looking at the role of public, private and related sectoral institutions in overcoming market failures in the canola sector.

The Role of Public-sector Institutions

7

Peter W.B. Phillips

Introduction

Agriculture has been one sector in almost all countries where the public sector has historically contributed a significant share of research resources and undertaken a large share of the research effort. Except for those agri-food products with effective hybrids (e.g. maize), most of the effort has been undertaken by governments, publicly funded universities or by private companies funded by public grants. That relationship held true in the canola sector, in Canada and globally, until the early 1980s. Between 1944 and the early 1980s all of the key activity in developing canola varieties was financed, undertaken or managed by government agencies. Since then, however, new, proprietary technologies have been developed and most of the resulting crop innovations have been commercialized by private companies. The transformation from a largely public industry to an increasingly private one has been precipitated by new, more cost-effective technologies, by significant industrial restructuring facilitated by large financial investments and by the introduction of legally sanctioned intellectual property rights for biotechnological processes, genetic discoveries/constructs and commercial varieties (Santaniello *et al.*, 2000).

As the germplasm, technologies, genes and seeds industries have been privatized, the public sector's historical role as proprietor or lead innovator has been challenged. In the past the public sector financed, undertook and commercialized all of the innovations for canola; now public institutions are testing a variety of new roles, ranging from partner to promoter. This has involved a shift away from doing all of the varietal development in public or university laboratories to doing more custom work and collaborations, often on pre-commercial or non-competitive projects. As the direct role for government has diminished, new public mechanisms have been introduced: capital grants for investment attraction (e.g. provision of public funds for investment attraction of multinational companies, such as Limagrain Canada and PGS); support for venture capital pools such as the Western Economic Diversification–Royal Bank knowledge-based lending facility in Saskatoon; fiscal support for private research through such mechanisms as research and technology tax credits and Industry Research Assistance Program (IRAP) grants; and provision of physical infrastructure such as Innovation Place and public greenhouses. Although it is next to impossible to determine explicitly the exact impact of these policies on private research, there are a number of examples that demonstrate that, at least at the margin, these policies can, and do, influence private decisions about location of research effort.

In this chapter we look at how the public sector has evolved from being the proprietor to being a partner in research, investing both manpower and funds in the industry. Chapters 10–13 contain a more detailed examination of the regulatory role of government.

The Rationale for Public Investment in Agri-food Research

Public involvement in agri-food research has historically been justified based on four specific market failures. First, governments argued that private firms were not doing the optimal amount of research to develop new varieties for farmers. The absence of an institution such as plant breeders' rights created a market failure in almost all agri-food research sectors. Numerous studies, reflecting that market failure, showed that research into agriculture provides extremely high returns to society. Alston *et al.* (1998) undertook a meta-analysis of the returns to agricultural research, identifying 1821 studies that examined the issue. Estimates of the returns ranged from −100% to +724,323%, with an average return of +62% (for 90% of the sample – dropping the upper and lower 5% of the sample). Nagy and Furtan (1978) estimated that public canola research up to 1979 yielded a 101% internal rate of return. If those rates of return are actually there, it is clear that there is too little investment in agricultural research. The traditional response has been for the public sector to step in to make up the difference. With the introduction of a more effective intellectual property rights regime for agri-food research (and more efficient research methods), both the incentives for and the level of private research into canola have accelerated. An assessment in Chapter 15 shows that the net result of this acceleration of research has been to lower the estimated return to research to below 10%, which puts in doubt continuation of public investment.

Second, the extensive market failures related to developing new products for new markets, as discussed in Chapter 6, are viewed as further reasons why private investment in products which could both develop and diversify the western Canadian economy is not forthcoming. The Canadian government had, from the beginning of its canola research effort, a goal to establish a new crop and income option for western Canadian farmers (NRC, 1992: 2). Through both the public and not-for-profit associations' efforts, canola has successfully provided a diversified option to traditional wheat, durum, barley and oats crops, now using on average 11 million acres annually and producing a crop worth on average more than Can$2.5 billion annually. Canola at times vies for wheat as the most valuable crop in western Canada. With the industry now developed, however, the continued role for the public and not-for-profit institutions comes into question.

Third, governments have always been concerned about equity, specifically how the gains to research are shared among producers, private companies and consumers. Recent economic research shows that imperfect competition in the input sectors and the monopsonistic nature of the food processing industry reduces the returns to farmers and possibly to consumers. The argument often made is that the presence of a public-sector seed developer that gives away its intellectual property effectively reduces the market power of oligopolies and potentially increases the returns to farmers and consumers. Now that the public sector breeders use IPRs to protect their intellectual property, it is not clear that this rationale remains valid.

Fourth, some governments have viewed public research, especially in the 1990s, as a factor in competitiveness. If knowledge spillovers (e.g. know-how related to genetic transformations) are limited to a specific location (perhaps because the diffusion of the knowledge requires face-to-face interactions), then any scale and scope economies that result will be captured by the region that undertakes that activity. That creates the possibility that countries engaging in technological competition may endogenously generate comparative advantage. In short, 'comparative advantage evolves over time' (Grossman and Helpman, 1991). Thus, if the final product (e.g. canola seed for crushing, oil or meal) is tradeable but the innovation-based knowledge is a non-transferable intermediate factor of production (e.g. seeds for germination), then the fact that innovation begins or is supported in one jurisdiction could indefinitely put that site on a higher trajectory of R&D and new product development (Grossman and Helpman, 1991). As a result, the high-technology share of gross domestic product (GDP) and of exports will be higher than otherwise, and farmers could realize higher incomes as they earn a premium for being early adopters.

The purpose of this chapter is not to examine whether these market failures both existed and were valid reasons for government action in earlier years, but to determine whether they remain valid in the face of increased private activity in more recent years.

The Public Sector as Proprietor

To get a full appreciation for the changing role of the public sector, it is instructive to look at the end points in the process. As recently as 1982, there were only six canola cultivars grown in the world, all bred by public-sector institutions in

Canada: the Agriculture Canada research stations in Saskatoon, Alberta and Ottawa, and the Universities of Manitoba, Alberta and Guelph, often working in collaboration with researchers at the National Research Council laboratories in Saskatoon and Ottawa and academics at the University of Saskatchewan. These six varieties were developed using largely non-proprietary technologies developed by those institutions, e.g. the half-seed method and special techniques using gas–liquid chromatography. All of the seeds produced and sold until then were in the public domain (Kneen, 1992). The rate of development of new varieties was also relatively slow, with an average of one new variety every 2 years, with the result that the average life span of a cultivar was about 10 years (Table 7.1).

Around 1985 there was a sharp acceleration of private-sector research and investment in canola development. Between 1982 and 1997, a number of new proprietary technologies replaced the publicly developed breeding methods and more than 125 new varieties were introduced. More than 75% of the new varieties were developed by private companies, so that by 1997 only about 10% of the seed sold in Canada had been developed by public institutions (this may understate the role of the public sector somewhat because many of the privately registered varieties were either developed using AAFC germplasm or were developed in collaboration with AAFC or NRC). The average active life span of a cultivar declined to only about 3 years by 1997.

The public research institutions – the universities and public laboratories – have been seeking a continuing role to play, now that the technologies and the marketplace have been largely privatized.

An Evolving Research Role for the Public Sector

The public sector no longer plans and invests as if it is the only actor developing canola. The significant investment by the private sector into the research of know-what knowledge and product development (e.g. varieties) has effectively met many of the public objectives for investment. The public sector has only had to set the environment for private efforts. It is instructive to look at the role of the state in the four elements of knowledge development and, more generally, in the product development process, as characterized by the chain-link innovation model (Fig. 2.2).

Product development

All of the early work on canola was done by the private sector, based primarily on a linear model of innovation (Fig. 2.1). Although researchers from Canada Packers were involved in and funded some of the search for a Canadian-sourced edible oil, the bulk of the product development effort was done by scientists in Agriculture Canada, the NRC and four Canadian universities (see Table 6.2).

Table 7.1. Canola varieties developed by institution and by year.

Years	1940–1959	1960–1969	1970–1979	1980–1984	1985–1989	1990–1994	1995	1996	1997
Number of varieties developed									
Public institutions	4	6	8	6	10	12	2	4	2
Private companies	0	0	0	0	9	32	18	28	29
Total all institutions	4	6	8	6	19	44	20	32	31
Market share by institution									
Public institutions	100%	100%	99.8%	99%	98%	49%	27%	26%	11%
Private companies	0%	0%	0.2%	0.7%	0.4%	43%	57%	61%	63%

Source: Canola Council of Canada, Canola Growers Manual (http://www.canola-council.org/manual/canolafr.htm); market share estimates by Nagy and Furtan (1978), Three Prairie Pools Varieties Survey and author (market shares do not add to 100% due to share of the acreage not being reported to specific varieties; private market share is likely under-reported as a result).

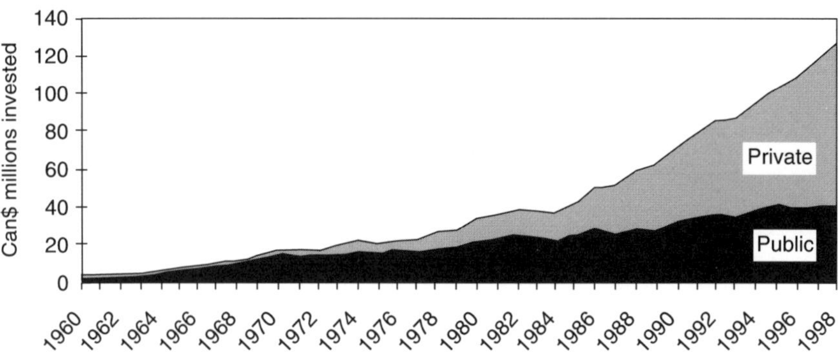

Fig. 7.1. Annual global public and private investment in canola (Millions 1997$).

Until 1985, all of the varieties in Canada were developed and registered by those institutions. Since the mid 1980s, however, a rapid expansion of private investment in canola research has supplemented and almost overwhelmed the public effort (Fig. 7.1).

After 1985, and especially after the introduction of plant breeders' rights in 1990, the public institutions have almost ceased producing new varieties on their own behalf. Almost all of the varieties registered by the public sector since then have been developed on behalf of others. For example, Canamera has contracted with the University of Manitoba to develop a selection of new modified oil varieties of canola; although these varieties are registered by the university, the private company owns the exclusive rights to use the variety. Similarly, Agriculture Canada has undertaken contract breeding on behalf of a number of companies. This new research effort conforms more to the chain-link model of innovation than the linear model that prevailed when the public sector undertook research on its own behalf. The development of Liberty-Link™ canola illustrates the new approach. AgrEvo in the mid 1980s identified the market opportunity for herbicide-tolerant canola, undertook much of the basic gene isolation work in Frankfurt and then partnered with Agriculture Canada to get access to its germplasm and to tap into its know-how and know-who to develop its Liberty-Link™ canola. It then partnered with the wheat pools in western Canada to prove-up the product and market it. Since 1985, more than 75% of the new varieties have been developed privately along these lines (often developed directly by public institutions using private funds) and the seeds market has shifted heavily towards private seed sales, accounting for between 63% and 89% in 1997, up from only trace amounts in the mid 1980s (Table 7.1).

Using the four public objectives as criteria, it is not clear that the state has any continuing justification for investing public funds in varietal development. Recent rates of investment and the resulting lower returns to research (see Chapter 15) suggest that instead of under-investing in varietal development, private industry may in fact be over-investing in research and development for canola varieties, at least partly because of public subsidies. Industry participants suggest that each new agronomic variety requires demand for seed for about 250,000 acres year^{-1} for 3 years to yield the targeted rate of return for biotechnology investments (i.e. a return that both covers the cost of the investment and provides a risk-adjusted return). Given an average of about 11 million acres planted in Canada, that would suggest that only about 45 varieties could be sustained over the long term. Currently well over 100 varieties are registered for sale. Furthermore, with an average life of 3 years, the optimal annual replenishment rate of new varieties should be about 15, not the current 30 or more per year. The potential to capture new market share for farm chemicals (upstream value added) or in the specialty oils market (downstream value added) may justify this rate of investment from a firm perspective, but clearly this accelerated 'creative destruction' is potentially socially wasteful. With the estimated return to research now below 10%, the entire sector as a whole may be losing money on the margin, as the average cost of capital for investment is currently more than 10%. Nevertheless, the public sector may provide one key service that is not and will not be provided by private companies. The conversion of the canola seed business from a public to private industry appears to have reduced the rate of advancement of seed quality. During the public development phase, the average yield was rising by between 1% and 1.4% per annum (Table 7.2). Yields dipped in the early 1980s due to the changeover to canola from rapeseed. Subsequently, yields have returned to previous peaks but appear to have stalled at that point. At least part of the reason is that private breeding efforts are less concerned about improving the yield of canola per acre and more focused on reducing input usage or modifying the oil and meal qualities of the output.

Economic development and diversification seem to be progressing apace, with new agronomic traits being rapidly bred into commercial varieties. The dispersion of canola into new growing areas has actually accelerated since 1985, at least partly due to the profit incentive of the seed merchants to develop and market new varieties for new growing areas. The total area seeded to canola in western Canada rose to a record 14 million acres in 1994, and canola's share of total seeded acreage has risen to 15% in 1986–1995, from 9%

Table 7.2. Changes in rapeseed/canola yields in Canada. (*Source:* Gray *et al.,* 1999.)

	Average yield for period (1980–84 = 1.0)	Average annual change in yield
1960–1964	0.89	
1965–1969	0.94	1.0%
1970–1974	0.99	1.2%
1975–1979	1.06	1.4%
1980–1984	1.00	−1.1%
1985–1989	1.07	1.3%
1990–1994	1.09	0.5%
1995–1997	1.08	−0.3%

in 1976–1985. All indications are that more new private varieties suited to new areas are imminent. Between 1994 and 1997, there were 97 field trials for transgenic varieties that involved introducing stress, insect, viral and fungal resistance into canola (CFIA, 1999), all factors that will permit growing canola in different areas of western Canada and the world.

Equity continues to be a key concern of the public sector. Recent research suggests that rapid innovation rates observed in the industry may at least partly reduce the market power resulting from imperfect competition in the input sectors – the four-company concentration ratio is between 40 and 67% in the canola seed industry (Phillips, 1998) and is 65% in the chemicals industry (Just and Hueth, 1993) – and the monopsonistic nature of the end users – the four-company concentration ratio is 100% in the oilseed processing industry in Canada (Wensley, 1996). There are two schools of thought about whether there is any continuing equity-based rationale for state investment. Green (1997) has argued that as innovation cycles shorten, the market power of oligopolies is reduced. Thus, the shorter breeding cycle (5–7 years) and the corresponding short life span of each new variety (approximately 3 years) may effectively limit the market power of oligopolistic seed companies. On the other hand, the advancing corporate concentration between chemical and seed companies through mergers and alliances (see Chapter 8) and the increasing complexity of the research methods, threaten to reduce the number of effective competitors in the seed business. Some have argued that the public sector and not-for-profit associations should devote some effort to developing and maintaining an open-pollinated, public variety of canola, with adequate quantities of seed available to effectively 'contest' the market. The existence of such varieties would reduce the capacity of private companies acting oligopolistically, thereby effectively driving the price of seeds closer to their marginal cost.

Competitiveness in the research and seeds business is more a local than a global concern, but it has in the past (and is currently) influenced Canadian public research policy. A number of governments (as is shown below) believe that a public presence in the industry creates the base to attract private investment. That certainly may be a reason to sustain know-how and know-who capacity in the public sector, but it does not necessarily justify investment of public funds for development of public varieties. It is hard to see how public development and marketing of seeds helps Canada attract private investment. Given that the rate of varietal development already exceeds the capacity of the industry to generate adequate revenues, the existence of pure public varieties would be more likely to squeeze out private seed, reducing the local potential market for private breeders and thereby reducing their incentive to locate locally.

Based on the above, the decision in the late 1980s by Agriculture Canada and most of the universities to reduce their efforts to develop their own varieties appears to have been the right one. Rather than re-enter the business, the public sector breeders may want to review their remaining support for breeding public commercial varieties. Rather than seeking more varieties, the public sector may wish to monitor the rate of creative destruction in the industry. Given the extensive public support for private activity (see below for details), there may be some potential to adjust the tax and policy system to reduce the incentives that are creating the excessively competitive private research.

Know-why knowledge

After knowledge of the marketplace, basic scientific knowledge is the second most critical factor in developing and sustaining a knowledge-based industry or economy. As one might expect, few private companies undertake know-why research. Without a financial incentive, there is little reason for this activity in the private sector. A search of the ISI special database of canola-related research published in academic or scientific journals between 1981 and 1996, shows that staff from about 130 private companies published at least one journal article in that period, accounting for about 6% of published articles. More than half of the companies had only one article credited to one of their staff over the 15-year period. Staff in only three companies – Allelix Crop Technologies (purchased by Pioneer Hi-Bred in 1990), Calgene and Unilever – published an average of more than one journal article per year. This may actually overstate the private sector effort as many of the articles were published by recent graduates in collaboration with their supervisors. Hence, much of their effort might more reasonably be allocated to the university setting. Theory confirms that private, for-profit firms are highly unlikely to increase their share of research to add to the stock of know-why knowledge, as there is little chance that they could ever capture any of the returns on their efforts. Thus, the creation of know-why knowledge would appear to be a critical, long-term role for the public sector.

The vast majority of know-why work on canola has been done by the public sector, 58% in universities and 36% in publicly funded research agencies (Table 7.3). Looking at Canadian sources alone, one can see that the 60 or so Canadian universities and public research laboratories were critical to the creation of knowledge in the canola sector. Over the period 1980–1996 they contributed 27% of the new global know-why knowledge.

Table 7.3. Sources of canola-related know-why knowledge.

	Number of entities		Number of articles		Citation rate	
	Total	Canada	Total	Canada	Total	Canada
Private companies	130	63	358	150	7.9	8.0
Universities	660	31	3616	1139	6.0	6.0
Public institutes and agencies	670	29	2305	558	3.4	5.7
Total institutions	1460	123	6279	1847	5.9	5.9

Source: Author's calculations using ISI special tabulation of academic publications related to canola.

Two recent trends in the public sector, however, may jeopardize the development of new know-why knowledge. First, the recent efforts by the public institutions, both universities and research institutes, to enter the know-what business by patenting their knowledge has directed much of the research effort at the universities and public laboratories toward research that can be patented and exploited commercially. All of the universities, AAFC and the NRC have actively moved their organizations to protect and exploit their intellectual property (IP), first by setting up IP offices and then by patenting and licensing their 'know-what' innovations. To support that drive, faculty and research scientists are now rewarded for their commercial innovations, both with a share of the financial returns and with patentable work providing credit toward merit increases and promotion. At the University of Saskatchewan, for example, since 1956 faculty have been subject to an intellectual property agreement with the university that appropriates all of the ownership rights to the university, but shares any resulting royalties or, if the university fails to commercialize the product, gives the researcher the right of first refusal. One recent change is that faculty have increasingly derived status and promotions based on the amount of resources they can attract from elsewhere, which are easier to attract if the faculty does patentable work. Hence, faculty increasingly look to patents as a means of attracting financial support for their work, and ultimately as contributing to advancement and promotion.

At the same time, public institutions have shifted some of their effort to develop and exploit their know-how, selling their services either on a straight fee-for-service basis or through 'collaborations'. Many of the faculty and research scientists at the universities working on canola have entered contractual or collaborative relationships with the private companies (Alberta with Alberta Wheat Pool, Calgary with DowAgroSciences, Manitoba with Rhône-Poulenc and UGG, and Saskatchewan with Saskatchewan Wheat Pool) while Agriculture Canada has undertaken extensive fee-for-service research with private companies (e.g. the AgrEvo example above). The NRC, in contrast, has engaged in less fee-for-service work and instead has favoured extensive collaborations where both cash and intellectual capital are exchanged (since 1989 the NRC has done collaborations valued at eight times the volume of straight fee-for-service work (source: NRC special tabulations)).

The evidence suggests that the shift of emphasis towards know-what research has diminished the public output of know-why research from the public institutions. Five Canadian universities (Guelph, Saskatchewan, Manitoba, Alberta and Calgary) and two Canadian public institutions (AAFC and NRC/PBI) produced more than 1223 papers, or 25% of the total global effort, between 1981 and 1996 (Table 7.4). While the output of those institutions in aggregate has been relatively steady, the contribution of any institutions has undergone significant shifts. One overriding trend, however, is that after each institution began to actively seek commercial arrangements, their share of papers declined. Most researchers contacted at the universities indicated that private collaborators often demand extensive periods of non-disclosure (up to

Table 7.4. Output of the top Canadian research centres relative to the canola research undertaken globally.

	% of total papers per period		
	1981–1985	1986–1990	1991–1996
AAFC	9.9	6.2	7.8
NRC/PBI	0.7	0.7	1.2
University of Alberta	3.6	3.9	2.3
University of Calgary	0.0	1.0	1.3
University of Guelph	5.5	6.4	2.6
University of Manitoba	2.5	5.1	3.8
University of Saskatchewan	5.0	4.9	3.2
Total 7 institutions	27.2	.29.5	23.6

Source: Author's calculations using ISI special tabulation of academic publications related to canola.

5 years in many cases) of any results that may have commercial value, in order for the private partner to have time to commercialize those innovations. Some university research offices attempt to resist this trend because, at its root, it reduces the public good flowing from the university. Research offices are also concerned that many of these public–private research projects employ graduate students and post-doctoral students to do the laboratory work and the non-disclosure provision can at times make it impossible for a student to submit their work for examination.

A more disconcerting trend is that the quality of the work being published publicly in most Canadian institutions has dropped relative to the rest of the institutions doing research. As each institution shifted resources to working with private partners, the quality of the published work dropped (based on the number of citations). Institutions such as Alberta, which partnered with the three prairie Pools, has always had a substandard citation rate. The others,

Table 7.5. Relative citation rate for pure agricultural research papers produced by the top Canadian research centres for canola (average for all canola papers in period = 1.0).

	1981–1985	1986–1990	1991–1996
AAFC	0.84	1.22	0.93
NRC/PBI	1.25	1.68	1.70
University of Alberta	0.94	0.82	0.96
University of Calgary	0.00	3.09	1.69
University of Guelph	0.98	1.14	0.87
University of Manitoba	1.11	0.92	0.85
University of Saskatchewan	1.35	0.68	0.80

Source: Author's calculations using ISI special tabulation of academic publications related to canola.

however, delivered lower quality when private objectives began to set the research agenda. Calgary had a very high citation rate in the late 1980s but that rate has fallen as commercial activities have grown at the university (Table 7.5).

The drop in the quantity and quality of output does not necessarily mean that the public institutions are less productive. For the most part it likely simply reflects the shift of focus away from public-good research towards private activities. Given that researchers have not in many cases changed, it is likely that most of them are still producing good amounts of quality work, but they now seek to position their results to raise new funding or exploit the benefits, both of which are inimical to publishing in academic journals. Some argue that this publicly developed, proprietary information ultimately becomes public either in patents or in commercial products, so even if effort is diverted to know-what research, the benefits accrue to the broader economy. Industry Canada estimates that about 70% of the information contained in patents does not appear in any trade journal for at least 5 years after the patent has been granted, and at least 50% of this information is never published in mainstream technical journals. Given that in the US the information included in a patent application is kept confidential until the patent is issued (often up to 2 years after application), this diversion of output to the proprietary route slows the dissemination of information, and the cost and difficulty of accessing full patent information at times may make the results of the research inaccessible to most academics.

Further analysis of these data suggests two other observations. First, the National Research Council appears to be following a different path. Both the quantity and quality of the NRC output has risen, both absolutely and relative to the total. Comparing that to the results in the other public institutions suggests that NRC may be pursuing a different strategy. The NRC had been involved from the early years in helping to develop gas–liquid chromatography for use on rapeseed but after that let its research programme wind down. In the early 1980s the NRC made a strategic decision to re-enter the canola research business. The institution's first key decision was to steer clear of varietal development. Instead, the NRC concentrated on technology development, which was often pre-commercial and non-competitive. In 1989 the NRC adopted a new objective of actively seeking to work with private-sector companies. Since then 90% of the work with private companies has been via joint-venture collaborations that involve the pooling of money, staff and intellectual property. Only 10% of the work with private companies was fee-for-service. In contrast, AAFC has tended to do a significantly larger share of its work with others via fee-for-service, which usually does not result in any other private contribution than money. In the NRC case, the greater exchange of non-financial information appears to support the development of more useful know-why knowledge. This difference of operation may help to explain why the NRC has seen a significant rise in its volume and quality of know-why work, while AAFC has seen stagnation in both.

Second, further examination of those Canadian public agencies with more extensive collaborations (e.g. more than five joint papers) revealed interesting results (Table 7.6). The literature often suggests that cross-fertilization between

Table 7.6. Collaborations among Canadian institutions.

	AAFC	NRC	CGC	U of M	U of S	U of C	U of A	U of G
AAFC	290	–	–	–	–	–	–	–
	5.2							
National Research Council	7	24	–	–	–	–	–	–
	15.4	5.0						
Canadian Grain Commission	7	0	25	–	–	–	–	–
	3.3	–	5.9					
University of Manitoba	30	0	11	137	–	–	–	–
	4.8	–	1.7	5.2				
University of Saskatchewan	13	8	0	13	164	–	–	–
	5.3	12.9	–	5.5	5.6			
University of Calgary	2	8	0	0	0	35	–	–
	29.0	20.6	–	–	–	8.1		
University of Alberta	13	0	0	0	7	2	131	–
	3.1	–	–	–	2.4	2.5	5.5	
University of Guelph	4	1	3	1	2	0	0	200
	1.8	10.0	4.0	28.0	4.0	–	–	7.0
% effort via collaborations	21	50	43	23	21	26	14	5

The first element is the number of joint papers and the second is the average citation rates for those papers; the intra-institutional figures may include an occasional collaboration with other institutions not identified.
Source: Author's calculations using ISI special tabulation of academic publications related to canola.

academic institutions improves the quality of the work. It would appear that, in the canola sector at least, that does not generally hold. AAFC collaborated on 21% of its papers between 1981 and 1996 but the citation rate for those papers was slightly lower than for papers produced solely in-house. The Canadian Grain Commission collaborated on 43% of its articles but saw no boost to their citations from that effort. The universities produced similar results. Manitoba and Saskatchewan both collaborated with others on about half of their articles, and similarly earned lower citations on those articles. The Universities of Alberta and Guelph, which collaborated on a relatively smaller number of papers, showed similar results. In contrast, the University of Calgary, which collaborated with others (often the NRC) on about 25% of its papers, reported higher citations on those papers. The main exception, once again, was the NRC. About half of its papers were collaborations – albeit with a smaller group of collaborators, drawn from AAFC, and the Universities of Calgary and Saskatchewan – and their citation rate for those papers was more than three times higher than the rate for papers without collaboration. This tends to confirm the earlier point that the operation of the NRC is qualitatively different and that, as a result, it is producing different results. Although based on limited data, this tends to suggest that not all collaborations are necessarily effective.

Collaborations, especially in the university setting, may actually act to stifle know-how research.

Know-what knowledge

All participants in the Canadian canola research community – both public and private researchers – are now focused on trying to protect and exploit new innovations. In spite of the increased interest in patenting by public researchers, it is not clear that there is any compelling reason for the public sector to invest in research for that purpose. So far, private companies hold patents on most of the key technologies involved in the transformation of canola (see Table 11.2) and, given the variety of technologies now available, it is unclear whether there is any shortage of supply. Furthermore, although many point to the anticompetitive potential of patents, in this case they appear to have facilitated the international transfer of technology. Although few of the canola-related technologies protected by patents acquired by private companies were developed in Canada, the resulting technologies have, for the most part, been transferred to Canada to be used.

In spite of this, there has been an increase in recent years of patenting by public laboratories and universities as they have sought either to justify their existence or have sought new sources of funding for ongoing research. Public breeders and scientists who have recently moved from the public to private sector appear to be the most concerned about the privatization of the technologies, partly because it is a radical change in the culture of the industry, which was open and collegial until recently, and partly because licence fees and Materials Transfer Agreements place strains on already tight public budgets. One response of the public agencies has been to seek to protect and exploit any intellectual property they develop in order to keep 'in the game' and have 'chips' for bargaining with the private companies (Lesser, 1988a). NRC and AAFC, for example, both stipulate in their collaboration agreements that any resulting innovations will be the intellectual property of the public institution, and that the only right the private collaborator would have is first refusal on commercializing the technology. Although the public laboratories and universities have thereby begun to develop a portfolio of intellectual property, so far many of the innovations have had relatively minor commercial application. An assessment of the NRC records, for instance, shows that the revenues up to 1998 from the licensing of the technologies have not repaid the cost of acquiring the patents, let alone the cost of running their intellectual property offices.

Furthermore, some private companies collaborating with NRC or AAFC to develop new technologies which the public agencies then patent, are concerned that public intellectual property rights at times reduce the private sector's ability to use the technologies. They say that potential private-sector partners want clear ownership of intellectual property before they will partner to use this new technology. The public-agency patents do not appear to provide this assurance.

Instead of gross output of research, the greater 'know-what' issue appears to be access to the supply of technologies, or in the jargon of the industry 'freedom to operate'. Over and above any legal requirements to use or lose their patented innovations, most companies provide some access to their proprietary technologies, at least partly because few, if any, of the companies are fully self-sufficient. Although a firm may control one or more patents, it usually needs to license or joint-venture with some other patent holder to get access to parts of the transformation process for which they do not have patents. This reciprocal dependence ensures at least some access. Nevertheless, consultations with private companies have revealed that many firms use their patents strategically. There have been suggestions from within the industry that some firms have at times strategically withheld access to the best-patented technologies in order to slow competitors, while others appear to restrict access to technologies if the resulting product would compete with the patent-holder's product line. At other times, the patents are used by their holders to negotiate an equity stake in follow-on inventions.

Rather than having the public sector duplicate private research to ensure open access to technologies, the state would be advised to use the powers vested in its intellectual property rights regime or competition laws to encourage greater dissemination of non-rival, patented innovations, which would enable more access and hence greater spillover effects (see Chapter 11 for a discussion of this).

As far as competitiveness is concerned, it would appear that knowledge of basic transformation processes (i.e. know-what patentable recipes) can flow relatively freely across borders or continents, given a basic level of prior learning (or 'know why'). Therefore, it is not clear that the transformation processes need to be developed within the local research community for the research centre to produce commercializable products that add greater value to the local economy. Rather, it may be sufficient for the technologies to be available. Looking at the chain-link model of innovation, the key to success is taking these new ideas and ultimately placing them in the market to earn a return. The question for a region worried about competitiveness, then, is how to assemble the various pieces in a way that the local economy and society benefit from the effort, which points directly to the know-how and know-who elements of knowledge generation.

Know-who and know-how knowledge

Potentially, the most important public policy role, and arguably the key to regional competitiveness in a knowledge-based world, is the generation and transmission of the non-codified knowledge that holds things together – the know-how and know-who types of knowledge. Economists usually assume that the human elements of a firm or industry operate in some black box, governed by either the 'invisible hand' or a Walrasian auctioneer. In reality, human relationships that go beyond economic transactions hold an economy together – people develop skills and have relationships, which together convert bits of information into operable knowledge. This tacit type of knowledge is learned almost exclusively through experience. Researchers learn how to do things and whom to work with through trial and error.

Most of the innovation literature assumes that this know-how and, perhaps more importantly, this know-who evolves within corporations or institutions (Lundvall, 1992). That may hold true in an industry or within firms that are largely self-sufficient but, as noted above, there are few firms in this industry that have the internal capacity to undertake all the research and development necessary to create a marketable variety. Some companies may have that capacity within their global operations (e.g. Monsanto/Calgene, Aventis, Dupont/Pioneer Hi-Bred, Dow AgroSciences/Mycogen, Advanta) but in many cases working through the geographically dispersed multiple layers of these multinational enterprises is more complex and less cost-effective than buying-in from a more accessible and timely local source. Hence, although Monsanto, AgrEvo and Dow have giant research 'universities' and laboratories at their headquarters in the USA or Germany, they all have collaborated extensively in Saskatoon with both AAFC and the National Research Council. Furthermore, in knowledge-based industries training and upgrading are critical, making it essential for private researchers to interact with the broader research community. For all these reasons, most of the firms in the industry have developed an extensive community of networks with both collaborators and competitors, involving other private companies, universities, AAFC and the NRC (Fig. 8.10 shows the arrangements that currently exist in the global canola industry).

Collaborations may be the key to the true impact of public institutions. Both AAFC and NRC have extensive arrangements with each other, with universities and with private companies. In 1995–1996 alone, NRC had more than 31 arrangements – ranging from research agreements to collaborative work agreements and licences – that brought more than 65 guest researchers from other institutions into the NRC laboratories (NRC, 1997). In 1997–1998, the NRC welcomed 109 guest researchers into their laboratories (Table 7.7) and set a goal to expand that effort by at least 15% by 1998–1999 (NRC, 1997). The key

Table 7.7. Relationships between private companies and the Plant Biotechnology Institute.

	Total number of guest researchers in each laboratory	
	1995–1996	1997–1998
Other government departments	8	3
Private companies	50	72
Universities	8	27
PhD students	–	7
Collaborative arrangements	2	–
Alliances	1	–

Source: http://www.pbi.nrc.ca/96annrpt/bus.html

feature of these arrangements is that the core research team at NRC, comprised of permanent scientists, learns from all of the collaborations, thereby adding to the know-how knowledge as well as obtaining a visible, efficient point of entry for know-who knowledge.

As with most communities, proximity matters. Formal and informal face-to-face meetings and working side-by-side on laboratory benches and in the greenhouses are critical elements of both developing the know-who and transmitting the know-how. It is highly unlikely that the community would have developed if there were only competitive firms in Saskatoon – the non-competitive environment offered by AAFC and NRC creates the platform for these relationships.

The training and recruitment role played by the two key public institutions helps to solidify the sense of community. Both institutions have achieved critical mass, with a significant number of full-time permanent scientists working within their operations and with a regular flow of young postdoctoral scientists who work in the public laboratories on the way to a permanent career. An informal review of NRC staff changes revealed that about 30 staff scientists have left the NRC in the past 5 years for other research appointments in the Saskatoon area (Keller, 1998). Furthermore, the NRC collaborations provide a handy recruitment and screening system for the companies. Once a collaboration is begun between the NRC and a private company, the NRC usually hires a recently graduated PhD scientist, most frequently on a 1- to 3-year contract, to undertake the work. The permanent staff at the NRC collaborate with the private-sector scientists to manage the work of the contract employee(s). At the end, or commonly before the end, of the collaboration, the contract research scientist may be offered a permanent appointment with the private-sector collaborator. In one sense, the collaboration provides a screening process for recruitment. If the contract scientist does not meet expectations, the private company is not obligated to hire the person. The process also benefits from scale economies – the NRC has at any one time, on average, 15–20 contract scientists on staff, which enables them to develop the special mentoring and assessment skills that both help the entering scientist and reduce the costs of screening and management. During summer 1998, more than one-third of AAFC and NRC employees responding to a survey were in non-permanent

Table 7.8. Relative importance of contract employees in industry.

	Full time permanent	Full-time contract	% of full-time staff on contract
Public sector	103	61	37
Private firms	94	16	15
University	19	18	49
Other	19	6	24
Total	235	101	30

Source: Canola industry employees survey, August 1998.

contract positions, while almost half of university respondents were contract employees. The use of contract employees by companies themselves was only about 15% (Table 7.8).

It is impossible to state conclusively whether an efficient amount of know-how and know-who knowledge would be generated without the public institutions. One gets a sense of their importance in the system when one examines the list of collaborators and their location. Even firms not resident in Saskatoon have developed extensive links to gain access to the knowledge in those two institutions. This suggests that any spillover benefits from the know-how and know-who located in Saskatoon are real and may not move far from Saskatoon, except through these collaborations.

In conclusion, the shift from proprietor to partner in the Canadian canola research community has revealed a number of aspects of knowledge-based growth. There would appear to be a spectrum of potential efforts and public returns, with the greatest returns to public effort possibly coming from the development of know-why, know-how and know-who knowledge. Working from the other end of the innovation–production chain, there appear to be few obvious incremental returns from varietal development and doubtful benefits from investing in the pursuit of know-what, patentable knowledge.

Promoting Growth

The Canadian and provincial governments have explored one additional role: promoter of canola development. Unlike in the role as proprietor or partner, where the public sector has actively participated in the creation or use of knowledge, in the role of promoter the various governments have invested time and money to create a climate for private initiative to flourish.

The recent challenge for economists and governments has been to determine what conditions and policies can spur private activity. Harvard Professor Michael Porter (1991) has undertaken the most extensive analysis of the market and policy structures that underlie localized, often knowledge-based industrial development. In the early 1990s he examined a number of regions of the world to determine the factors that led to localized growth. He characterized them as: (i) factor conditions (land, labour, capital, and technology); (ii) related and supporting industry; (iii) competitive demand conditions; and (iv) firm strategy, structure and rivalry. This part of the study examines those factor conditions provided directly by the public sector, while later chapters will look at others.

Four specific public interventions have been undertaken to encourage and support private investment into the industry. First, the public sector in Canada and western Canada has signalled to both potential investors and to related and supporting industries that the biotechnology sector, and the canola sector in particular, are high-priority growth targets. In 1983 the government of Canada developed and introduced a National Biotechnology Strategy to ensure that

Canada realized the potential benefits of biotechnology. To achieve this goal, the Strategy established several objectives: to focus on research and development in certain strategic areas; to ensure an adequate supply of highly trained personnel; and to create a climate conducive to investment. Another component of the Strategy has been to provide a science-based regulatory system for the protection of human and animal health and the environment. This strategy led to restructuring in the operations of public agencies, to new opportunities with many funding sources and to renewal of the regulatory system. In 1998, the federal government reviewed and brought forward a new Canadian Biotechnology Strategy. Meanwhile, all of the provincial governments introduced technology policies that supported and encouraged knowledge-based growth. This policy focus led to tangible investments by many of the governments in infrastructure and various programmes and services that were designed to increase private investment into the industry.

Infrastructure investments are the most visible efforts. AAFC, NRC and the Saskatchewan government all invested heavily after the mid 1980s. AAFC added incrementally to its greenhouse and laboratory capacity in Saskatoon throughout the 1980s and early 1990s. After the decision in 1995 to consolidate all of its canola research in Saskatoon, AAFC began construction on a Can$37 million expansion and retrofitting of the Prairie Regional Research Centre on the University of Saskatchewan campus. The brassica germplasm collection was also moved to Saskatoon as part of the consolidation. With the re-entry of the NRC into canola research in the 1980s, significant investments were made in the Plant Biotechnology Institute in Saskatoon. By 1998, the research programme had groups working on brassica biotechnology, seed-oil modification, growth regulation, promoter technology and gene expression. These investments by the public agencies provided the physical space to undertake more extensive collaborations and contract research for private firms.

Equally importantly from the perspective of promoting more private-sector activity, the province of Saskatchewan invested heavily in the development of Innovation Place as a research park to attract biotechnology companies. Innovation Place was initially developed, beginning in 1976, as a research park for information technologies, and in 1980 the first two buildings were completed. In the late 1980s management in the park identified biotechnology as a potential growth area. By then the university and the park were already home to AAFC, NRC, the Saskatchewan Research Council and two biotechnology operations, the Saskatchewan Wheat Pool research programme and Philom Bios. In 1987 a consultant identified a demand for greenhouses for research firms and by late 1989 a Can$3.5 million, 26,000 ft^2 research greenhouse complex was opened, with six laboratories and 16,000 ft^2 of growing areas in ten glass units. Although the NRC Plant Biotechnology Institute was a foundation tenant, precommiting to rent 60% of the space, it took 5 more years for the rest of the space to rent, resulting in a Can$1 million loss for the Innovation Place enterprise. Since then, Innovation Place has invested heavily in infrastructure to support biotechnology development. In late 1990, Innovation

Place opened the Atrium Building and made a strategic decision to put all agricultural biotechnology firms there. Saskatchewan Wheat Pool was the first tenant. To encourage others to come, Innovation Place management offered to pay for companies to move their enterprises into the building, for phone installation and for new stationery. In addition, the centre offered reduced or free rent to commodity groups, such as the Saskatchewan Canola Growers Association, which was then used to encourage companies, such as Monsanto, to locate in the building. In 1998 the Atrium had 23 biotechnology companies on site and 16 other research facilities or companies (e.g. Saskatchewan Wheat Pool and AgrEvo) located in the park in purpose-built facilities. In 1997 the greenhouse complex was expanded by 2975 m^2, with 1860 m^2 of glass growing areas and 1115 m^2 of laboratory facilities. Unlike in the earlier period, all of the new space was pre-sold. In 1998 the 6500 m^2 Atrium fermentation support facility was opened.

The key to the role of these investments is not simply provision by the public sector. Rather, the greenhouses, laboratories and Atrium complex provide an extremely flexible operating environment for start-up or expanding enterprises. Unlike other centres, where firms are required to build and own their own facilities, which requires both financial resources and time, Innovation Place owns almost all of the buildings (the province owns the POS Pilot Plant and the Canamino building) and rents them. Hence the province has effectively reduced the risk and capital costs of operating in Saskatoon. In addition, a number of other public agencies (e.g. Saskatchewan Opportunities Corporation and Ag-West Biotech Inc.) have offered investment capital or grants to a number of companies to locate at Innovation Place (e.g. Limagrain, Plant Genetics Systems and Mycogen accepted investments, while Pioneer Hi-Bred and Cargill/Intermountain Canola declined).

Although the multi-tenant approach has realized some efficiencies (potting sheds, computer and internet facilities, dining room, common areas and professional and social programmes), the rents at Innovation Place are significantly above rental rates in other parts of the city. A survey of firms both operating at Innovation Place and those operating elsewhere suggested that the offerings of the park were valued most at the point of entrepreneurial start-up or expansion. The flexibility offered by the rental system was valued by these firms and justified the higher rents. In contrast, established companies have expressed less interest in Innovation Place and a number of firms have relocated their operations away from the park to save expenses.

A second approach to encouraging private investment has been monetary incentives. Incentives have included a variety of fiscal measures that are both generally available and targeted, joint public–private-sector venture loan funds, and one-off capital investments for selected projects. A study undertaken by ABT Associates of Canada (1996) for Finance and Revenue Canada on the performance of the federal income tax incentives for scientific research and experimental development (SR&ED) surveyed 501 firms undertaking research, 10% of which were in agriculture, forestry and fishing. The survey asked investors to

Table 7.9. The importance of different forms of government scientific research and experimental development support – by rank order of mean scores.

Means of support	Mean scores
The federal SR&ED tax credit	
basic operation of system	7.8
refundability of the federal tax credit	7.7
expensing of capital expenditures	5.9
Intellectual property rights	6.3
Provincial tax incentives for R&D	5.8
Government grants and contracts for R&D	5.1

Note: these are averages; each firm was asked to rank the measures from 1 to 10; the items were rotated randomly in the survey to avoid order bias in responses.
Source: Survey of Scientific Research and Experimental Development Claimants – project report prepared by ABT Associates of Canada for Department of Finance and Revenue Canada, June 1996 (http://www.fin.gc.ca/toce/1998/resdev_e.html).

rank different forms of government support. Table 7.9 shows the results. Firms clearly valued the basic federal tax credit more highly than intellectual property rights, provincial tax credits or government grants and contracts.

No specific data are available on the flow of funds to private companies in the canola sector as a result of the general SR&ED tax credit system or the provincial tax systems, but estimates can be made on the domestic investment in the canola industry. The basic 20% SR&ED tax credit (for large corporations) can be applied to the sum of the estimated investment in canola by private companies, plus the investment made by the commodity associations (e.g. Canodev, Canola Council of Canada), which have structured their operations to take advantage of the tax credit. Given that a total of Can$330 million has been invested by private companies and commodity associations in Canada between 1989 and 1998, the SR&ED credit at even the lowest (large company) rate could have been worth Can$66 million.

The provinces also provide significant tax relief on R&D expenditures. Provincial and territorial governments provide full deductibility for current and capital expenditures on qualifying research and development. Provincial rules generally adopt federal rules relating to the definitions of qualifying work and expenditures, and the treatment of government assistance, non-government assistance and the federal SR&ED tax credits. A number of provinces with canola-related R&D also provide tax credits. Manitoba introduced a 15% non-refundable Research and Development Tax Credit in its 1992 budget. Ontario provides three provincial income tax incentives for SR&ED. The Super Allowance is an additional (or bonus) deduction of 25–52% for research and development expenditures over past levels; this allowance is not considered to be government assistance and, therefore, does not reduce the amount of expenditures for federal and provincial tax credits and deductions. Effective from

1 January 1995, Ontario introduced a 10% refundable Innovation Tax Credit available to those smaller Canadian companies eligible for the enhanced rate of the federal SR&ED tax credit. Saskatchewan introduced in its 1998 budget a 15% R&D tax credit. Alberta does not have an R&D tax credit.

A recent study by the Conference Board of Canada (Warda, 1997) provides an international comparison of income tax support for R&D in Organization for Economic Co-operation and Development (OECD) countries. The R&D tax systems were ranked by comparing the minimum benefit–cost ratio at which an R&D investment becomes profitable, given a country's income tax treatment for corporations performing this work. That study showed that, after taking account of both federal and provincial incentives, Canada's income tax treatment for R&D investments was the most favourable among the G-7 countries, with an implicit subsidy of about 28% compared with second-place Australia, with a 10% implicit subsidy. The US, France and the UK were the only other OECD countries with positive subsidies. Italy, Japan and Germany were the least favourable.

The other generally available programme is the Industrial Research Assistance Program which offers small and medium-sized businesses financial assistance for projects at the precommercialization stage. These repayable contributions can be used to support later stage development of products, especially those near-market development and demonstration projects for new or significantly improved technological products or processes. The programme, offering support to private companies and public–private collaborations, provides up to 50% of the funds to a maximum of Can$250,000 per project, for up to 3 years. Approximately 28% of the commercial companies in the canola industry surveyed in 1998 confirmed that they had used the programme in recent years.

In addition, Saskatchewan has two specific programmes designed to support agronomic research. The Agricultural Development Fund, created in 1985, invested Can$5.7 million in canola-related research up to 1998 while the Agri-food Innovation Fund has invested both in infrastructure (Can$3 million or one-third share of the fermentation plant at Innovation Place) and Can$0.3 million in 1997–98, its first year of operation (SDAF, Special Tabulations, April, 1998).

Meanwhile, the federal government and some of the provinces have engaged with private companies to establish venture loan funds, targeted either exclusively or partially at the biotechnology industry. Western Economic Diversification has collaborated with the Farm Credit Corporation and two commercial banks (the Royal Bank of Canada and the CIBC) to offer venture loans to developing companies (Table 7.10). Meanwhile, the Saskatchewan Opportunities Corporation (SOCO) partnered with the Bank of Nova Scotia to create an 'opportunities fund' to invest in knowledge-based firms. Sources at the three funds confirm they have invested in canola-related research enterprises but decline to give details.

Finally, all of the provinces with interests in canola-related biotechnology

Table 7.10. Public sources of funds for canola research for the 1989–1998 period.

Source	Estimated Can$ investment	% private firms using	Application
Federal tax credit	$66 M	100	All firms doing canola research
Provincial tax credits	$7 M	100	1992, Manitoba; 1995, Ontario; 1998, Saskatchewan
IRAP precommercialization assistance	$7 M	28	Grants of up to 50% or Can$250,000 for 3 years
Public–Private Venture Loan Funds			
WED–RBC	$7.5 M	–	$30 M fund makes loans of Can$50–500 K
WED/FCC–CIBC	$25 M	–	$100 M fund makes loans of up to Can$1 M
SOCO–BNS	$2.5 M	–	$5 M fund makes loans of Can$100–500 K
AFIF	$3.3 M	16	
ADF	$5.7 M	8	

Source: Author's estimates based on programme information, discussions with programme managers and Canola Survey.

have invested directly in new start-up or relocating research companies. Saskatchewan, Alberta, Manitoba and Ontario all have invested in at least one venture. In total, the four provinces have contributed an estimated Can$10 million in nine companies.

In total, there was an estimated Can$520 million of canola-related research over the 1989–1998 period, with about Can$190 million undertaken in public universities and public laboratories. The remaining Can$330 million of investment was directed by private companies. The sum total of public support through tax credits, grants, loans and public infrastructure programmes is estimated to have contributed at least Can$110 million towards funding the private research, or about one-third of the total private expenditures. In other words, the public sector directly or indirectly funded at least 57% of the canola research activity over the past decade, yet acquired ownership or rights to only a small portion of the resulting technologies or varieties being commercialized. Chapter 16 contains a more detailed analysis of the economic and policy implications of this large public subsidy to private research.

It is next to impossible to determine the exact impact of these policies and incentives on private research. Finance and Revenue Canada, which undertook a review of the SR&ED tax credit system in 1996, concluded that the system generated a 1.38 cost-effectiveness ratio in 1994, which means that every dollar of tax revenue foregone generated approximately Can$1.38 of incremental

R&D. The study also reviewed other recent estimates of cost effectiveness and showed that their estimate was about mid range, with Asmussen and Berriot (1993) estimating only a 0.26 ratio for France in 1985–1989 and Hall (1993) estimating a 2.00 ratio for a US incremental tax credit in 1981–1991. Nevertheless, the tax system is only one of the partnership options used by governments. The combination of direct public research expenditures, infrastructure investments, general tax incentives, industrial grants, loan funds and targeted industrial investments has touched virtually every actor in the canola research system operating in Canada. Although it is unclear how much of the public funds provided to private enterprises may have leveraged incremental investment, almost all of 25 companies doing canola research surveyed in 1998 confirmed that without the support discussed above, they would not have started, would not have relocated to Canada or would not have expanded their research programmes as aggressively as they did. In that sense, governments in Canada have been prime promoters of canola development.

Conclusion

The combination of a transformation in the innovation process – away from a supply-driven linear model and towards a demand-pull, chain-link system – and the introduction of a new intellectual property rights regime have both worked to shift the impetus in canola development towards private-sector investment and away from public involvement. As a result, the public sector in Canada has been challenged to find a new role. The government has, to the discomfort of many public-sector scientists, relinquished its role as proprietor and instead become a partner and significant promoter of development, often at high cost. In Chapters 14–16, possible returns realized by this investment are discussed.

The Role of Private Firms

8

Peter W.B. Phillips

Introduction

The recent changes in the industrial structure in the canola sector amply illustrate the impact of new technologies on hitherto stable industries. The opportunity presented by biotechnology to manage the research process to deliver custom products has presented an attractive investment opportunity for some private companies while creating a threat or risk to many existing businesses.

In the past decade the industry has been transformed from a disjointed group of small, medium and a few large chemical and food companies producing a quality oil product for sale to a few markets. Now it is becoming an integrated sector where the production system from basic genetics right up to the final consumer is increasingly tightly managed and controlled within a small group of industrial networks.

Even in the period before 1980, when there were no private investments in research or seed development, the private sector was an integral part of the canola business. The seed was developed by the public sector but the private firms and cooperatives multiplied the seed for sale, marketed the seeds to producers, produced and sold the chemicals and herbicides used in the production of canola, purchased and marketed the crop to domestic processors or to the offshore market and processed the seed into oil products.

The emergence of the new biotechnologies in the 1980s precipitated a shake-up in the industry. As Just and Hueth (1993) point out, the chemical companies quickly realized that the new ability to manipulate the genetic coding of crops changed both their markets and their concept of their business. The new technologies created the opportunity for them to develop new varieties tolerant to their patented herbicides, thereby opening new markets for their product. In the canola industry, Monsanto and AgrEvo did just that, investing in research to isolate the genes that were tolerant to their patented herbicides and then inserting them into commercial canola varieties. Canola previously was not a target crop for those broad-leaf herbicides; the result of these new herbicide-tolerant varieties is a sharp shift in the market shares of chemicals in the canola sector. With approximately 50% of the acreage planted in 1998 with one of the three main herbicide-tolerant varieties, the majority of the chemicals market has been captured by new chemicals.

At the same time, some companies decided that the biotechnology-based parts of the production system had the potential to produce significantly higher profits than many other areas. Many companies that were upstream or downstream of the farmer restructured to capture this business, often divesting themselves of lower-return activities and replacing those activities with direct investments or acquisitions of knowledge-based parts of the production system.

Looking at the organization and structure of industry can help to determine whether innovative activity supports or discourages globalization of research and production across a variety of markets. The evidence from other studies suggests that innovative activity is, as the theory suggests, inherently indivisible at certain levels and will tend to be located in relatively few discrete locations rather than in every market. Patel (1995) used US patent data to examine whether large, global firms globalize their technological activities. He concluded that the most internationalized firms 'are not in the "high-tech" product groups that are normally labeled as having a world mandate. On the contrary they are in product groups where adaptation for serving local markets is important.' He argued that this is consistent with early analysis by Vernon (1966) and Porter (1990) and the 'national systems of innovation' literature that what happens in home countries is still vitally important in the creation of global technological advantage for firms. The following analysis confirms that this is equally true in canola. In short, individual firms appear to have created global research networks where specific steps in the transformation are undertaken in discrete locations, and then assembled the elements in the key markets. As Canada is the primary market for transgenic canola varieties, the bulk of the assembly has occurred in Canada.

This chapter provides an examination of the transformation of the industry from a stable situation, with largely arm's-length relationships between different producers in the production system, towards a highly volatile and rapidly evolving, vertically coordinated global production system.

The Production Chain

The core canola production chain has traditionally involved input manufacturers and marketers, farmers, processors and marketing and logistics companies. Figure 8.1 demonstrates the scale and distribution of value in this traditional chain. Farmers purchased, often through arm's-length transactions, seeds and other inputs worth about one-third of the value of their output. Downstream of the farm gate, marketers and processors added another 50–100% to the value produced on-farm as the product flowed towards consumers. Although there was a public research system devoted to canola development upstream of this chain, it was difficult to say that it was integrated into the supply chain. Grounded on university research and developed through public research programmes, the resulting innovations were often less driven by market needs and demands than by government or producer interests. Downstream of the oil processors in the core chain were a large number of food-processing companies, wholesale and retail operations and, ultimately, millions

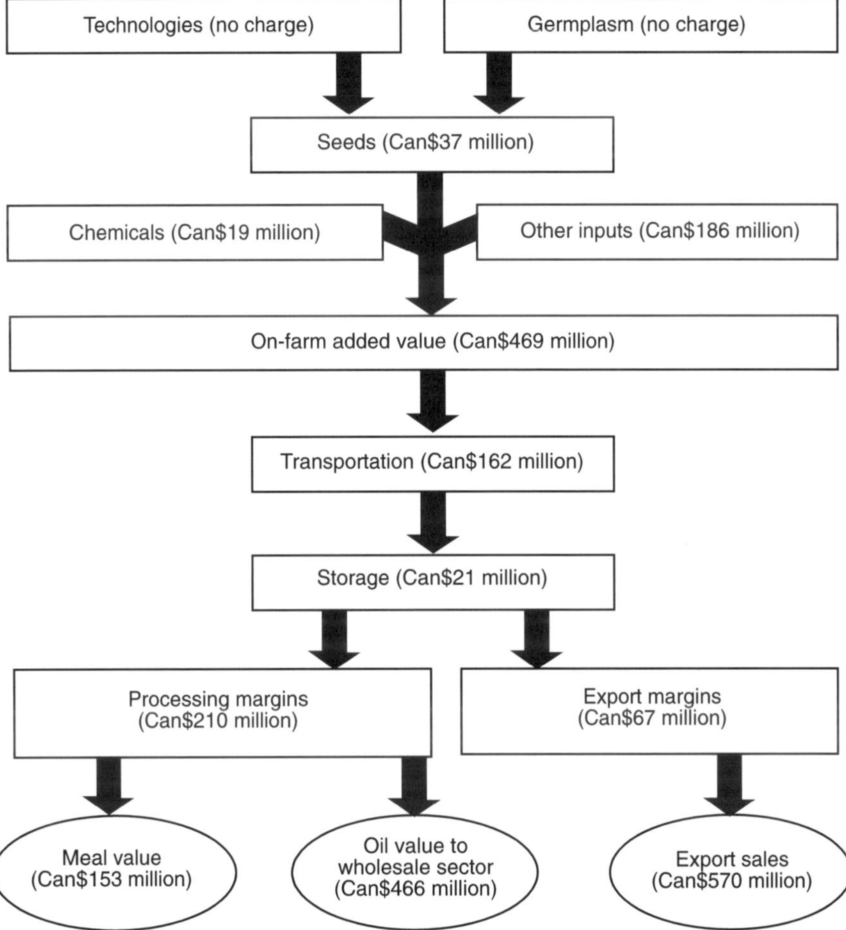

Fig. 8.1. Saskatchewan's canola supply chain in 1990. (*Source:* Weights based on Statistics Canada economic impact evaluation undertaken using the interprovincial input–output model, 1990 (1997).)

of consumers domestically and abroad. These could not be viewed as part of the core industry, as once the seed was crushed into oil, the subsequent marketing and use of the oil was largely unrelated to the Canadian industry. Rapeseed and canola oils were often only one of many potential oils for marketing or processing and were simply placed into the global commodity markets. Consumer demands only episodically fed back explicitly into the system.

With the advent of private capital in the research and seed-development industries, the chain in the 1990s clearly extended further upstream into research, while the new ability to manufacture custom oil products forged links downstream into the processing, food and other value-added industries. Now, there is a more clearly defined and increasingly coordinated production system that links the consumer to the basic genetics business.

Theory

Traditional, neoclassical microeconomic theory does not provide a rationale for institutional structures because it assumes that there is full and transparent pricing of all factors of production and output features, in such a way that prices provide complete information about consumer wants and needs. This may have made some sense in a world of physiocratic farmers, who produced true commodities that were homogeneous and virtually indistinguishable from other, like produce. In practice, however, as noted before, the supply chain involves many actors, and produces output with natural heterogeneity and many output traits, each with widely different market values. In addition, many of the inputs in the chain are provided by incomplete markets, with little transparent pricing available. In the canola case, the bulk canola crop exhibits wide variations in oil quantity, oil profiles, meal nutritional values and, historically, in percentage of unwanted qualities (e.g. erucic acid and glucosinolates). The grading system developed and administered by the Canadian Grain Commission provides part of the discovery mechanism, grading based on percentage of oil, physical shape of the hull and presence of erucic acid and glucosinolates, but the system is not fine enough to provide full price discovery for the full value of the oil or the meal. Hence, some of the potential value in the crop is dissipated in the system. Furthermore, many of the new varieties (e.g. with industrial oils, nutraceutical and pharmaceutical properties) are not at all addressed in the current grading system. Meanwhile, the input markets have increased in importance, with patented technologies and proprietary germplasm now underpinning the industry. The markets for these factors are incomplete, if not absent, with no transparent prices or standardized auction system. Hence, the industry cannot rely on the impersonal price system – it requires some form of institutional structure to manage the potential failures in the marketplace.

As discussed in Chapter 6, the theory of 'institutional' economics helps to define the potential for industrial structure to assist firms to overcome the limitations of the price discovery system and to capitalize on knowledge-based market opportunities. The theory suggests that firms exist to manage risk – namely those risks and uncertainties resulting during price discovery, negotiation and monitoring of transactions. This 'transactions-cost' approach posits that firms integrate vertically to avoid the cost of contracting. Contracts are costly because markets operate with 'bounded rationality' and individuals and companies act opportunistically. The cost of those contracts varies with the relative bargaining position of an actor, which is determined by the specificity of the assets each party has invested. The firm with assets that have little alternate use is most at risk of having its returns bid away by other actors in the production system, and it therefore has the most incentive to coordinate its activities vertically. A complementary approach assumes that firms ('principals') will contract with 'agents' to avoid market risk. Once again, there is a concern that 'opportunistic' agents will take advantage of any imbalance of power, in this case resulting from the inability to measure either their contribution to the total output (called non-separability) or the quantity of effort invested in the task (called task programmability). In short, the greater the measurement problems, the higher the cost of buying-in relative to the cost of doing-in, with the result that vertical coordination is more likely to be pursued.

Putting together the two approaches to institutional economics, one can predict the organizational form of vertical integration based on the degree of asset-specificity, task-programmability and non-separability (Table 1.2). If these factors are low, spot markets are optimal. However, as these factors increase in importance, a greater degree of vertical organization is required to manage the resulting higher risk of opportunistic behaviour. At the extreme, a formal hierarchical relationship may be necessary to manage the risk.

Looked at in this way, one can see that if the agrochemical and seed businesses simply produced the inputs and left farmers, marketers and processors to produce the value and pay the input suppliers their due, there is significant potential for either the value to be lost in the production system or for downstream actors to bargain to capture the returns. Hence, almost all of the companies that have targeted the canola sector for investment have moved to vertically coordinate their production system, spanning from genetics all the way to the consumer. Thus, these companies have sought to manage the opportunistic activities of other actors that would either diminish total return to the product or redistribute shares of the return.

Institutions evolve for other reasons than simply to overcome the limitations of the price system. Risks in the financial system have tended to push investors to invest through intermediaries (traditionally through conglomerates), which encourages further corporate concentration. It is likely that many of the investments in the agri-food chain have been supported, if not encouraged, by the availability of capital to the expanding companies. In short, many of these ventures are relatively good places to invest in, to earn an above-average return. Looking at the enlarged supply chain, spanning from genetics to the end consumer, one can anticipate the potential interest of profit-seeking firms in the various stages of the production system by looking at the ability of the owner of the assets to exclude others, or by looking at the structure of the market and the extent of competition. Extraordinary profits (i.e. economic rents) can be realized in two specific circumstances. One would expect profit-seeking firms to be attracted to those parts of the production system where there are legal barriers to entry – namely clearly defined, defensible property rights that can be exploited – or where there are few competitors due to scale economies. Looked at from that perspective, one can hypothesize that profit-seeking firms would invest as core business activities in germplasm collections, patents on applied research and patented gene constructs, all of which now exhibit clear property rights which facilitate exploitation. In contrast, there is little economic incentive flowing from these characteristics to invest in basic research, many other production inputs, farm production, marketing or exporting. Furthermore, it is unlikely that primarily seed companies would move downstream into the oligopoly agrochemical business, but for those companies with agrochemicals, moving upstream into the research business and downstream into the seeds business (which also offers access to proprietary germplasm) makes good economic sense (Table 8.1).

This approach implies that firms would pick and choose to enter only a few disjointed parts of the supply chain. In practice, however, companies seldom allow their search for gains from innovation to be left up to the chance operations of the marketplace. Although excludability is defined initially as the result of the attributes of the innovation, by the way they structure their operations firms can improve the odds of gaining a larger share of non-excludable benefits. This framework is used to examine the evolution of the private sector activity in the canola industry over the past 20 years.

Commercializing Knowledge

The development of the commercial agricultural biotechnology industry has been described by Robert Swanson, one of the founders of Genetech, as 'a kind of marriage between science and business' (Kenney, 1986). In the first instance, the scientist was key. Novel ideas that have commercial application were developed in a large number of widely scattered locations almost, but not quite, randomly spaced about the world. Key developments were made in laboratories in Saskatoon, Calgary, Davis, Ghent and New Jersey, to name just a few. Many of these innovations were made in academic or public research agency laboratories. The prime means of commercializing those ideas has been for the scientist or academic to work with their university to patent and then license the technology, or alternatively, for the scientist to get the rights to the innovation and work with a team made up of the scientist, an entrepreneur, a manager and a venture capitalist (Kenney, 1986). Each brings their skills and experience to the commercialization of the product. In the canola case, many of the early commercializations of technology occurred in that way.

In other knowledge-based industries, such as computer programming and

Table 8.1. Factors influencing market organization.

Industry stage	Barriers to entry	Market characteristics	Potential for rents
Germplasm	Via trade secret	Finite supply	High
Basic research	None	None	None
Transformation processes	Via patents	S monopoly[a]	Medium–high
Genes	Via patents	S monopoly[a]	High
Seeds	Via plant breeders' rights	Oligopoly (4C ratio = 60% but volatile)	Medium
Herbicides	Via patents/trademarks	Oligopoly (4C ratio = 65%)	Medium–high
Other inputs	None	Generally competitive; contestable	Low
Farm production	None	Competitive (25,000 growers in W. Canada)	Low
Marketing	None	Competitive; contestable	Low
Processing	Via trademarks/brands	Oligopoly (4C ratio = 100% in Canada)	Medium

[a]S monopoly, Shumpeterian monopoly acquired by being the first to innovate.
4C, four company concentration ratio.
Source: Author.

advanced instrumentation, that process led to a large number of small, entrepreneurial firms that provide the new innovations and new products to the marketplace. In practice, this has led to a pyramidal distribution of firms, with many smaller start-up firms, fewer medium-sized firms and a few large firms. For a number of reasons, the current distribution of firms in the canola sector is reversed. There are only a few small, mostly service-providing firms remaining in the industry. Most of the independent, small- and medium-sized research firms (e.g. Calgene, Mogen and Mycogen), as well as most of the smaller seed companies that marketed their products (e.g. Newfield Seeds), have been acquired by a few large multinational seed and agrochemical conglomerates. Only a few large actors remain and they are seeking to develop webs of relationships that control the development and flow of product even more tightly. As a result, entrepreneurs are few and far between. The industry is now dominated by large actors, bankrolled by profits either from agrochemical or seed product lines.

Private activity in canola began to grow about 1985, with a sharp acceleration of private-sector research and investment into canola development. Between 1986 and 1998, the private sector invested more than Can$730 million (inflation-adjusted to 1998) globally in canola research and development (Canola Research Survey, 1997). Four key factors contributed to the invasion of private money into the industry.

First, and perhaps most importantly, in 1985 the USA affirmed low erucic acid rapeseed oil as a food substance 'generally regarded as safe' (GRAS) and in 1988 the use of 'canola' on food labels in the USA was approved. Rapeseed oil had been used as a human oil for a long while, but medical studies had shown recently that in young laboratory animals fed diets high in rapeseed oil, the animals lacked an enzyme that would break down the 22-carbon chain of erucic to an 18-carbon chain fatty acid that the body could use. As a result, erucic acid tended to accumulate around organs that use fat as a source of energy, such as the heart and adrenals, until the animal developed the enzyme and the fat was utilized. However, after the fat deposits were gone, researchers found muscle lesioning of the heart muscles (Downey, 1998). When Agriculture Canada bred a rapeseed variety with less than 5% erucic acid, the potential for the oil market was expanded. Early in the 1980s the market position of canola improved as longitudinal health studies revealed that the unsaturated fats in canola pose less health risk than animal oils and many other vegetable oils, such as palm and coconut (Malla, 1995).

Second, breakthroughs in breeding methodologies improved the economics of private-sector breeding. The general practice of shuttling seeds between northern and southern climates and the application of computers as aids in the laboratories significantly shortened the traditional breeding period. This was also the period when a number of companies recognized that biotechnology innovations and processes (i.e. cell fusion, genetic recombination, polymerase chain reaction and genome maps) could significantly reduce the average time to develop new canola varieties. The average variety took about 10–15 years to develop in earlier periods. If a company used all the innovations aggressively, it could reduce the period to develop a sister variety without novel traits down to 3 years and the period to develop one with novel attributes (e.g. tolerance to a new herbicide) to 7 years. The length of time until the company gets a return on an investment often determines which investments will get made. A company would need to expect a 50% increase in the return for a 7-year project compared to a 3-year investment, in order to get the same present value on that project. If the average period of development was 12 years, the return would need to be 250% higher than that for the 3-year project to yield the same net present value. Clearly, as the period of time for development shortened, more private interest was generated. Related to this was the different reproductive characteristics of the two varieties. *Brassica napus* is self-incompatible, which makes it relatively easy to control the pollination process and get pure breeding lines. In contrast, *B. rapa* is self-pollinating, which makes it much more difficult to control the breeding process. For this basic reason, most of the private research has focused on *B. napus*. Only 22 *B. rapa* varieties were developed by private companies between 1985 and 1996, compared with 87 *B. napus* varieties; recently, with the development of an effective hybrid system which controls pollination, there has been increased interest in *B. rapa* by some firms.

Third, financial deregulation in the early 1980s in North America led to a large pool of capital seeking new investment opportunities, which coincided with the budget crunch in universities and public institutes and new pressures to commercialize new technologies for profit. As a result, the biotechnology industry became a focal point for private investment. County NatWest WoodMac tabulated the scale of agricultural biotechnology R&D spending in 1990 and identified more than US$390 million of reported spending on plant and animal health products in that year alone. Hodgson (1995) estimated that the total spending on all research by the key multinational companies also involved in canola research totalled about US$148 million. Global canola research in 1990 was estimated at about Can$70 million. The private portion of that investment, totalling approximately Can$35 million, accounted for approximately 16% of the crop biotechnology investments by these companies.

The fourth, and perhaps most crucial, factor was the introduction of intellectual property rights for biological inventions. In 1980, a US Supreme Court decision (Chakrabarty vs. Diamond) explicitly allowed patents for living organisms, and in 1985 US plant patents were explicitly allowed (Lesser, 1998b). Meanwhile Canada, in 1985, modified the Seeds Act to allow varieties to be introduced that were 'as good as' rather than 'better' than reference varieties, and then in 1990 Canada assigned intellectual property rights to private developers via the Plant Breeders' Protection Act. These regulatory changes, combined more recently with new hybrid technologies and new contracting methods, helped to secure the excludability of innovations, assisting firms to capture the economic rents of innovation.

These four factors set the stage for intensive innovative activity in the sector in the 1990s.

The Emergence of Private Effort

The first visible evidence of private investments in canola was the rapid introduction and effective privatization of the seed stock. Before 1985 there were no private varieties registered for planting. With the change in incentives and the application of new technologies, private companies have dominated the lists of new varieties released since 1985. Between 1985 and 1998, 175 new varieties were registered in Canada, 82% of them by private companies. Most of the varieties registered by public breeders were also funded and supported by private companies. In the past 4 years, private breeders registered almost 90% of the new varieties.

Two features of the evolving industry have come to light. First, as shown in Table 8.2, there has been significant growth in the number of independent breeding programmes, with an average of more than 18 different entities undertaking research in the decade. Although the number of breeding programmes has expanded, there is significant change within the industry. As the larger companies (AgrEvo, Pioneer, Dow, Cargill, Monsanto) have bought smaller independent operations, new ventures have sprung up to fill the gap. At the same time, there has been a marked tail-off of public activity, with both AAFC and the universities withdrawing from certain breeding activities.

As recently as 1985, virtually all of the seeds planted in Canada were developed by one of the public institutions (Table 8.3). In just 12 years, the market share of publicly bred varieties dropped to about 10% of the acreage seeded. Global agrochemical and seed companies such as BASF/Svalof AB (Swiss/Swedish), Zeneca (UK), AgrEvo (German), Limagrain (French) and Pioneer Hi-Bred (US) captured more than half the seed market (in 1997, the latest year with data, there was more than one-quarter of the market unallocated, which was likely made up of bin-run seeding and undisclosed purchases of private seed). Although this precipitated some modest restructuring of the industry, it heralded a much more fundamental change that was moving from universities to private laboratories.

Table 8.2. Independent breeding programmes producing at least one commercial variety per period.

	B. napus		*B. rapa*	
	Private	Public	Private	Public
1950–1959	0	1	0	1
1960–1969	0	2	0	2
1970–1979	0	2	0	1
1980–1984	0	3	0	1
1985–1989	5	4	2	1
1990–1994	13	3	3	3
1995–1998	14	3	3	1

Source: Author's tabulations of CFIA variety registration data.

Table 8.3. Canola market share (% of acreage seeded to varieties developed by selected institutions).

	1960–1969	1970–1979	1980–1984	1985–1989	1990–1994	1995	1996	1997
AAFC	88.9	89.0	53.9	93.8	38.1	10.8	5.7	0.6
U. of Manitoba	11.1	18.5	27.3	1.7	9.8	15.3	8.0	3.4
U. of Guelph	0	0	0	2.2	0.3	0.2	0.1	0.5
U. of Alberta	0	0	0	0.2	1.1	0.2	12.3	6.1
Total public	**100.0**	**99.8**	**99.3**	**98.3**	**49.3**	**26.5**	**26.1**	**10.6**
Svalof AB/BASF	0	0.2	0	0.4	41.9	42.9	20.6	24.8
Zeneca	0	0	0	0	1.3	12.7	21.6	10.5
Pioneer HB	0	0	0	0	0	0.6	11.8	9.8
Limagrain	0	0	0	0	0	0.4	4.2	4.5
AgrEvo/PGS	0	0	0	0	0	0	2.4	11.4
Total private	**0**	**0.2**	**0**	**0.4**	**43.2**	**56.6**	**60.6**	**63.1**
Unallocated	0	0	0.7	1.3	7.5	16.9	13.3	26.3

Source: Nagy and Furtan, 1978; Prairie Pools Varietal Survey 1963–1992; Manitoba Crop Insurance Corporation varietal data weighted by historical shares for 1993–97.

Until the 1980s the technologies used to breed new varieties of canola were largely developed by public universities or laboratories and then were placed in the public domain. Beginning with the ground-breaking 1973 patent by Cohen and Boyer, there was a rush to perfect and patent the rapidly evolving new biotechnology methods. During the 1980s, companies brought these new technologies to the market. Table 8.4 shows the extent of the new processes developed and commercialized since then. Now almost all of the key elements required to apply biotechnology processes to canola are held privately, mostly by one of the large agrochemical companies.

The results of application of these new technologies in private laboratories have just emerged in the marketplace. Beginning in the mid 1980s, a number of chemical companies, led by Monsanto and AgrEvo, began using these new transgenic technologies to insert patented herbicide-tolerant (HT) genes in canola. The first HT varieties were released in Canada under identity-preserved production in 1995, with unconfined release approved in 1997. There are now four basic herbicide-resistant packages available and others are being developed. Given the success of HT varieties, seed and chemical companies have gone to work on other agronomic factors. Between 1994 and 1997, there were 97 field trials for transgenic varieties that involved introducing stress, insect, virus and fungal resistance into canola (CFIA, 1999). This targeted search for commercially attractive seeds has precipitated a massive restructuring of the supply side in the industrial structure. The ultimate phase of development is already under

Table 8.4. Key technologies related to canola breeding processes.

	Key technologies (and owner, if any)
Genomic information	*Arabidopsis* genome project
	Amplified fragment linkage polymorphing for gene mapping (patented)
	Molecular markers
Germplasm	Public gene banks in Canada, USA, Germany, Russia, India, Pakistan, Australia, Japan and others
	Private collections
rDNA strands/genes	HT genes (Calgene and Monsanto, USA; AgrEvo, Germany; American Cyanamid, USA; and Rhône Poulenc, France)
	Antifungal proteins (Zeneca, UK)
	Antishatter (Limagrain, France)
	Fatty acids (Calgene, USA)
	Pharmaceutical compounds (Ciba Geigy, Switzerland)
Transformation systems	
General	*Agrobacterium* (Mogen, The Netherlands; PGS, Belgium)
	Whiskers (Zeneca, UK)
	Biolistics (Cornell University and Dupont, USA)
	Chemical mutagenesis (public domain)
Brassica specific	*Agrobacterium* methods for *Brassica* (UC Davis and Calgene, USA)
Selectable markers	Large number of privately patented markers for selecting specific transformants (Monsanto, USA; PGS, Belgium; others)
Growth promoters	
Constitutive	Constitutive promoters (e.g. for HT, disease, drought, salt resistance) to express genes in all cells in plants, including 35S (Monsanto, USA)
Tissue specific	Pod/shatter control (Limagrain, France)
	Floral morphology (AgrEvo, Germany; multiple others)
	Oil traits (Agriculture Canada; others)
Hybrid technologies	InVigor™ (PGS, Belgium)
	CMS System (Zeneca, UK)
	Ogura CMS Systems (INRA, France)
	Lemke (NPZ, The Netherlands)
	Kosena system (Mitsubishi, Japan)
	Polima (China; public domain)
Oil-processing technologies	Oleosin partitioning technology for separating and purifying recombinant nutraceutical or pharmaceutical proteins (SemBioSys, Canada)
	Other oil-processing technologies (various)
Traditional breeding technologies	Double haploid process (Canada)
	Backcrossing (public domain)
	Gas–liquid chromatography (public domain)

Source: Personal communications with canola researchers and patent searches.

way. End users, such as Procter & Gamble, Frito Lay, Nabisco, Campbell Soup, Mobil Oil, Shell, Lubrizol and Novartis (Ciba-Geigy), to name but a few, have invested in relationships with R&D companies to modify canola oil for its nutritive value as well as for its value as a vector for other transgenic features. In 1994–1997 in Canada alone, there were 184 field trials of transgenic varieties of canola that have been manipulated to modify the oil composition, change the nutritional balance of the seed or to produce nutra- or pharmaceutical products (CFIA, 1999).

The advent of these new technologies has led to significant restructuring in the industry.

The Impact of Innovation on Industrial Structure

Before the mid 1980s, canola was a genuine commodity exhibiting low task-programmability, low non-separability and low asset-specificity. This explains the reliance on spot markets observed during the period. But as innovation yielded new seeds, the characteristics of the industry changed. As the production technologies have become more linked (e.g. herbicide-tolerant canola, such as Round-Up Ready™ or Liberty-Link™), task-programmability has risen. This has been matched by new vertical coordination (packages of production inputs) between the seed merchants, chemical companies and farmers. Meanwhile, recent efforts to breed in new product characteristics (e.g. modified oils, high erucic acid content, low linolenic oils, nutraceuticals and pharmaceuticals) targeted to serve niche markets have increased both the asset specificity of farmers producing these canola varieties (because their production only has value if sold to that specific market, which is often a monopsony or oligopsony) and the non-separability of the production (because the new varieties only have value if produced to the specifications of the end user, which requires greater quality assurance throughout the production system). As a consequence, there are a number of new vertical management systems that span the supply chain from seed developer to the wholesale industry.

The commodity-based industry (1940–1985)

Traditionally, the canola industry can be divided into two key components in Canada: approximately 60–70% of the rapeseed/canola produced annually is exported as seed, primarily to six big oilseed mills in Japan, but more recently to the EU, Mexico and the US (Wensley, 1996). The remainder is processed into oil in western Canada or Ontario for domestic consumption or export. The residual meal was used as fertilizer until glucosinolates were removed in the 1970s; since then the meal has been sold as feed to nearby cattle producers or feed lots.

The commodity seed trade dominates the business (Fig. 8.2). Western Canada has planted between 5 and 14 million acres of canola annually since

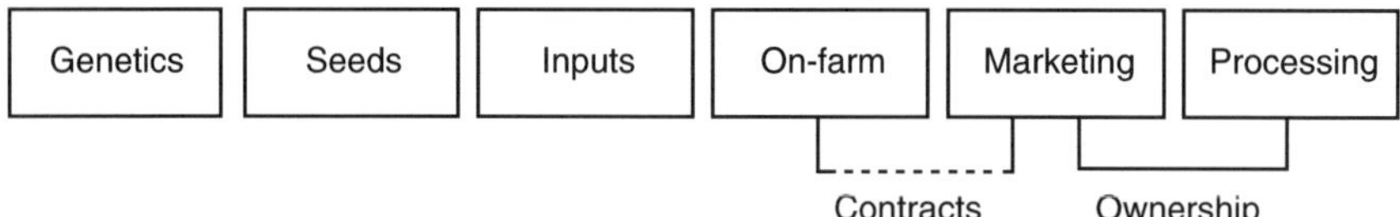

Fig. 8.2. The commodity seed trade and oil-crushing supply chain (1940–85).

1980, producing between 3.2 and 7.2 million tonnes (Mt) of canola seed, while annual exports have ranged from 1 to 4 Mt, accounting for approximately 15–20% of world production and 60% of world trade. Although some firms operated in more than one link in the seed-export supply chain, few operators developed more than spot-market relationships. In essence, the low asset-specificity, low task-programmability and low non-separability made unnecessary any special industrial structures to manage risk.

With the genetics and seed industry in the public domain, task-programmability and non-separability remained low throughout the chain. Furthermore, there was little prospect of strategic investments in research (Just and Hueth, 1993), which minimized the asset specificity of the patented chemicals and machinery, or the corresponding investments in their production. The only area where asset specificity was moderate was for farmers themselves. Once they had made a decision to produce canola, they had a relatively specific asset – the resulting seeds. Given that rapeseed/canola is not marketed by the Canadian Wheat Board, farmers sought other ways to manage the risk of asset specificity. In September 1963, the Winnipeg Commodity Exchange introduced rapeseed trading and broadcast daily prices (McLeod, 1974). Farmers used a variety of stategies to manage risk. Some used the exchange to hedge, some contracted forward with one of the multinational grain companies, while other farmers encouraged their locally owned cooperatives (the Pools) to develop a marketing presence to 'look out for their interests' (e.g. XCAN Grain). Each case provided a form of vertical management of the risk between the farmers and the marketers.

The canola oil industry, although smaller than the seeds business (between 1 and 2.7 Mt of canola seed are crushed annually in Canada), is still a major part of the industry and has exhibited more industrial integration because of the greater degree of asset specificity in the oilseed-processing link. As in the seed business, there was little task programmability and no separability problems. The prime area of asset specificity was between the marketing link and the processing industry. Given that canola seed has limited value except as an oil, marketers faced asset specificity of holding canola seed. Only, in their case, the risk was a limited international market. Meanwhile, crushing facilities in the Canadian prairies only had canola as a commercially viable oilseed to crush. This mutual dependence created an opportunity that the three Wheat Pools pursued to develop their own processing system. The resulting company, CSP Foods (now Canamera) developed plants in Manitoba and Saskatchewan, while a number of groups of local investors and a few non-resident firms invested in other plants. Over time, some of these plants vertically integrated, adding product refining to the basic crushing facilities.

The private, product-based canola sector (1985–date)

As the seed stock was progressively privatized after 1985, asset specificity, task-programmability and non-separability all increased, so that the pricing system was not able to fully accommodate the new elements. The extent of that change can be observed by looking at the supply-driven development of herbicide-tolerant (HT) canola, and new demand-driven varieties with characteristics modified to end-user specifications.

On the supply side, research to develop herbicide-tolerant varieties would appear to have been precipitated by the four factors identified as leading to the privatization of the seed stock. Clearly, once the seed industry was open to private development, the chemical companies, with patents on a variety of herbicides and insecticides, had most to gain (or lose) from changes in the seeds. Agriculture Canada and the University of Guelph had by 1984 already bred a HT (triazine) canola. The chemical companies could see that whichever could develop a variety tolerant to their patented chemicals could either protect their market or gain market share.

This opportunity precipitated a number of developments (Fig. 8.3). Perhaps most dramatic was the industrial restructuring that occurred in the chemical sector itself. All of the large chemical companies moved to partner their agrichemicals divisions with genetics and seeds units. AgrEvo initially developed an in-house genetics capacity in Frankfurt, then partnered with Agriculture Canada to get access to a base of canola varieties and, in 1996, purchased 75% of Plant Genetics Systems of Belgium (now wholly owned), an early leader in transgenics in canola and the owner of the InVigor™ hybrid technology. Meanwhile, Cargill bought InterMountain Canola from Dupont, Dow AgroSciences bought Mycogen and a minority interest in SemBioSys, Monsanto purchased Calgene in 1996 and has since acquired 49% of Limagrain Canada, and Dupont purchased Pioneer Hi-Bred for US$9.4 billion. In addition, most of the companies partnered with NRC, Agriculture Canada or one of the universities to bolster their research capacity. As a result, the genetics and seeds businesses became almost fully vertically integrated.

Once the HT varieties were in the fields, other vertical links began to develop. The HT varieties exhibit strong task-programmability, as farmers would only purchase the seed if they were going to use the appropriate chemicals. Most of the companies chose to control the distribution of their seed and sell a package of seed, herbicide and agronomic advice. AgrEvo's Liberty-Link™ system, which includes a gluphosinate-tolerant variety and Liberty™ (a gluphosinate herbicide) sold as a package to farmers either through an in-house marketing arm called InterAg or a consortia of the Wheat Pools, priced the seed in 1997

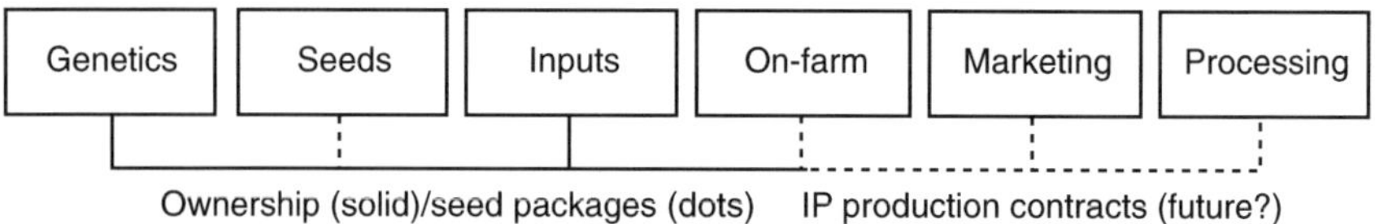

Fig. 8.3. The HT-canola supply chain (1985+).

at about Can$2.51 per lb, compared to AC-Excel (a reference public variety) at Can$1.60 per lb. In addition farmers must buy the patented herbicide Liberty™ at Can$198 per 10 litres. The average cost of the package per acre was then about Can$40. Monsanto, the owner of Round-Up™, has significant competition in the herbicide market due to the end of its patent (causing the price per 10 litres to be only Can$88.25), and so has adopted another approach. It has developed and patented a Round-Up Ready™ (*RR*) gene, which it licenses to a variety of breeders (so far to Alberta Wheat Pool, Pioneer Hi-Bred, Agriculture Canada, Svalof and Limagrain). Instead of taking its return on the seed, Monsanto levies a 'technology fee' of about Can$15 per acre for use of the *RR* gene. It manages this system by requiring farmers to attend informational sessions and to sign a technology-use agreement before it will allow them to purchase the seed and chemical package. Although farmers were required to use Monstanto's glyphosate herbicide in 1995–1999, Monsanto concluded agreements with other glyphosate producers and, as of 2000, allows farmers to apply herbicide produced by other companies. Meanwhile, Pioneer Hi-Bred has bred an imidazolinone-tolerant canola and sells the package with Pursuit™, a herbicide produced by Cyanamid, and Rhône Poulenc worked with Svalof and University of Manitoba to develop a bromoxynil-tolerant variety in 1999. The three HT varieties available in 1996 captured only about 4% of the area seeded, but were estimated to have accounted for 4 million acres or 33% of the area in 1997, approximately 50% of the acreage seeded in 1998 and up to 75% of the acreage in 1999.

So far, the HT varieties have only affected the market structure at the farmer level. The one imponderable at this point is whether rising non-separability will force a downstream identity-preserved production (IPP) and marketing system to evolve. The European Union has introduced a number of regulations that would require 'genetically modified organisms', such as transgenic canola, to be labelled in the EU food system. Some breeders and marketers fear that this action may cause undifferentiated canola from Canada in that market to be banned, boycotted or discounted. This could drive the Canadian industry to develop identity-preserved production systems to allow the industry to segment its production for different markets. If this were to happen, non-separability would rise sharply, as it is impossible visually to distinguish between transgenic and non-transgenic varieties. Developing an effective IPP system should not be an insurmountable problem as, during 1995, HT varieties of canola authorized for 'confined release' used an identity-preserved production system to segment the new varieties from the traditional varieties. The system entails a variety of rules, including: all seed to be grown under contract and growers purchase new seed each year; the identity-preserved and traditional commodity crops are not produced on the same farm; individual fields of identity-preserved crop are routinely checked and the yield estimated to reduce the risk of growers acting opportunistically and adding commodity seed when delivering the specialty crop; and separate elevators, unloading locations and storage are used at the crushing plants to prevent co-mingling of commodity seeds and product (Del Vecchio, 1996). Such a system, as applied in 1995 and 1996 in Canada for the new herbicide-tolerant varieties, cost an estimated Can$30–40 per tonne (Manitoba Pool Elevators, AgrEvo), which would not be sustainable in the long run unless the market value of the IPP product rises.

Meanwhile, some firms have recognized that there is uncaptured value in the production chain, due to the incomplete pricing of oil and meal qualities in the canola seed. Cargill, for example, offered contracts for the 1999 year for a specific variety of canola with desired oil traits and Cargill's Saskatchewan-based crushing plant contracted to purchase the resulting crop at a premium.

The next phase of demand-driven development, already moving from the laboratory to test fields, has been to use biotechnological techniques to modify the output traits. That development has raised asset-specificity, task-programmability and non-separability significantly, with the result that all of this production is likely in close-looped, identity-preserved production systems (Fig. 8.4).

Novel varieties of canola are not new. As far back as the 1960s, a number of industrial users demanded varieties with novel characteristics, such as high erucic acid or low linolenic oils. Each of these varieties produced oils destined for industrial rather than food markets and so they used a closed, identity-preserved production system. Both the seeds and the resulting crop exhibited high asset-specificity, because they had incremental value (or at times only had value) when sold to a limited industrial-oil market (often a single buyer). As a result, contracting between the processor, breeders and farmers has at times been necessary to get production started. Furthermore, novel characteristics increase the non-separability of the product, as its marketability requires rigorous quality assurance. The change has been that biotechnology has now created the opportunity of adding significantly more novel traits, such as engineered fat chains, industrial-oil genes (e.g. laurate), nutraceutical characteristics or pharmaceutical proteins and enzymes. The possibilities appear almost endless. Canola, for example, can be modified to produce a wide variety of pro-

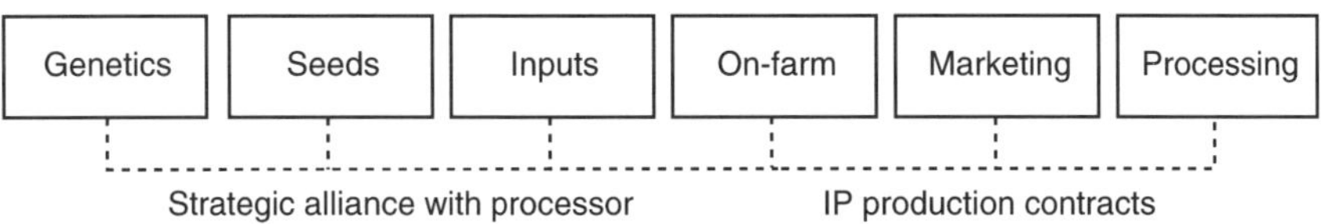

Fig. 8.4. The supply chain for novel canola oil products (1985+).

teins: low-value, commodity, end-of-the-scale, proteins for improved nutritional value of the seeds; intermediate-value, bulk proteins such as industrial and food enzymes; and high-value proteins, mainly of interest to the pharmaceutical industry. The high asset-specificity for the genetics, seed and unprocessed crop, due to a monopsonistic or oligopsonistic end-market, requires some measure of contracting or vertical integration. Furthermore, many of these crops require tailored agronomic practices to ensure the quality of the resulting oil. For example, certain pesticides and herbicides can only be applied at set times or the resulting seed and oil would have unacceptable residues and would be worthless. It is not possible to allow farmers and others in the production chain to exercise their own judgement in the production; they require specific instructions to ensure the integrity of the output. Hence, task-programmability has also risen. This combination of high asset-specificity, high non-separability and high task-programmability necessitates more than simple one-off contracts. Given that the Canadian legal system only allows civil courts to assess claims to cover real and reasonable losses from the actions of farmers and others who might act opportunistically, and not punitive damage claims, there are limited legal disincentives to act opportunistically. The losses to the greater supply chain of a contaminated product, however, would be larger. At the extreme, all the value in the product could be lost. Therefore, many firms have judged that they need long-term contracting, strategic alliances or interlocking ownership and management to increase the incentives to act appropriately.

This can be observed in a number of the novel canola products currently being produced. Procter & Gamble invested US$5 million in Calgene in the late 1980s to finance development of an alternative source of high-quality laurate to use in detergents. The traditional source is the second crush of coconut and palm oil from lesser-developed countries, which often exhibits wide variations in quality. Calgene isolated the laurate gene in the bay tree and transferred it, using genetic engineering techniques, into a variety of *B. napus*. The resulting canola was produced initially in Georgia under an identity-preserved production contract and crushed by Cargill's subsidiary, Stevens Industries of Georgia. More recently the production of Laurical™ canola has moved into the northern plains states, and in 1997 into Saskatchewan, where the Saskatchewan Wheat Pool has a contract to arrange both the crop production and oil processing (through Canamera). Other examples abound. Mycogen has a contract with Lubrizol to develop lubricants from canola; Lubrizol funds the research at Mycogen laboratories and the resulting crops are produced under an identity-preserved production contract in the US. SemBioSys of Calgary, partially owned by Dow Agrosciences, allied with Novartis (formerly Ciba-Geigy Canada) and linked to the University of Calgary by joint appointments, has successfully developed transgenic canola that produces the anticoagulant, hirudin, which has been traditionally obtained from leeches and is used to treat patients undergoing major surgery. SemBioSys has also produced a human cytokine, interleukin-1, which is used experimentally as a modulator of the immune system (Moloney, 1995).

Recent estimates suggest that at the commodity stage of novel attributes, each product could generate between 20,000 and 50,000 acres of demand, and some speculate that there could be as many as 20 different specialty enzymes under production in coming years (Moloney, 1995). At the niche market level, the high-value but low-volume pharmaceutical proteins would likely require only 2000–3000 acres of production per product. In total, the novels market could require 400,000 to 1 million acres of cropland, representing as much as 10% of the total cropland currently being cultivated in canola. Almost all of the firms have research programmes targeted on this opportunity.

The Impact of Biotechnology on Firms in the Canola Business

The result of the new technology on the industrial structure can be observed readily. At the commercial level, the increased interest in canola research and development has precipitated and contributed to a number of corporate mergers and alliances. Table 8.5 illustrates the diffusion of activity between the early 1980s and the 1990s and the interconnections that are developing. A survey in 1998 of 28 public and private institutions around the world involved in breeding new canola varieties illustrated the array and range of new arrangements. Of the 28 entities, 13 had produced at least one variety for public release in the 1985–1997 period while the other 15 provided services, technology or germplasm to the process. Although 14 companies produced at least one variety at some point in the period, mergers and acquisitions and public cut-backs have reduced the number of independent programmes to about half that number. The key programmes now revolve around Monsanto, Aventis, Pioneer Hi-Bred, BASF, Svalof, Dow AgroSciences and Syngenia. The 28 organizations surveyed identified 118 different arrangements they had pursued in recent years to expand their capacity, involving research contracts or alliances with universities, public laboratories or private companies, distribution and

Table 8.5. Strategic alliances and contracts in the canola research community.

Purpose	Number of arrangements	% of total
Total	118	100
Research	80	68
university	47	40
public laboratories	33	28
Distribution	19	16
Marketing	19	16
Mechanism		
Exchange of equity	7	11
Strategic alliance/contracts	57	89

Source: Industry survey conducted between March and June 1998.

licensing arrangements with farm service companies to control the sale of their innovations, and research or marketing alliances with end users to lock in markets and returns. Although 11% of the deals were sealed with equity exchanges, the favourite method of collaboration, especially for distribution purposes, was through strategic alliances or contracts. This approach is consistent with recent surveys of technology companies in other sectors (e.g. Gibbons, 1995).

The research community has changed dramatically over the years (Figs 8.5–8.10). St Louis-based Monsanto has probably gone the furthest in developing a biotechnology base to its business. Beginning in 1960, the American chemical company targeted agriculture as a core market, with a biotechnology unit beginning in the early 1980s. The first major commercial breakthrough related to canola was the isolation and patenting of the Round-Up Ready™ gene. Beginning in 1987 the company started an agronomy and breeding programme to insert that gene into canola. Round-Up Ready™ canola varieties were granted approval for unconfined release in Canada in 1995 and accounted for about 15% of the Canadian canola acreage in 1997. The Monsanto research laboratories have also isolated and patented a number of growth promoters, which give it a competitive edge in breeding. Throughout the late 1980s and into the 1990s, Monsanto acquired a number of seed companies around the world to gain access to germplasm and markets. In 1996 Monsanto also purchased 75% of Calgene (since raised to 100%), the owner of the patent for the

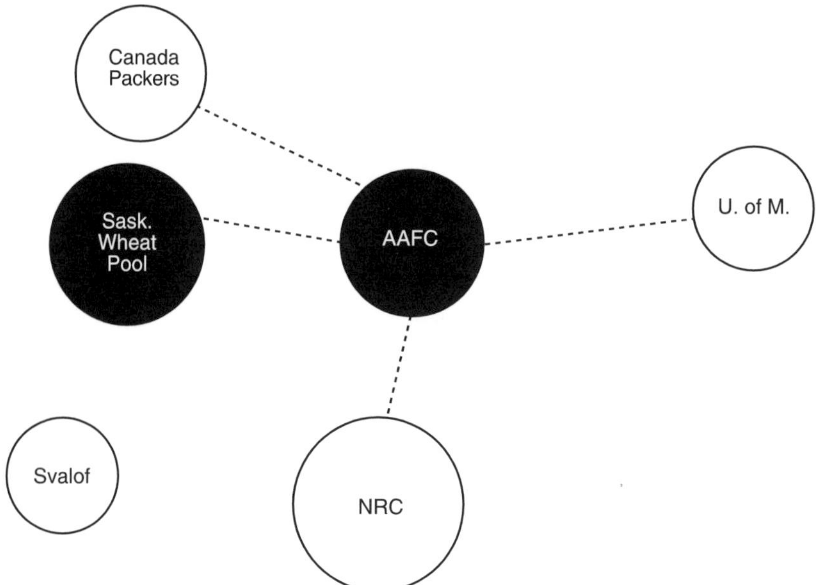

Fig. 8.5. Key relationships in the global canola research and marketing community, 1950. The solid lines and percentages denote ownership. The dotted lines denote other relationships (only some of the relationships are illustrated). The shaded circles represent a Saskatchewan location.

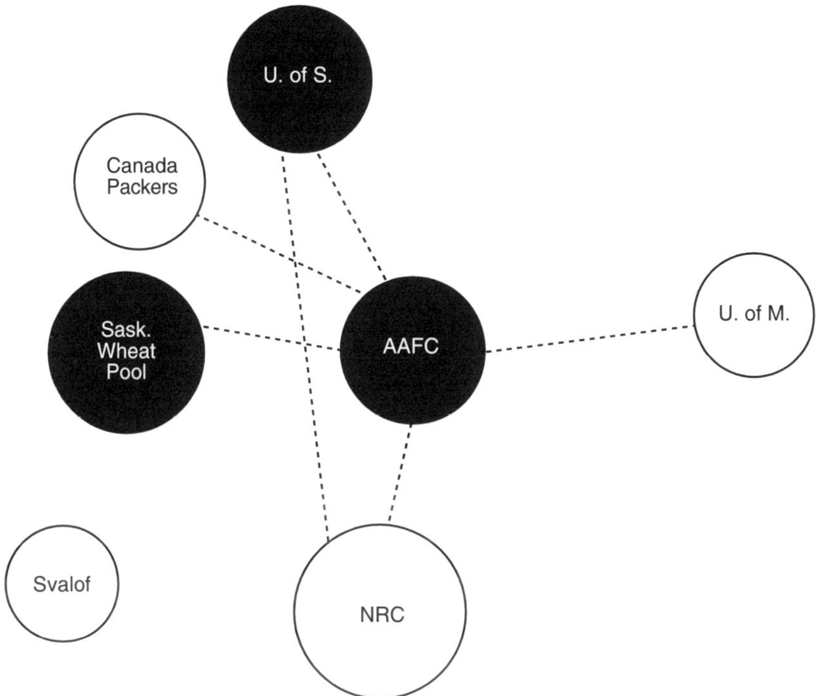

Fig. 8.6. Key relationships in the global canola research and marketing community, 1960. The solid lines and percentages denote ownership. The dotted lines denote other relationships (only some of the relationships are illustrated). The shaded circles represent a Saskatchewan location.

main transformation technology for canola and the developer of Laurical™ canola. Calgene, in turn, has actively courted arrangements with end users of their oil products, including Mobil Oil and Procter & Gamble and has targeted to develop a number of new rapeseed-based modified oil products. Calgene also has a strategic alliance with the Saskatchewan Wheat Pool, under which SWP will adapt Calgene's modified oil varieties for Canada and then will manage their marketing and processing through SWP's oil processing subsidiary, Canamera.

Late in 1997, Monsanto restructured its operations, divesting itself of its US$3 billion industrial chemical business as Solutia Inc. The remaining company, which in 1997 earned about US$3 billion from agricultural products and US$3 billion from pharmaceuticals and food products (*The Economist*,1997) has since pursued growth through acquisitions. Over the past few years Monsanto has acquired a large number of seed and germplasm companies, including Holden's Foundation Seeds, Sementes Agro, Asgrow, Dekalb Genetics, Carlo Erba, CIAGRO, Plant Breeding International Cambridge Ltd, Delta and Pine Land (Monsanto abandoned its offer to purchase in January 2000 and paid an US$81 million termination fee) and in 1997, 49% of Limagrain Canada, a

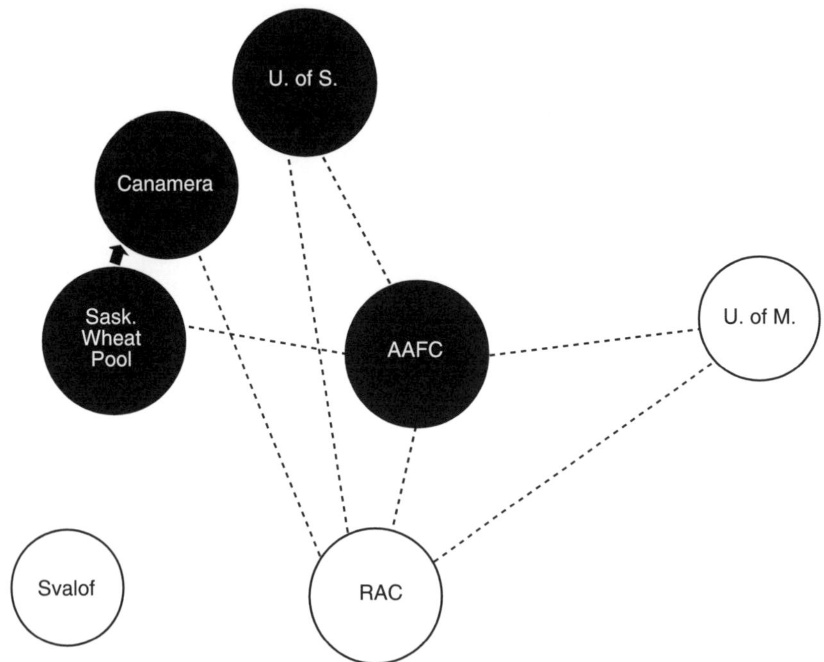

Fig. 8.7. Key relationships in the global canola research and marketing community, 1970. The solid lines and percentages denote ownership. The dotted lines denote other relationships (only some of the relationships are illustrated). The shaded circles represent a Saskatchewan location.

major developer of canola seeds. Limagrain's parent company, a cooperative with 680 farmer owners in France, has been in the biotech business since 1984 and is currently the third largest seed producer in the world. In June 1994 Groupe Limagrain purchased Kingroup Inc. (Price Brand Seeds) from Elf Anofi and created King Agro in eastern Canada and Limagrain Canada Seeds in western Canada. Limagrain has extensive breeding programmes in Australia, England, France, Germany and Canada. Monsanto has also acquired a variety of technology companies which own potentially valuable proprietary technologies, including Cereon, Incyte, Mendel, ArQule, Agrecetus, Ecogen, Millenium and Calgene.

Since then, Monsanto has aggressively pursued greater control of its downstream market. In May 1998, Monsanto entered into a major strategic alliance with Cargill, the world's largest oilseed crusher, and owner of InterMountain Canola, a breeding company concentrating on modified oil products. Together Monsanto and Cargill are pursuing value-added oil markets under the operating name Renesson. Within a few weeks of that announcement, Monsanto and American Home Products announced a US$35 billion merger to broaden

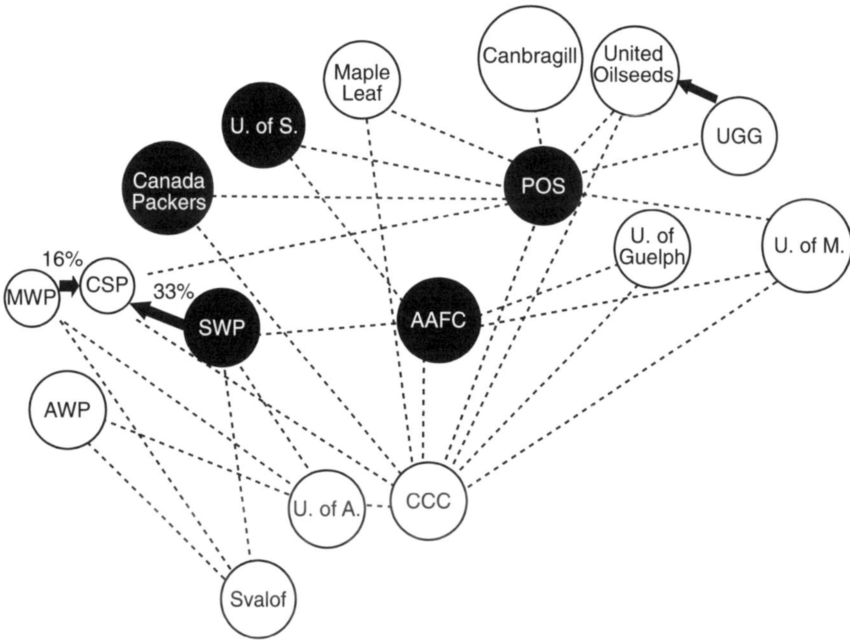

Fig. 8.8. Key relationships in the global canola research and marketing community, 1980. The solid lines and percentages denote ownership. The dotted lines denote other relationships (only some of the relationships are illustrated). The shaded circles represent a Saskatchewan location.

Monsanto's potential market (incidentally it would also have brought access to the research that American Cyanamid had done with competitor Pioneer Hi-Bred); that merger ultimately was abandoned. In 1999, Monsanto and Pharmacia and Upjohn, a global pharmaceutical firm, concluded an agreeement to merge their operations, creating a pharmaceutical venture and spinning off and recapitalizing the agricultural operations. Outside of its rapidly expanding core business group, Monsanto has developed close links with the public research agencies in Canada, with Saskatchewan Wheat Pool through a strategic alliance with Calgene, and with Pioneer Hi-Bred, Svalof and Alberta Wheat Pool through the licensing of the Round-Up Ready™ gene to them for insertion in their varieties. In addition, Monsanto has, through its technology use agreements for Round-Up Ready™ varieties, acquired significant information about a majority of canola growers in western Canada. The company reports that approximately 30,000 producers had attended their meetings by mid 1999, representing virtually all farmers producing commercial quantities of canola in western Canada. So far the company has not translated that information into a tighter relationship, but the potential exists.

Most of the other research-based producing networks have evolved in

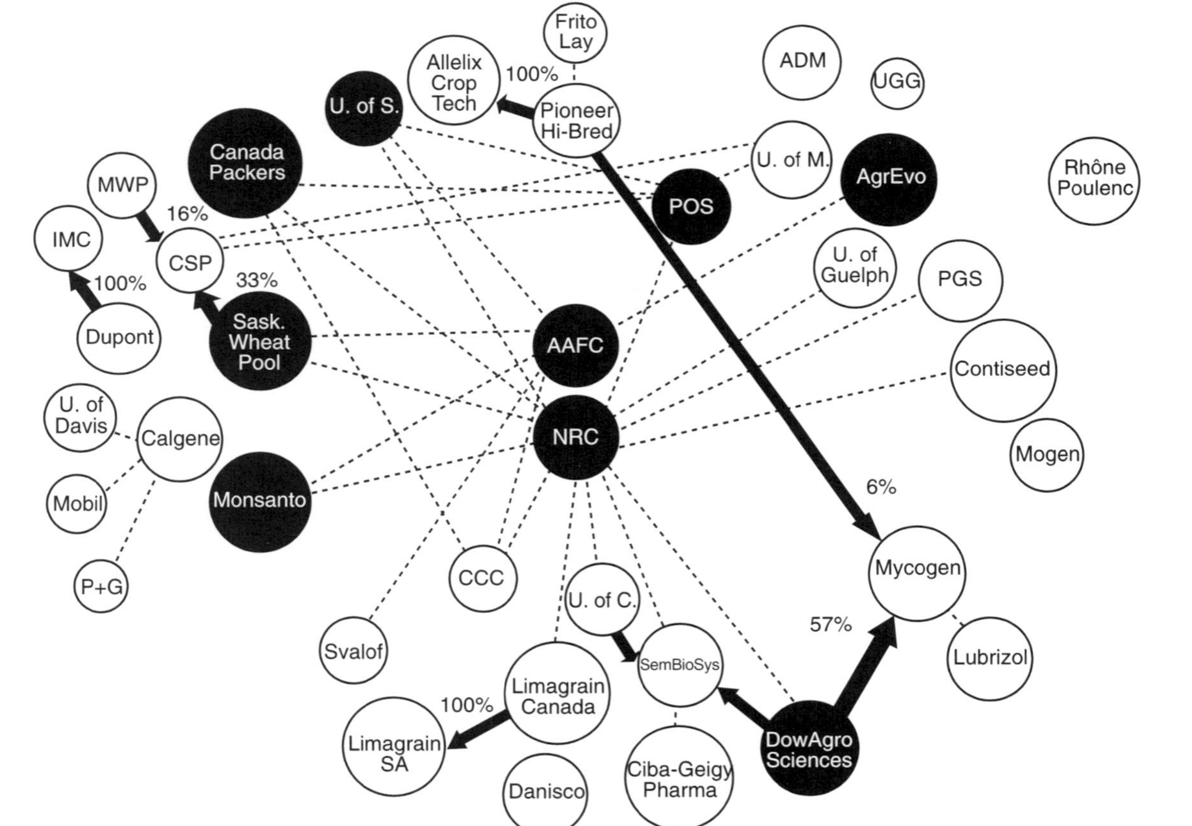

Fig. 8.9. Key relationships in the global canola research and marketing community, 1990. The solid lines and percentages denote ownership. The dotted lines denote other relationships (only some of the relationships are illustrated). The shaded circles represent a Saskatchewan location; some multinational firms have both Canadian and foreign programmes.

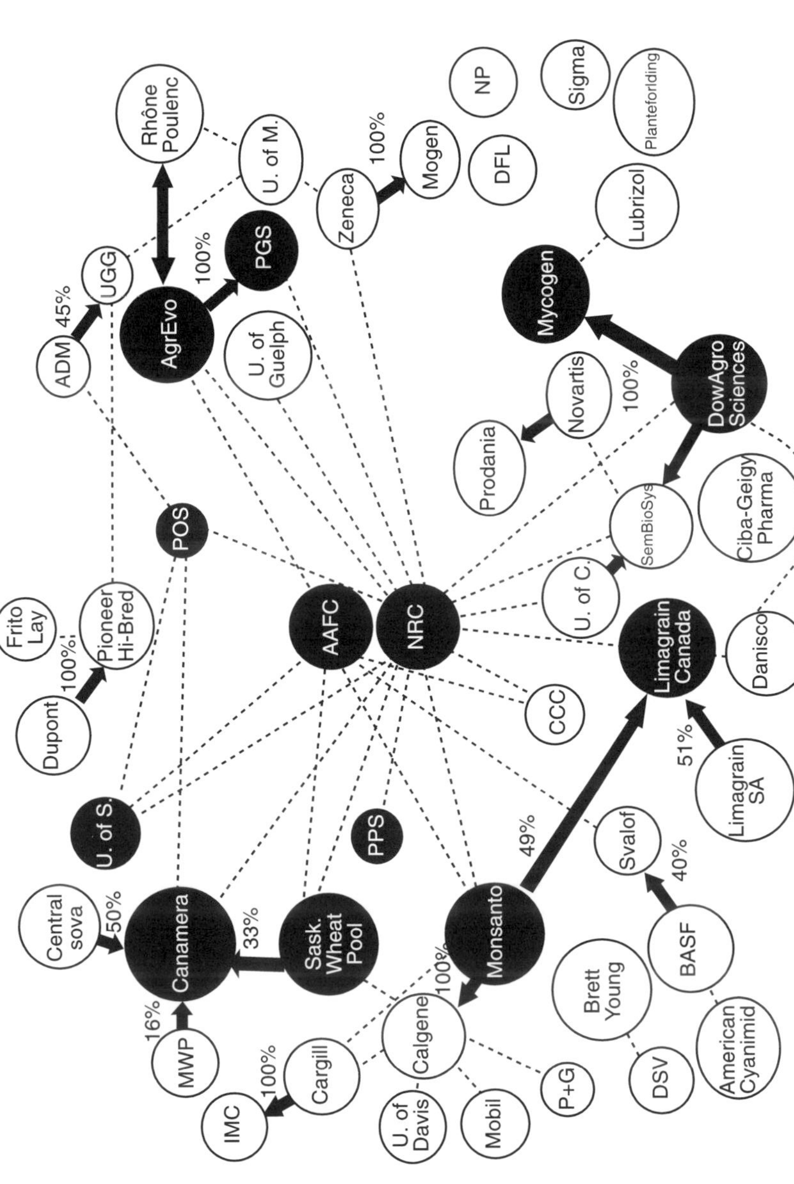

Fig. 8.10. Key relationships in the global canola research and marketing community, 2000. The solid lines and percentages denote ownership. The dotted lines denote other relationships (only some of the relationships are illustrated). The shaded circles represent a Saskatchewan location; some multinational firms have both Canadian and foreign operations.

similar ways. AgrEvo, based in Berlin, was created in 1984 by the merger of the crop protection businesses of Hoechst AG and Schering AG, and in 1998 was one of the five largest crop protection companies in the world. In 1996 AgrEvo earned DM232 million on gross receipts of DM3639 million, yielding a return on equity that ranges from 19% in 1996 to 26% in 1994. AgrEvo employed 8057 workers worldwide in 1996. AgrEvo invested DM436 million in 1996 in research and development. Similar to Monsanto, AgrEvo has pursued growth through acquisition and mergers. Over the past few years AgrEvo acquired ownership of Cotton Seed, GeneLogic, Nunhems and Kimeragen. Most important for canola, in August 1996 AgrEvo purchased 75% of Plant Genetics Systems (PGS), headquartered in Ghent, Belgium, for US$550 million (it has since acquired the rest of the company). PGS holds the patents on a number of relevant proprietary technologies, including InVigor™ hybrid technology and a number of selectable markers useful for canola. Combined with AgrEvo's patents on the Liberty-Link™ gene and growth promoters, the combined research effort, linked as it is with AAFC and NRC, is leading in the development of hybrid transgenic canola varieties. The Liberty-Link™ system, introduced in 1994, accounted for approximately 20% of the Canadian acreage in 1998. On the marketing side, AgrEvo has collaborated with the three prairie pools to produce Innovator™, a Liberty-Link™ product. Other AgrEvo products are marketed by InterAg, the company's seed division. In 1998 Hoechst announced a merger with Rhône Poulenc that created Aventis, a new US$20 billion life science company. Aventis CropScience, which comprises AgrEvo and Rhône Poulenc Agro, focuses on crop protection and crop production. Rhône Poulenc had been working with the University of Manitoba and Svalof to develop a new bromoxynil-tolerant variety, which was finally approved and registered under the Seeds Act in 1999.

Zeneca Seeds was created in 1993 when ICI, which had been in the seeds business since 1927, broke up its pharmaceutical, agrochemicals, seeds and specialties businesses. ICI had previously acquired Contiseed in 1988, a subsidiary of Continental Grain Company, and thereby acquired a canola breeding programme in Australia. This effort was expended through investments in the UK, Canada, USA, India and Belgium, with the result that Zeneca now has a significant global breeding programme. Although Zeneca Agrochemicals is a major agrochemical company, the separation of the two entities in 1993 effectively eliminated any synergies between the two companies. Zeneca's only herbicide-tolerant canola variety as of 1998, Hyola 357™, uses Monsanto's Round-Up Ready™ technology. Zeneca recently acquired Mogen, a Dutch company with expertise in disease resistance and value-added modifications. Combined with Zeneca's proprietary technological innovations, the combined company has taken an early lead in the development of hybrid canolas, with five varieties in the market. Zeneca markets all of its own varieties through Zeneca Seeds. In 1996, Zeneca Seeds formed a joint venture with the Dutch seed company Royal Van der Have, creating Advanta, the fifth largest seed company in the world, with more than 2500 staff in 20 countries. In 1999 ASTRA (a Swedish pharmaceutical company) and Advanta merged, and in 2000 Novartis and Advanta merged, creating Syngenta, the world's largest agrochemical firm and third largest seed company.

Dow AgroSciences, formed in 1989 when The Dow Chemical Company and Eli Lilly and Company combined their agricultural, specialty and plant sciences businesses, had annual sales in 1998 of US$2 billion, representing about 7% of the global agrochemical market. Dow has a substantial programme in biotechnology, with R&D expenditures in excess of 10% of sales. Dow has been a long-time majority investor in Mycogen, the fourth largest seed company in the USA, and in 1998 agreed to purchase the remaining shares. Dow also has investments in Morgan Seeds (Argentina), Dinamilho (Brazil), Verneuil Semences (Europe) and Phytogen (USA). In addition, Dow has made a minority investment in, and developed a long-term contract with, SemBioSys from Calgary and has announced a US $1.2 million research and marketing alliance with Canadian-based Performance Plants Inc. of Kingston, Ontario. Using these investments and relationships, Dow has targeted to capture a share of the value-added oil, designer meal, nutraceutical and pharmaceutical markets. Mycogen holds one of the patents for the *Bt* gene, while SemBioSys holds patents on a new oil-processing technology. Through Mycogen and SemBioSys, Dow also acquired an extensive set of relationships with industrial oil and pharmaceutical companies, including Novartis (Ciba-Geigy) and Lubrizol. Dow also has a strategic alliance with Natunola Health Inc., an Ottawa-based biotech company concentrating on modified oil derivatives for the cosmetics and toiletries industry (Natunola). In 1998, Dow announced a number of additional efforts to build capacity, including the creation of Advanced AgriTraits LLC, a new subsidiary to act as a clearing house for companies seeking to market their genetic traits, germplasm or biotechnology capabilities. This new venture, a first of its kind, may provide a new market for proprietary research elements. In October 1998, Dow AgroSciences announced a worldwide biotechnology research alliance with Rhône Poulenc Agro, to modify six crops, including canola. Given that shortly thereafter AgrEvo and Rhône Poulenc Agro announced they were merging, this alliance may not be realized. Then, Dow announced a strategic alliance with the Danisco group. Established in 1989 by a merger of Aktieselskabet Danisco (Danisco), Aktieselskabet De Danske Spritfabrikker (Danish Distillers) and Aktieselskabet De Danske Sukkerfabrikker (Danish Sugar), Danisco is a leading international supplier of sugar, food ingredients and packaging to the industrial market, and a supplier of selected foods and beverages to consumers in the northern European markets. Danisco earned DKK1952 million on gross sales of DKK17,002 million in 1996/97, posting a 14.8% return on equity. The company invested DKK321 million in R&D in 1996/97. The company employed 12,937 people in 1996/97. In the canola area, Danisco by itself has issued a few selected varieties and has collaborated with Limagrain on a number of others.

Pioneer Hi-Bred, headquartered in Des Moines, Iowa, is one of the largest seed companies in the world, with a market capitalization of US$7.1 billion in 1996 and earnings of about US$223 million on gross sales of approximately

US$1.7 billion, yielding a 20% return on equity. Pioneer produces, markets and sells seed (especially maize) in nearly 100 countries worldwide. Pioneer entered the canola business in 1990 with the purchase of Allelix Crop Technologies (a joint public–private venture in Ontario). Pioneer was an early innovator in breeding, applying computers to reduce the time to produce conventional varieties from 6 years to less than 3 years. Pioneer now has canola research operations in Canada (Edmonton, Saskatoon and Georgetown, Ontario) and in India. Recently Pioneer has been working with upstream and downstream partners, such as the chemical company American Cyanamid and the end user Frito Lay, to develop specialty oil products. In 1995 Pioneer bought 13.5% of Mycogen, with which it collaborated on *Bt* technologies, but sold its shares to Dow in 1998. Pioneer meanwhile developed a strategic marketing relationship with United Grain Growers (UGG), building upon a pre-existing deal between Allelix and UGG. Recently, Pioneer contracted with UGG to be the exclusive distributor of Pioneer varieties of canola in Canada. In 1997 Dupont, which had earlier sold its investment in InterMountain Canola to Cargill, purchased 20% of Pioneer for US$1.7 billion. Together, the two companies developed an alliance to create one of the world's largest private agricultural research and development collaborations. The equally owned joint venture company, Optimum Quality Grains, brought together Dupont's and Pioneer's research effort, which together invest more than US$400 million in agricultural research annually. A key target of this initiative is collaborative research in genetic modification of maize, soybeans and other oilseeds to improve their oil, protein and carbohydrate composition. This alliance was solidified in 1999 when Dupont purchased for US$7.7 billion the remaining 80% of Pioneer that it did not own. Although Dupont Agricultural Products, a leading world supplier of crop-protection products, was not part of the Optimum Quality Grains effort, the new merger may begin to bring greater integration in the company.

Svalof Weibull Seed Ltd is a subsidiary of the Swedish Farmers' Supply and Crop Marketing Association (Svenska Lantmannen, SLR), a farmers' cooperative with approximately 75,000 owner/members, which has the longest history of private rapeseed/canola seed development, going back to the 1940s. Svalof has introduced the largest number of new varieties in Canada since 1985 and has at times commanded a lion's share of the Canadian seed market. Svalof AB and W. Weibull AG, both Swedish companies with more than 100 years' experience, merged in 1993. Their merger involved a number of subsidiaries, including SvSeed and Newfield Seeds in Canada and Semundo GmbH in Germany. Svalof, which does the majority of its research in Sweden, has smaller operations in Germany, France and Canada. Svalof developed a strategic alliance in the 1980s with the three prairie Pools, developing seeds for sale in their networks. Although Svalof has relationships with other players (including Monsanto, AgrEvo and Rhône Poulenc), their arrangement with the Pools continues. Svalof moved into the biotechnology business in 1995 with the start-up of DNA LandMarks Inc. of St Jean sur Richelieu, Quebec, an R&D company specializing in DNA technology, development and services, which has evolved into a world leader in gene markers and mapping in canola. Svalof sells its varieties through a variety of outlets, including the Pools, wholly owned subsidiaries Newfield Seeds and Wheat City Seeds, and a number of other companies (e.g. Cargill Seed, UGG/Proven Seeds, Brett Young Seeds and Value Added Seeds). In 1999, Svalof and BASF negotiated a partnership, with BASF acquiring 40% of Svalof's equity. In 2000 the companies spun-off their combined plant research efforts into BASF Plant Science GmbH, with research programmes in Germany, Sweden, Canada and the US. BASF also purchased American Home Product's Cyanamid Agricultural Division, which previously had research and marketing links with Pioneer Hi-Bred and UGG.

A number of smaller breeding ventures remain, including among others the Pools (Canada) and a number of smaller breeding programmes in Europe. The three prairie Pools first entered the canola development business in the late 1940s, when both Manitoba Pool Elevators and the Saskatchewan Wheat Pool assisted with crushing ventures (McLeod, 1974). Both companies slowly developed and expanded their crushing capacity over the years. Eventually they merged their ventures with others, to form Canamera. Manitoba Pool currently owns 16.7% of that venture while Saskatchewan Wheat Pool owns 33.3% of the company. In the early 1970s when plant breeders' rights legislation was being discussed in Canada, the three Pools agreed to cooperate on new crop development: Saskatchewan Wheat Pool was to concentrate on cereals; Alberta Wheat Pool was to lead on *B. rapa* and *B. napus*; and Manitoba Pool Elevators was to work on specialty crops. Much of the effort on canola that followed was done in collaboration with Svalof; the Pools tested Svalof's breeding lines from Sweden and got first choice for marketing the new varieties. The Pools began direct research activity in 1981, when the Alberta Pool hired a breeder to work with breeders at the University of Alberta. That effort was replaced in the late 1980s with an agreement between the Pools and Agriculture Canada, the Canola Council and the government of Alberta, which jointly funded a transgenic breeding programme at the University of Alberta. Quantum was one of the resulting varieties. The Pools got exclusive rights to all resulting varieties. In 1992, Saskatchewan Wheat Pool expanded its efforts in developing *B. juncea*, a close relative to canola, and opened a research programme in and around Saskatoon to develop new varieties. Saskatchewan Wheat Pool has also used its one-third ownership in Canamera to pursue new value-added markets. Along that line, Saskatchewan Wheat Pool and Calgene formed a strategic alliance in December 1996 by which SWP will use its breeding programme to develop Canadian-adapted specialty canola varieties, combining Calgene's genetically modified oil traits with SWP's germplasm. SWP will then use its distribution and crop-handling systems to produce crops in Canada for Calgene using these varieties (Calgene, 1998). As the Pools have drifted apart in policy terms, their relationships through both the joint research and with Svalof have changed somewhat. The key difference is that each Pool now only gets access to the primary breeding line in their own province. The second lines, which previously were often contracted to other companies (e.g. UGG) are now offered to each of the Pools for sale in other jurisdictions. For example, Saskatchewan Wheat Pool may, in the future, choose to acquire marketing rights for second lines in Alberta and Manitoba to sell through AgPro, Saskatchewan Wheat Pool's subsidiary. Similarly, AgriCore, the company resulting from the merger of Alberta and Manitoba Pools, may choose to compete against the Saskatchewan Wheat Pool in Saskatchewan.

The past few years have seen the arrival in Canada of new varieties developed by a number of smaller breeding programmes in Europe. DLF-Trifolium, formerly Prodana, has three varieties in the market and an arrangement with Bonis and Co. to market the varieties in Canada. Other European companies include DSV Lippstadt, from Germany, which has a development and marketing arrangement with Brett Young Seeds in Manitoba to import European germplasm and adapt it to the Canadian Market. It registered one variety for sale in Canada in 1998. Norddeutsche Pflanzenzucht from Germany has developed with Svalof one variety for Canada, while Sigma of France, through SERASEM and in collaboration with Ringot, Semences Vertes and Semences de France, develops and markets rapeseed and canola varieties in Europe and has research under way in the EU to develop new varieties. Similarly, Planteforlding D.P. of Germany has developed and introduced a new variety in Canada in 1999. It is not clear yet whether these new entrants will be able to act as niche breeders or whether they will simply be absorbed by the increasingly concentrated life-science companies.

The public universities and research laboratories, which initially developed canola and supplied all of the new varieties for the market until 1985, have significantly changed their efforts, in most cases ceasing to develop independent varieties and increasingly working with private companies to provide custom breeding services. AAFC, for instance, produced no new varieties of *B. napus* for registration after 1992, and only two new *B. rapa* varieties after 1993, while most of the varieties brought forward by the universities of Manitoba and Alberta were sponsored by private companies. Guelph has released a few varieties in recent years but their market share has remained small. Instead, these institutions have, in many cases, become 'partners' in the breeding process, providing the know-how and know-who lacking in most private companies.

In short, the industry no longer resembles a traditional commodity-based system. New entrants and new objectives have revolutionized the playing field, leaving little room for public breeders and often only temporary room for smaller private breeding programmes.

Future Trends

Canola is potentially the prototype or template for the evolution of the rest of the agri-food industry. The infusion of private capital into the sector since 1985, the development of proprietary breeding technologies and the privatization of the seeds industry fundamentally altered the industry. Individual actors in the supply chain, either farmers, input providers or output processors and marketers, are now influenced by the effort of previously remote firms as they endeavour to optimize the wealth creation potential of their product. This increased management has worked to integrate the industry vertically but also to segment the industry horizontally, with farmers, input companies and processors watching for signals from within their supply chain rather than from among their peers in the industry. As discussed in the next chapters, this push–pull result presents significant opportunities for regions to win or lose, and creates new pressures for regulations to adapt.

The Impact of Location on Production

Peter W.B. Phillips

Introduction

'Location, location, location', the battle cry for estate agents everywhere, is increasingly becoming the focal point for discussion of the dynamics and benefits of knowledge-based growth. The ultimate question in the minds of many public-policy makers and companies is where is the best location in which to undertake research and commercialize the results? Economics has had a great deal to say in the past about how firms will organize themselves to research and produce products, but has had relatively little to say about where firms will locate their activities.

A survey of most of the companies and public-sector research and development efforts in the canola industry in 1997/98 revealed that since the advent of private capital in the sector, the canola research effort has been slowly but steadily concentrating in Canada and, specifically, Saskatoon. Table 9.1 shows that Canada's share of the global total research effort, which was approximately 75% in the 1945–1966 period, dropped to almost 40% in the late 1980s, but has since rebounded to approximately half of the global effort. This concentration has been almost exclusively due to the relocation or development of private research and development programmes in Canada. The public share of the Canadian effort has dropped from almost 100% in the 1944–1975 period to only about 44% in the late 1990s. Meanwhile, the effort in Canada has been concentrating in Saskatoon. Although companies like Zeneca, Pioneer Hi-Bred and Cargill/InterMountain Canola have located their efforts in other provinces, extensive private research has located to Saskatoon, supplemented by an increasing concentration of public research in AAFC and NRC in the city since the late 1980s.

Table 9.1. Canada's and Saskatchewan's share of the total canola research effort globally, and proportion of share produced by public sector.

| | Global employment | | Saskatchewan | | Canada | |
	Annual average	Public effort as % total	Share of global total (%)	Public effort as % share	Share of global total (%)	Public effort as % share
1980–84	191	68.3	32.2	96.8	48.1	94.6
1985–89	287	56.1	25.8	96.8	42.1	85.3
1990–94	499	44.1	25.3	81.3	41.3	65.9
1995–98	702	36.4	29.5	59.5	48.3	44.5

Source: Canola industry survey, 1997–1998; based on person-years of employment devoted to research and development.

This chapter examines the pressures that have led to the increasing concentration of research and development in the canola industry in Canada and Saskatoon.

The Theory of Local Economies

Spatial economics and central place theory, developed over the past 150 years (Von Thunen, 1826; Weber, 1909; Christaller, 1933; Isard, 1956), have studied the location of production within a region but did not conclusively provide a basis for explaining international location decisions, or even all regional decisions. Recently, a number of economists have been examining the impact of increasing returns to scale on a large number of theories, including location. Much of that work returns to the basic observations by Marshall (1890), who identified three clear and straightforward sources of external economies (Krugman calls them 'centripetal forces') that explained the location of some industry. His first of three explanations, that external economies develop from informational spillovers, is the most complex and difficult to measure:

> When an industry has thus chosen a locality for itself, it is likely to stay there long: so great are the advantages which people following the same skilled trade get from near neighbourhood to one another. The mysteries of the trade become no mysteries; but are as it were in the air, and children learn many of them unconsciously. Good work is rightly appreciated, inventions and improvements in machinery, in processes and the general organisation of the business have their merits promptly discussed: if one man starts a new idea, it is taken up by others and combined with suggestions of their own; and thus it becomes the source of further new ideas.

Secondly, he noted that some centres developed an extensive local and regional network of related and supporting industry (often called 'backward linkages') as well as access to large, sophisticated markets ('forward linkages'), which makes that location more attractive for investors:

> And presently subsidiary trades grow in the neighbourhood, supplying it with implements and materials, organising its traffic, and in many ways conducing to the economy of its material. Again, the economic use of expensive machinery can sometimes be attained in a very high degree in a district in which there is a large aggregate production of the same kind, even though no individual capital employed in the trade be very large. For subsidiary industries devoting themselves each to one small branch of the process of production, and working it for a great many of their neighbours, are able to keep in constant use machinery of the most highly specialised character, and to make it pay its expenses, though its original cost may have been high, and its rate of depreciation very rapid.

Finally, he argued that as the size of the labour market grows, it begins to specialize and support further development:

> Again, in all but the earliest stages of economic development a localised industry gains a great advantage from the fact that it offers a constant market for skill. Employers are apt to resort to any place where they are likely to find a good choice of workers with the special skill which they require; while men seeking employment naturally go to places where there are many employers who need such skill as theirs and where therefore it is likely to find a good market.'

Clearly, these are not the only forces working on industrial location. If they were not offset by 'centrifugal forces', all industries would tend to specialize and there would be extensive trade between locations. Krugman (1998) proposes three offsetting forces (Table 9.2). First, agglomeration tends to create congestion, which is a pure diseconomy of scale. Second, immobile factors – land, natural resources and, internationally, labour – at times force production to go where the resources are. Finally, as economic activity in a region expands, it increases demand for land, which is relatively fixed in supply, tending to drive rents higher.

Recently a number of economists have looked at the role of regional dynamics in creating competitive industries and sectors. Krugman (1998) and Audretsch (1998) have both examined the influence of the Marshall–Arrow–Romer pure external economy, arguing that it explains a significant share of location, especially for innovative firms and sectors. Meanwhile, Porter (1990) and others have looked at the role of linkages and market size on production,

Table 9.2. Krugman's categorization of forces affecting geographic concentration (modified from Krugman, 1998).

Centripetal forces	Centrifugal forces
Pure external economies	Pure external diseconomies
Market-size effects (linkages)	Immobile factors
Thick labour markets	Land rents

while Zucker *et al.* (1998) have examined the role of the labour markets in geographical concentration. Each develops compelling arguments. The purpose of this chapter is to examine the centripetal and centrifugal forces in the canola industry to determine which ones are driving the observed trend and to examine how they can be, or have been, influenced by government action (Phillips, 2000).

Economies and Diseconomies of Scale

As discussed earlier, there appear to be limited economies of scale in basic research (Chapter 4) but there are significant economies of scale in the know-how and know-who stages of development, if only because imperfect markets make smaller ventures invest relatively more of their resources in price discovery and contracting, ultimately leading to larger firms (Chapters 7 and 8).

Increasing economies of scale are usually the result of high fixed costs. With a high fixed cost to enter, firms in such industries face declining average costs over the feasible operating range. As a result, each unit of input yields a successively higher volume of output. This is especially true in a knowledge-based world where the high cost to enter is not the purchase of lumpy capital equipment but rather is the result of an investment in 'learning by doing'. This investment is a real barrier to entry as there is significant uncertainty that the resulting output from the research effort will have any commercial value. It is not like making and selling a standardized product. In crops-based agriculture, for example, breeders must first convince the regulators that the product is both safe and conforms to the industry standard. Monsanto and AgrEvo, as the first developers of HT canola, learned that convincing the regulators of the safety of their product imposes significant costs. Furthermore, Rhône Poulenc learned in 1998 that although it had a variety that it believed added value and conformed to industry standards, the Western Canadian Canola/Rapeseed Recommending Committee (WCCRRC) did not, delaying introduction of its bronoxynil-tolerant varieties for 1 year. Although total costs have risen more than tenfold in real terms over the intervening period, the real average cost per variety has dropped sharply since the early 1980s and the marginal cost has remained relatively small.

Beyond the basic economies of scale in the industry, there is significant potential for externalities to influence the industry. On the positive side, the sharp drop in the average cost per variety is at least partly due to the mysteries 'being in the air', disseminating advancements and improvements throughout the increasing number of breeding entities. A survey of canola firms in Canada, and undertaken globally, in early 1998 suggests that the flow of knowledge has a real impact on their operations. The survey asked companies about the proximity of competitors and/or collaborators as factors in locating their research efforts. Half of all the respondents, representing the majority of private companies responding, acknowledged the importance of proximity to either collabo-

Table 9.3. How important are the following to your decisions to both undertake research and to locate the research in laboratories in Saskatoon or elsewhere?

	N = 28	%
Proximity to competitors or collaborators	14	50
Proximity to collaborators	11	39
Proximity to competitors	8	29
Access to local pool of skilled labour	7	25
Access to large and accepting farm market for seeds being produced	6	21
Key scientists either in your company or in partner organizations	5	18
Role of government agencies (federal, provincial, regional, SREDA) related to hospitality, red tape (or lack of)	5	18
Access to laboratories, greenhouses and test fields	4	14

rators or competitors. About 40% recognized the importance of being close to their collaborators, particularly the NRC and AAFC in Canada and key research universities in the US (Table 9.3).

In spite of the pull to concentrate in one centre caused by spillovers, positive externalities have not been strong enough to concentrate the entire industry, or even the majority of the industry, in Canada or Saskatoon, as the theory would tend to suggest. Interviews with firms that have collaborations in Saskatoon but are not resident in Saskatoon suggested four other factors that are strong at the firm level, thereby overcoming the pull of the positive externalities. First, some companies identified the increased risk of losing commercially valuable intellectual property by being close to competitors. At least one company said it chose an out-of-the-way place to make it less likely that other companies will find out about what they are doing. Although they recognized that they were giving up any locationally specific economies of scope, they judged them as being of less importance than the potential losses of intellectual property. Second, larger companies identified economies of scale within 'discovery' laboratories that are working on multiple crops. If canola is only one of many crops that the private laboratory is working on, the spillovers resulting in canola research in Saskatoon and Canada are often too small to overcome the scale economies generated within a global research laboratory. Third, some companies identified locked-in, undepreciated capital stock as a factor impeding mobility. Given that the NRC and AAFC only recently shifted to collaborations as the base for much of their work, the pull to Saskatoon has likely grown strong enough to attract investments only in the past few years. Firms that made investment decisions before the pull became strong have undepreciated research capacity elsewhere (e.g. in Ontario, Manitoba, Alberta or other countries). At the same time, employees put down roots in existing communities, often making it difficult to get them to move; losing key personnel in a move could more than offset any other gains. Fourth, the length of the Saskatchewan growing season and the variability of the weather can combine with other

Table 9.4. The potential for spillovers from canola to other biotechnology areas (179 of 390 research employees indicated they were working on other crops).

	Number	%
Wheat	48	24
Peas, lentils and legumes	25	12
Flax	23	11
Forage crops	19	9
Herbs	15	7
Other oilseed products	14	7
Other cereals	13	6
Other crops	47	23
Total	204	100

factors (e.g. prior investments and internal economies of scale) to make other locations more competitive. At least one respondent said the diseconomies of operating in Saskatchewan were rising. Popularity of Saskatoon is creating both congestion and higher land costs. Given the requirement to do extensive field trials for new innovations, it is getting more difficult to find easily accessible, inexpensive land to conduct test plots close to Saskatoon.

Over the longer term, the potential for spillovers generating more consolidation of research depends on the opportunity for the work on canola to be applied to other areas. A survey of agri-food research employees in Saskatoon in summer 1998 demonstrated that 179 of the 390 employees responding, or about 46%, were working on other crops in addition to, or instead of, canola. Table 9.4 shows the extent of that work.

Market-size Effects

The location theories suggest that firms locating where there are extensive backward linkages into supporting industries and forward linkages into the market can realize economies of scope. This has been tested in Saskatoon by examining the extent of linkages that exist and their impact on location and production decisions.

Looking backwards from the private research efforts, there are three types of inputs that help to sustain and improve their productivity: (i) knowledge; (ii) know-how; and (iii) supporting infrastructure and services. As examined in Chapter 4, the germplasm and basic know-why and know-what knowledge are increasingly being imported from the USA and other countries. Few of those inputs are uniquely available locally. Rather, as discussed in Chapter 7, the presence and operation of the university, the NRC and AAFC all provide the platform for successfully importing and using those technologies and inputs (e.g. they contribute to the absorptive capacity of Saskatoon as an R&D centre).

The third element is the related and supporting local infrastructure and firms that provide inputs to the operators in the firms. This involves facilities, finance, accounting and legal support, among others. As discussed in Chapter 7, the Saskatchewan Opportunities Corporation (SOCO), through its investments in Innovation Place provides a relatively attractive site for commercial development. Nevertheless, only about 14% of the respondents to the survey indicated that this was a major reason for choosing a location. Furthermore, a number of firms indicated that the relatively high cost of renting in Innovation Place partially or, for some, fully offset the benefits of flexible leases and prime location. A number of other public investments in Saskatoon also support the industry. In the past, the Saskatchewan Research Council did some of the technical work on oil and meal properties of canola, and has since launched Genserv, a commercially directed laboratory to undertake genetic testing for breeders, and has opened a Can$6 million, scale-up fermentation facility. More important, however, has been the federal investment in the POS Pilot Plant, which undertakes scale-up work on oil and meal properties for companies with new varieties, and consults on oil-processing technologies. More than 68% of its resources in the late 1990s came from fee-for-service work for local, national and global companies. The recent addition of Good Management Practices/Hazard Analysis Critical Control Points (GMP-HACCP) programmes has increased demand for its services; with the plant operating 24 hours a day, 7 days a week, 52 weeks a year, the average waiting time stretched out to 6 months in late 1998 (Carr, 1998).

As part of this study, the authors surveyed the financial, accounting and legal communities to determine how they have structured to accommodate and service the growing private research effort in Saskatoon. The surveys were conducted by telephone and fax between July and October 1998, and many respondents, citing concerns about client and commercial confidentiality, were unwilling to divulge any data regarding how or in what form they serviced the industry, if in fact they did at all. Nevertheless, the results suggest that only limited accommodation has been made in Saskatoon to support and service the developing biotechnology industry (Table 9.5).

Each of the major banking institutions, as well as those companies which provide venture capital, were contacted. The Royal Bank of Canada and the Canadian Imperial Bank of Commerce (CIBC) are the only institutions that have specific knowledge-based lending facilities, both in partnership with Western Economic Diversification. The Royal Bank programme is run out of Saskatoon while the CIBC programme is administered from Winnipeg. The other banks and the credit union have clients in the industry but do not have special facilities targeted on the sector. One factor limiting the provision of this service is the limited market for financing. In a related survey of firms undertaking canola research, all of the multinational agrochemical firms responding indicated that they did their banking as part of the corporate effort, with much of the service coming from their Canadian or global headquarters which were always located in another city or province. The public lending facilities are more focused on

Table 9.5. The financial industry and the canola sector.

Institution	Location and service	Canola clients
Royal Bank of Canada	Special loan pool in Saskatoon; office at Innovation Place	Yes
Canadian Imperial Bank of Commerce (CIBC)	Office at Innovation Place; special loan pool in Winnipeg	Yes
Bank of Montreal (BOM)	Two knowledge-based investment loans officers in Saskatoon; access to BOM Capital Corp in Calgary	Yes
Toronto Dominion Bank	No special loans officers; Saskatoon	Yes
Saskatoon Credit Union	No special loans or officers; Saskatoon	No
Business Development Bank	Saskatoon office	Yes
Western Diversification	Saskatoon office	Yes
Saskatchewan Opportunities Corp. (SOCO)	Innovation Place office	Yes
Agri-food Equity Fund	Innovation Place office	No

Source: Telephone survey, August 1998.

Saskatoon, with Western Economic Diversification, the Business Development Bank of Canada, SOCO and the Agri-food Equity Fund all located in, or having offices in, Innovation Place or Saskatoon.

The accountancy profession has similarly been challenged to service the new knowledge-based firms. The Chartered Accountants of Saskatchewan, the Certified Management Accountants of Saskatchewan and the Certified General Accountants of Saskatchewan, when contacted, indicated that no specific listings are available for practices that deal specifically in the area of biotechnology. To check on the industry, a fax-back survey of biotechnology companies was undertaken in August 1998, asking firms whether they had a local auditor or accountant. Too few responses were received to get statistically valid data. Nevertheless, the multinational firms responding confirmed that their accounting services were purchased through their Canadian head offices while the smaller firms responding indicated they used local accountants and auditors. None of the accounting practices were identified as having a speciality in accounting for knowledge-based enterprises.

A third key service required by knowledge-based firms is legal support for protecting intellectual property that may have commercial value. The first point made by many of the legal partnerships in Saskatoon is that there are no resident patent agents in Saskatchewan. Local firms that have clients requiring patents refer the work to patent agents operating in Calgary, Edmonton, Winnipeg or Ottawa. Most of the patent agents have relationships with local legal firms or have liaison offices in Saskatoon to link to the local demand for the service. Nevertheless, a number of practices in Saskatoon have capacity to support firms with protecting their intellectual property (Table 9.6). Of the more than 50 multi-partner practices operating in Saskatoon, nine indicated that they had some capacity in the area of intellectual property rights. These firms were surveyed to determine whether that expertise or practice provided service to canola companies (e.g. patent applications, plant breeders' rights applications or related litigation). Five firms indicated that at some point they had done some work with the canola industry but only two firms indicated that they had an ongoing practice. The rest of the firms and the lawyers concentrated on trademarks and copyright law. Only about five lawyers indicated any ongoing activity with the canola industry, with none of them working exclusively on that business. In addition, the focus of the local legal community is on the Canadian law, which limits their application as most of the innovations of multinational ventures are protected under US law. In short, firms are being serviced, but the presence of many multinationals, which buy these services internationally, combined with the absence of any registered patent agents in the province and few full-time lawyers specializing in intellectual property, limit the scope of benefits that could accrue to the industry.

Apart from publicly provided infrastructure, there is little evidence of any existing or developing critical mass of other specialized industry that strengthens the bond between the canola research industry and the Saskatoon location.

Moving downstream from the research effort, there are some indications that Saskatoon and western Canada provide forward linkages that increase their attractiveness as research sites. The presence of ample supplies of test plots for field trials makes western Canada a significantly more attractive location than parts of the US and almost all of Europe. In 1997, more than 530 confined field trials of canola were undertaken in Canada, with more than 230 in Saskatchewan alone (CFIA, 1999). Although only about 14% of the firms indicated in the survey (Table 9.3) that this was a positive feature for location, all companies responded that without available space Saskatoon and western Canada would be less attractive. Two companies indicated that the popularity of western Canada as a breeding and growing region is beginning to constrain their testing programmes. One company suggested that the demand for test plots within a narrow radius of Saskatoon has forced many firms to drive for up

Table 9.6. Saskatoon's legal community and intellectual property practices.

	Number	%
Total number of multi-partner practices	50+	
Firms with IP practices	9	<18
Firms with ongoing capacity related to agri-food research	2	<4
Estimated number of lawyers with IP practices	15	<5
Estimated number of lawyers with IP practices related to agri-food research	5	<2

Source: Telephone survey, August 1998.

to 2 hours to get to their sites, which is unpopular. Another company indicated that the widespread dispersion of canola production in western Canada is making it more difficult to find areas far enough removed from production areas to test varieties with some novel characteristics. These companies have been forced to try tests underground, in tents or in non-canola growing regions in the interior of British Columbia.

Equally important for companies developing canola varieties is access to a large and accepting market. Given that the supply chain is only now integrating downstream to the farm gate, the real concern to date has been proximity to a large-enough production area that optimal adoption can be realized. As discussed in Chapter 4, the rapid introduction of new varieties in the late 1990s (i.e. 30 or more varieties each year) and the shorter product life cycle (i.e. averaging 3 years in the late 1990s, down from 15 in earlier years), have accelerated the optimal adoption rate (i.e. with maximum market share being reached in 1–2 years rather than 4 or more years). More than one-fifth of all companies surveyed (Table 9.3), representing most of the larger breeding operations, indicated that rapid access to a receptive seeds market was critical to their decision to locate in Canada. They confirmed that undertaking the research and commercializing the resulting varieties under the same regulatory system was a key feature in their location decisions. Heller (1995) estimates that a regulatory delay of 1 year decreases the rate of return for a biotechnology product by 2.8%, while a 2-year delay decreases the rate of return by 5.2%.

Farmers, in the three prairie provinces combined, seeded an average of 10.7 million acres to canola and rapeseed varieties in the 1992–1996 period, with a record of 14.2 million acres in 1994. Production has been tending to concentrate in Saskatchewan. In 1981 there were more farmers planting canola in Alberta than in Saskatchewan (defined as farms seeding at least 100 acres to canola, which in 1996 represented 90% of the total canola acreage seeded). Since then, the number of farmers planting canola in Saskatchewan rose by 184% while the number rose by only 92% in Alberta; by 1996, Saskatchewan had almost half of all the Prairie farmers planting canola, and the highest percentage of its farmers in canola (23%) (Table 9.7). By the late 1980s, Saskatchewan's share of acreage planted was slightly less than 40% but rose to approximately 46% in the 1992–1996 period.

Table 9.7. Canola farmers in western Canada.

	Manitoba	Saskatchewan	Alberta	Prairie provinces
Number of farms 1981	1,932	4,524	4,687	11,143
Number of farms 1996	4,862	12,838	8,987	26,687
% change 1996–1981	152%	184%	92%	139%
% all farms planting canola	20%	23%	15%	19%

Source: Statistics Canada (1997), Agricultural Census, selected operator data for farms with rapeseed greater than 100 acres.

Table 9.8. Distribution of production by type of farmers and age, 1996 (%).

	Manitoba		Saskatchewan		Alberta	
	All	Canola	All	Canola	All	Canola
Under 35	18	32	16	29	16	30
35–54 years	51	53	49	51	51	51
Over 55	31	15	35	20	32	19
Total	100		100		100	

Source: Statistics Canada (1997), Agricultural Census, selected operator data for farms with rapeseed greater than 100 acres.

A number of reasons are behind the increased share of canola in Saskatchewan. Deregulation of grain handling with the end of the Western Grain Transportation Act (WGTA) has increased the incentives to plant non-Board crops while the significantly higher research effort on *Brassica napus* relative to *Brassica rapa* has tended to benefit Saskatchewan producers, who tend to plant the majority of their acreage to *B. napus*. Alberta historically planted about 20% *B. napus* while Manitoba plants only about 30% *B. napus*.

Another feature about the canola producers in western Canada is that they are relatively more innovative than the general farm population. In 1996, there was a strong tendency for canola producers to be younger. Almost one-third of western Canadian canola producers were less than 35 years old, compared with less than 20% of all farmers (Table 9.8).

Canola producers are also more likely to have structured their enterprises to expand and bear more risk. The average canola producer has between 44% and 100% more land than the average producer in the prairies and uses between 61% and 113% more capital than other farmers. To manage the increased risk of the larger ventures, a significantly larger percentage of canola producers have adopted some form of informal or formal partnership or corporate structure to manage their ventures (Table 9.9).

Canola producers involved in this database treated canola not just as a small-niche crop but as a major component of their operation. The average canola producer in the prairies planted between 266 and 316 acres of canola

Table 9.9. Farm structure by type of farmer, 1996.

	Number of acres		Partnership		Farm capital	
	All	Canola	All	Canola	All	Canola
Manitoba	785	1571	37	47	518	1101
Saskatchewan	1152	1663	29	33	523	843
Alberta	881	1776	39	44	680	1329

Source: Statistics Canada (1997), Agricultural Census, selected operator data for farms with rapeseed greater than 100 acres.

Table 9.10. Tillage and agronomic practices by type of farmer, 1996.

	Manitoba		Saskatchewan		Alberta	
	All	Canola	All	Canola	All	Canola
% Farms reporting						
Conventional tillage	74	71	63	48	74	41
Conservational tillage	31	37	39	48	30	41
No tillage	11	17	21	25	9	16
% Farms using						
Fertilizers	82	97	70	95	74	95
Herbicides	73	91	75	88	63	86
Insecticides	21	90	14	53	9	20

Source: Statistics Canada (1997), Agricultural Census, selected operator data for farms with rapeseed greater than 100 acres.

in 1996, a rise of between 16% and 35% from the 1981 census records. Given the recommended agronomic rotation for canola (rest 3 years between plantings) and the 16–18% of potential acreage that was being planted to canola in 1996, farmers would appear to be growing almost as much as their land resources will permit.

Canola producers are also adopting more innovative production techniques. A significantly larger proportion of canola producers in the three prairie provinces use conservational or no-till tillage practices and significantly less conventional tillage (Table 9.10). Partly due to the nature of the crop and partly due to the sophistication of the producers, canola producers use fertilizers, herbicides and insecticides much more intensively than the general farm population. A 1997 survey sponsored by the Canola Council of Canada (Zatylny,

Table 9.11. Distribution of non-traditional crop production by type of producer, 1996 (%).

	Manitoba		Saskatchewan		Alberta	
	All	Canola	All	Canola	All	Canola
Buckwheat	0.32	1.60	0.01	0.06	0.01	0.03
Dry field beans	0.31	1.54	0.07	0.32	0.04	0.26
Dry field peas	2.47	12.38	5.65	25.11	2.08	13.63
Maize for grain	0.47	2.37	0.00	0.00	0.01	0.07
Maize for silage	0.27	1.34	0.01	0.03	0.02	0.16
Potatoes	0.23	1.15	0.08	0.36	0.17	1.09
Soybeans	0.02	0.10	0.00	0.00	0.00	0.01
Sunflowers	0.49	2.47	0.11	0.47	0.01	0.04

Source: Statistics Canada (1997), Agricultural Census, selected operator data for farms with rapeseed greater than 100 acres.

1998) showed that 35% of the herbicide-tolerant canola acres were direct seeded, compared with only 19% of conventional canola.

Canola producers are also more innovative in their choice of products. In Manitoba, they are five times more likely to produce non-traditional products (especially buckwheat, maize, dry field peas and sunflowers) than the average producer (Table 9.11). In Saskatchewan, canola producers are 4.4 times more likely to produce a mix of non-traditional products, while in Alberta, they are 6.6 times more likely. Although small – 1.4% of total acreage cultivated by Alberta canola producers and 2.9% of the acreage in Saskatchewan and Manitoba – this focus on non-traditional products indicates that these producers are seeking new production and income opportunities more intensively.

The extensive cooperative farm service networks have supported this large, receptive and relatively sophisticated farm market for new seeds in the prairies. The Prairie Pools, in particular, with an historical delivery share of about 60% and a membership including the majority of farmers in the West, have aggressively positioned their organizations as wholesalers for new varieties, partnering extensively with Svalof in earlier years and more recently collaborating with AgrEvo and Monsanto to deliver their proprietary herbicide-tolerant seeds to their farmer members.

On net, it would appear that the linkages to upstream industries are adequate but not a determining factor in location, whereas the downstream capacity to move the product into the market is quite extensive and likely to be one of the factors contributing to the expansion of activity in Saskatoon and western Canada.

The Labour Market

The third centripetal force coinciding with industrial agglomeration, identified by both Marshall and Krugman, is the development of a 'thick' labour market, with an extensive supply of specialized skills. Such a labour market first enables employers to find employees more easily and secondly helps employees find employers more efficiently. The slow but gradual agglomeration of canola research in Saskatoon and Canada suggests that there should be some evidence of a speciality labour market evolving. More than one-quarter of companies responding to the canola industry survey (Table 9.3), and the clear majority of private companies, indicated that access to a deep, local, skilled labour pool was important.

A survey of canola research employees in Saskatoon was undertaken in summer 1998 in order to assess the dimensions and dynamics of the local labour pool. Approximately 1000 surveys were distributed to 40 research companies, public agencies and laboratories, service companies and related university researchers. Approximately 500 of the surveys went to entities operating at Innovation Place and the rest to employees working in public laboratories on the campus. A total of 390 responses were received, representing 169 person-

Table 9.12. Distribution of survey based on percentage of time working on canola.

	Number	% Distribution
<10%	152	39
10–25%	42	11
26–50%	40	10
51–75%	21	5
76–100%	135	35

Source: Survey of canola industry employees, August 1998.

years of canola-related employment in 1998, compared with the 248 person-years of related employment in Saskatchewan identified in the company survey (Table 9.1). Approximately 40% of the respondents only did a little work on canola, while 35% of the employees put most of their effort into canola-related research (Table 9.12).

As one might expect with such a rapidly expanding research area, few of the workers have been working on canola for a long time (Table 9.13). About 43% of the respondents said they had been working on canola for less than 1 year while 32% said they had been working for 1–4 years on canola. Only about a quarter of the workers had been working on canola for more than 5 years. The average time worked on canola was 3.5 years, suggesting that any agglomeration benefits may be relatively new and may not have had time to be factored into investment decisions by companies. In contrast, the average worker in the province has been employed by the same employer for almost 10 years.

As with any new sector with highly trained employees, the workers in this industry are relatively young, compared with the rest of the provincial labour force. Few workers enter the industry with less than a technical diploma and many of them will have at least an undergraduate degree. More than 60% of the respondents were aged 25–44, compared with only 49% for the provincial labour force, and only 19% were older than 45, compared with almost one-third for the labour force (Table 9.14). The average age of the employees in the industry is approximately 36, compared with 39 for the provincial labour force. This skew towards a younger age partially explains the focus in the Innovation

Table 9.13. Work experience of canola workers.

	Number	% Distribution
<1 year	169	43
1–4 years	125	32
5–9 years	49	13
10–19 years	33	8
>20 years	14	4
Average years	3.5	

Source: Survey of canola industry employees, August 1998.

Table 9.14. The Saskatoon canola workforce compared with the provincial workforce.

Age (years)	Canola employees	Total Saskatchewan
19–24	20%	19%
25–44	61%	49%
45–64	18%	29%
65+	1%	3%
Average age	35.6 years	38.9 years

Source: Canola industry employee survey, July 1998; Statistics Canada, Labour Force Survey, August 1998.

Place programming towards social events (e.g. softball) rather than more sedentary or contemplative activities.

From the employers' perspective, the benefit of a thick labour market is the ability to find appropriately skilled employees to undertake the work required. To gauge this in Saskatoon, the employees were asked about their academic and work background. Approximately 23% had technical diplomas, 60% had undergraduate degrees, 17% had master's degrees and 19% had PhDs (there is double counting as persons with PhDs also have undergraduate degrees). When the degree-granting institutions are examined, the dynamics of the industry become more pronounced. The vast majority of the workers with only technical diplomas or undergraduate degrees got their training either in Saskatoon or Canada. In contrast, an increasingly larger share of the employees with higher degrees were trained in other countries. Table 9.15 shows the results for the totals. When the data are segmented by public and private sector employers, there appears to be little difference in the hiring practice between the two groups of employers.

Table 9.15. Distribution of employees in the Saskatoon agri-food research community, by degree, 1998.

	Technical diploma	Undergraduate degrees	Masters' degrees	PhD degrees
Total	88	235	67	73
Saskatchewan as % of total	82	66	43	27
Canada as % total	98	91	84	64
USA as % total	1	2	6	10
Europe as % total	0	4	7	18
Other as % total	1	3	3	8

Source: Canola industry employees survey, July 1998; Statistics Canada, Labour Force Survey, August 1998.

All but two of the employees with a technical diploma were trained in Canada and 82% of the employees with technical diplomas got their training from the Saskatchewan Institute of Applied Science and Technology, mostly through the 2-year Biotechnology Technology Program in Saskatoon. Since its inception in 1969, the programme has produced 335 graduates (SREDA, 1998). The graduates have comprehensive training in biology and chemistry with emphasis placed on laboratory practice in the areas of plant, animal, chemical and microbial work, including molecular biology. Employment surveys of those graduating from this course show that 88% are employed in their field within 6 months. For the 1987–1996 period, almost half of the graduates had attained employment in Saskatoon in a related field. With the acceleration of activity in Saskatoon in the late 1990s, the enrolment rate in the programme rose to approximately 15 per year, up from under 10 per year in the 1980s. The same trend is seen at the undergraduate level. All but 22 of 85 of the respondents with a bachelor's degree were trained in Canada. There has been a bit more intra-country mobility at this level of training, however, with only 66% of the respondents getting their training in Saskatchewan. The Saskatoon labour market draws extensively from Ontario, Alberta and Manitoba.

The labour market becomes significantly more mobile at the graduate degree level. Less than half the employees with master's degrees and only about one-quarter of the employees with doctorates are trained in Saskatchewan. As shown in Table 9.15, the more advanced the degree, the greater mobility and cross-national movement of employees. At the PhD level, more than 35% of the workers were trained offshore, in Europe, the US and other countries. Nevertheless, it is important to note that the single largest source of both master's and doctoral level employees is the local university in Saskatoon. Table 9.16 shows that in the past 35 years Saskatchewan has granted 42 PhDs for research in areas related to canola, which represents approximately 7% of all the canola-related degrees granted globally. Although Saskatchewan's share of the global training effort at this level has dropped from 25% in the early years to only 5% in the current decade, the absolute volume of output has risen steadily.

Table 9.16. PhD dissertations published between 1925 and 1998 in Saskatchewan and globally.

	Saskatchewan	Global	Saskatchewan as % of global
1925–1969	2	8	25
1970–1979	7	33	21
1980–1989	11	117	9
1990–1998	22	422	5
Total	42	581	7

Source: Keyword search of the UMI Dissertations Abstracts on September 17, 1998 for the terms canola, rapeseed, *Brassica napus*, *Brassica rapa* and *Brassica campestris*.

The emergence of a 'thicker' labour market in the 1990s is beginning to show in the data. Of the 390 respondents, almost 40% indicated that they had worked at universities before working for their current employer and a further 36% indicated that they had worked with other private companies earlier in their career (Table 9.17). Given the sheer size and the know-how and know-who roles played by the public agencies, it is somewhat surprising that only 21 or 11% of the private-sector employees indicated that they had worked for either agency before entering their current employment. Discussions in the industry suggest that this probably understates the role these two agencies play in the labour market. Both agencies have more than 15 postdoctoral appointments annually, many who ultimately move to work for their private-sector funder, suggesting that the survey either missed these people or that they attributed their time in postdoctoral research in the agencies either to the university or to the private funding company. This clearly is an area for further research and investigation.

The data shows that the local labour market is able to supply much of the labour required, but that some of the higher skilled employees must be recruited from elsewhere. This is especially true for employees with higher degrees. All employees were asked what features of the job and community affected their willingness to move to or from Saskatoon. Both labour market features and personal factors were tested. Somewhat surprising, given the conventional wisdom in the industry, the 'thickness' of the labour market was the key consideration mentioned by respondents. All employees with graduate degrees that responded to this question ranked proximity to other companies or agencies that could hire them was in their top five considerations and 87% of the respondents put it as the most important consideration (Table 9.18). The second most important feature was the type of work in the job, another feature of a thick labour market. Salary and benefits came third, followed closely by career prospects. Almost all the other factors, either related to the job or related to the community, were ranked well below these four factors. Somewhat surprising to some may be that few of the respondents were concerned about taxes, cost of living, or community amenities. The key message appears to be that if the labour market and the

Table 9.17. Labour market mobility in the Saskatoon agri-food research community.

	Current	Past employment experience			
	Employer of respondent	University	Other companies	AAFC	NRC
Companies	189	45	81	13	8
AAFC	162	42	50	–	4
NRC	39	19	9	3	–
Total	390	151	140	16	12
% total		39	36	4	3

Source: Canola Employees Survey, 1998.

Table 9.18. Relative importance of job attributes and community features as they affect highly mobile employees.

	PhD ($n = 25$)					Masters ($n = 45$)				
	1	2	3	4	5	1	2	3	4	5
Proximity to other companies/agencies hiring	22			1	2	39	2	1	2	1
Type of work in the job	17	2				13	12	1	1	1
Salary and benefits		9	4	2	1	5	9	11	2	
Future career prospects within the company		6	5	5	1	4	3	8	5	1
University links (adjunct appointment; collaborations)	1		2	4				1	2	2
Workplace setting (e.g. research park)			2		1			1	2	2
Personal income tax levels			2		1			1	2	3
Commuting distance to work			1		7		1			4
Cost of living (excluding housing)			3	1				4	2	
Cost of housing			1	2				3	3	
Sales tax levels								1		
Proximity to friends and family				1	1	6	1	3	3	3
Community facilities (e.g. cultural, sports)	1	1	1	2				1	2	

Survey questions: If you have moved from elsewhere, have considered employment opportunities elsewhere or are actively considering a move elsewhere, what factors are most influential to your decision? Rank top five (1 = most important; 5 = least important).

job are attractive, all other factors can be ignored. The survey does not demonstrate categorically, however, that the labour market is thick enough yet.

One test of whether the labour market is thick enough to be attractive to highly skilled, mobile employees is the incidence of unemployment in the industry. Each respondent was asked to identify whether he or she had experienced any unemployment in Saskatoon. More than three-quarters said they had never had any unemployment (Table 9.19). Among those who had been unemployed, 40% said the period was less than 3 months, 52% said their period of joblessness ranged between 3 months and 1 year and only 7% said they had experienced unemployment totalling more than 1 year. At first glance both the incidence and duration of unemployment for canola workers appears to compare unfavourably both to the Saskatchewan and Canadian labour markets during 1996. More than 24% of respondents reported periods of unemployment, compared with 6.6% in the general population, while the average period of unemployment was 31 weeks among respondents, compared with about 20 weeks in the general labour market (Statistics Canada). Follow-up interviews in the industry suggest that most respondents answered the question as relating to the unemployment during their entire career. If the survey results are simply adjusted by the average tenure in the industry (3.5 years), the results fall in line with expectations, with the average incidence of unemployment becoming 6.8% and the duration of unemployment per year dropping to 8.8 weeks, compared with 20 weeks for the province. Nevertheless, the distribution of the incidence of unemployment is harder to explain away. Approximately 60% of the unemployment in the sector lasts longer than 3 months compared with only about 42% of the unemployment in the province. If this reflects true market conditions, then it may actually reflect agglomeration as unemployed workers with speciality skills are willing to remain searching for jobs in the local market rather than leave and find employment elsewhere.

The evidence suggests that the Saskatoon labour market related to canola is becoming thicker as the level of activity increases. Data are not strong enough, however, to indicate that the scope of benefits arising from the more specialized labour force are by themselves the key or even a major reason for the agglomeration of activity in Saskatoon.

Table 9.19. Incidence of unemployment in the Saskatoon agri-food research community.

Period of unemployment ($n = 390$)	Number of persons	% total	% of unemployed
No periods of unemployment	296	75.9	–
Total reporting some unemployment	94	24.1	100
0–3 months unemployed	38	9.7	40.4
3–12 months unemployed	49	12.6	52.1
Over 1 year unemployed	7	1.8	7.4

Zucker *et al.* (1998) go one step further in attempting to determine more explicitly how the labour market contributes to agglomeration. Their study examined the role of human capital in the birth of US biotechnology enterprises by looking for causalities between the location of research stars and the creation of new firms. They defined stars as scientists who had discovered 40 or more genetic sequences or scientists who wrote 20 or more articles on genetic-sequence discoveries. They concluded that the presence of active stars in a region were strongly positively correlated with the start-up of new ventures, stating that 'at least for this high-tech industry, the growth and location of intellectual human capital was the principal determinant of the growth and location of the industry itself'.

Using this approach for the canola industry mirrors the earlier trends noted in Chapter 8. If we take stars to be those who publish at least 20 articles, and borderline stars as those who publish 15–19 articles, we find that 68 individual scientists worldwide fit the criteria. About 45% of the stars are in Canada, 45% in Europe and 10% in Japan. Approximately 63% of the borderline stars are in Canada and the rest scattered in Europe, Japan, the US and Australia. In total, the 69 stars and near stars, which represent just less than 1% of all the scientists working on canola, produced 1523 articles, or about 31% of all the articles produced over the period (Table 9.20). (Zucker *et al.*, 1998: 292, found that the stars in their study represented 0.75% of all scientists but 17% of all articles.)

The largest single geographical concentration of stars and near stars in the world is in Saskatoon, where 11 (or 16%) of the scientists live and work. If the stars and near stars are then assessed by their citations rates, Saskatoon has six out of 40, or 15% of the total, and about one-third of all the Canadian stars and near stars. The next nearest is Winnipeg, which has eight stars and near stars, but only two stars or near stars when the citation rate is used to eliminate high producers of largely unnoticed work. Using the methods of Zucker *et al.* this would suggest that Saskatoon is a logical place to have entrepreneurial growth.

Leadership

Over and above the centripetal and centrifugal forces, there is an intangible role played by individuals. The aggregate data on research trends, economies of scale and labour market dynamics can only explain some location decisions. Krugman has argued that much of the existing pattern of industrial location is simply a matter of 'historical accident.' Looked at on another level, one can just as convincingly argue that the 'historical accident' was the uneven distribution of people who make things happen – the leaders.

As presented in Chapter 3, the history of the development of canola, researched and published in two stages by the Saskatchewan Wheat Pool (McLeod, 1974) and by the NRC in Saskatoon, highlighted the importance of leadership. The effort began with a team consisting of J. Gordon Ross, a former

Table 9.20. The location of research stars by country, 1981–1996. (*Source:* ISI, 1997.)

	Stars		Emerging/borderline stars	
	All scientists with 20 articles or more	At least 20 articles; >5.0 cite rate	All scientists with 15–19 articles or more	at least 15–19 articles; >5.0 cite rate
Australia	0	0	1	1
Canada	13	5	25	13
Saskatoon	3	2	8	4
Europe	13	8	9	6
France	3	1	1	1
Germany	4	2	2	0
Poland	1	0	0	0
Sweden	2	2	1	1
UK	3	3	5	4
Japan	3	2	1	1
USA	0	0	4	4
Total	29	15	40	25

Liberal Member of Parliament who became president of Prairie Vegetable Oils of Moose Jaw, Henry (Hank) Sallans, an oilseeds chemist with the NRC, and William White, a plant breeder with the Dominion Forage Laboratory of Agriculture Canada. With Ross as the promoter, Sallans and White began the search for a new oilseed crop for western Canada. As the effort progressed and grew, the three brought in and were ultimately replaced by a new generation of scientists and promoters. Sallans brought Burton Craig to Saskatoon and the Prairie Regional Laboratory, where he ultimately developed techniques for using gas–liquid chromatography that provided the means to test seeds for desired oil characteristics. Keith Downey, an in-law of White and the scientist credited with breeding the first canola variety of *Brassica napus*, began his career as a student labourer in the Dominion Lab in the 1940s and returned to the Agriculture Canada laboratory in Saskatoon in 1957, after working at the Lethbridge research station for 5 years. Meanwhile university-based scientists joined the team, at the encouragement of Sallans. Milton Bell, at the University of Saskatchewan and Baldur Stephenson at the University of Manitoba played key roles, respectively, reducing glucosinolates and breeding a canola-quality *B. rapa* variety. The leadership mantle for promoting the crop during the 1960s and 1970s shifted initially to R.K. Larmour, a scientist who left the NRC to become research director for Maple Leaf Foods and ultimately to the Rapeseed Association of Canada.

As the original leaders began to retire, new leaders emerged. Increasingly, leaders need to have peer status in more than one of the three key fields – business, government or university. As in the rest of the story, the leadership has tended to shift away from public-sector scientists toward industry. Brent Kennedy, the general manager for AgrEvo Canada, emerged in the late 1980s as the most articulate spokesperson for the industry and its needs. Many credit him for arranging for relocating to Saskatoon the NRC canola research effort, and particularly Wilf Keller, a plant biotechnologist who has been key in developing new transgenic varieties. Although Gerhard Rakow assumed the mantle as the senior oilseeds scientist at AAFC, concentrating on conventional breeding technologies, and AAFC hired Derek Lydiate, a molecular biologist with extensive experience in genomics, public-sector scientists have not been recognized as the leaders, except in narrow scientific discussions. The university-based scientists, similarly, have been unable to assume the leadership role they had earlier, as now they are largely financed and tied to private breeding programmes. Some, such as Ian Grant from Guelph, ultimately moved to the private sector and assumed leadership roles within the business world.

More recently, the industry has collaborated through Ag-West Biotech Inc., a government-funded, industry-led sectoral association (Phillips *et al.*, 1999). Ag-West delivers a wide variety of services to the Saskatchewan biotechnology industry, including mediation between business and government, project facilitation, project financing and visible leadership and direction for the biotechnology sector. On the investment side, Ag-West (combined with the International Centre for Agricultural Science and Technology, which operated between 1992 and 1997 and has since been rolled into Ag-West) invested Can$9.1 million in 49 projects in the agri-food sector, involving 35 companies, creating an estimated 437 person-years of R&D employment in Saskatchewan over the decade, at an estimated average gross provincial public cost of Can$18,550 per person-year of employment. Ag-West also has a facilitation mandate, which it delivers through five specific activities. First it produces and distributes to 1700 subscribers around the world *The Agbiotech Bulletin*, a monthly newsletter that provides timely information on the latest developments in Canadian agricultural biotechnology. Second, Ag-West runs the Saskatchewan Ag-Biotech Information Centre (SABIC), a one-stop resource centre for ag-biotech information, opened in 1997. The operation of the centre is designed to enhance public awareness and understanding of the ag-biotech industry. In the first 6 months of 1999, the centre conducted 75 different tours for 1000 visitors, including 35 groups of school students and teachers, 14 international groups, eight farm groups, five business groups and four government groups. Third, Ag-West established the Saskatchewan Agbiotech Regulatory Affairs Service (SARAS) in 1997 to provide companies moving products through the regulatory process with expert advice, mentoring and assistance to achieve product registration. Fourth, Ag-West staff participate in trade shows around the world, helping to increase the awareness of Innovation Place and Saskatoon as a leader in agricultural biotechnology. Fifth, Ag-West developed a continuing relationship with the University of Saskatchewan to build academic capacity in Saskatchewan to support commercialization of new products.

Ag-West has also developed an array of mechanisms to enhance networking in the industry. It provides a variety of seminars and conferences annually in order to disseminate technical, regulatory and market information to Saskatchewan's agricultural biotechnology community, and to create an environment for closer co-operation within and between private sector agricultural organizations. For the past 4 years, Ag-West and the Saskatoon regional economic development authority jointly sponsored major networking opportunities in Saskatchewan. In 1995 and 1997, they sponsored Strategic Partnering Events, where 15–20 biotechnology companies seeking additional financing were connected to over 100 investors. Ag-West has also enhanced the level of technological and information transfer needed for collective action by hosting Agricultural Biotechnology International Conferences (ABIC) in both 1996 and 1998. ABIC '96 and ABIC '98 each attracted more than 400 participants from over 30 countries.

The 1998 industry survey of companies undertaking canola-related initiatives revealed strong recognition of Ag-West in nurturing the Saskatchewan agricultural biotechnology community and significant kudos for the past and current presidents of the company. Beyond the financial support, respondents recognized the importance of a visible leader and spokesperson for the sector, both when dealing with governments and when responding to public concerns. Dr Murray McLaughlin, the first President, and Mr Peter McCann, the current president, both have become recognized leaders and advocates for the business, and gain significant respect from the private companies and public researchers in the industry.

This and other analyses suggest that the successful development of canola and the increased activity in the sector in the 1990s was not simply good luck, but was in large measure the result of a small articulate set of leaders. The lineage of leadership can be traced fairly directly, with often clear and direct hand-offs between generations. Busch *et al.* (1994) concluded from a sociological perspective that this leadership was instrumental in the effort to create the canola industry.

Conclusions

Harvard Professor Michael Porter's national diamond of competitiveness provides a framework for examining the extent of private-sector efforts (Fig. 9.1). In short, none of the factors examined above, by themselves, provide clear, causal reasons for the agglomeration of canola research within Canada or Saskatchewan. Taken together, however, they indicate the importance of locational features, be they economies of scale or scope, specialized supply and mar-

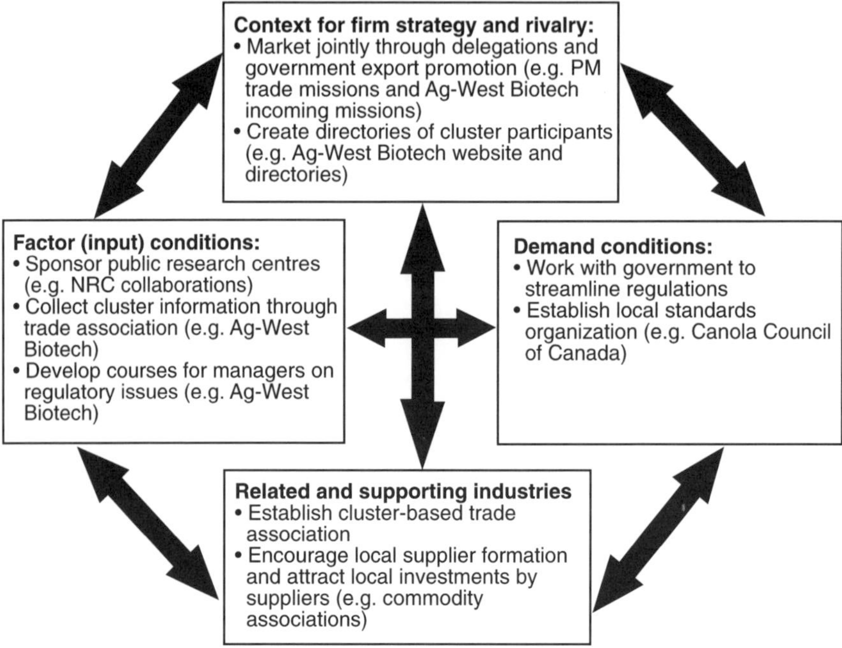

Fig. 9.1. Private-sector influences on canola research and development cluster in Canada (modified from Porter, 1998).

keting networks, thick labour markets or the presence of leaders. The lesson, if any, is that one-off industrial incentives are unlikely by themselves to generate a sustainable industrial development. Rather, concerted action on a number of fronts is necessary to achieve long-term, sustainable development. It must also be remembered that the benefits of agglomeration are not simply limited to the growing region. Ultimately, increasing returns and market efficiencies are factored into the product prices and will be passed along to consumers, in the form of both lower prices and greater choice.

Regulating Biotechnology-based Growth 

Why Regulate the Market?

10

Peter W.B. Phillips

Introduction

The state often plays multiple roles in an industry. Up to this point in this study, governments have been examined as partners and promoters in the discovery and development of new canola technologies and varieties. Governments in Canada and abroad have other objectives than simply generating wealth. As a regulator and custodian, the state often balances its efforts to pursue wealth creation with policies designed to address its fiduciary responsibility for public health and safety, the environment and equity considerations.

The canola industry is no different than many other sectors. The state has been, and continues to be, fundamentally involved in its creation and management as a partner, promoter and regulator. As shown in the next few chapters, various governments in Canada and abroad have intervened in the industry and its marketplace in order to achieve a mixture of economic and non-economic objectives. Harvard Professor Michael Porter, using his diamond of competitiveness model, provides a framework for identifying the potential role of governments in developing and nurturing clusters of economic activity. Figure 10.1 identifies the key national and provincial government contributions to development and management of the canola cluster in Canada.

This chapter proposes a short answer to the basic question of 'why and how does that state intervene in the market?'

The Rationale for Public Involvement

The public sector has become involved in the canola industry at a number of

© CAB *International* 2001. *The Biotechnology Revolution in Global Agriculture* (eds P.W.B. Phillips and G.G. Khachatourians)

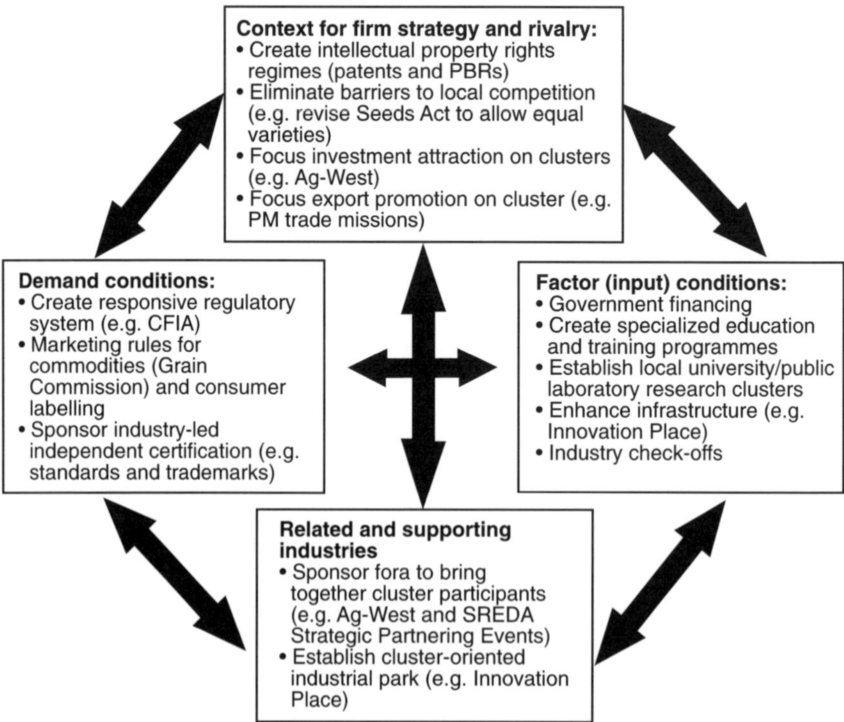

Fig. 10.1. Government influences on the canola research and development cluster in Canada (modified from Porter, 1998).

levels, including the research system, the production system, the product markets and ultimately at the consumer level. Table 10.1 shows the array of regulatory and related industry strategies used to manage the development of canola products in Canada. In each case, the state has pursued a variety of objectives. In Chapter 7, it is demonstrated that at the research level, the state has at times acted as the proprietor and, more recently, as a partner and promoter, creating the basic economic structure for public and private investment in R&D through direct investment (e.g. via AAFC or NRC) and through a selection of fiscal measures (e.g. subsidy and tax relief programmes) targeted on the industry. The state has been equally active on the legislative and regulatory side in recent years, using its powers to create an intellectual property rights regime to provide greater private-sector incentives for discovery (Chapter 11), to manage the domestic relationship between production and demand (Chapter 12) and ultimately to influence the international marketplace (Chapter 13).

Table 10.1. The array of regulatory and industrial strategies designed to manage development of canola products in Canada.

	Pre-1985 conventional varieties	Post-1985 conventional varieties	Post-1985 varieties with novel attributes
Know-why knowledge	All knowledge in the public domain; published in journals	All knowledge in the public domain; published in journals	
Know-what knowledge	All knowledge in the public domain	All knowledge in the public domain	Patents available after 1980 for technologies and after 1985 for plants in USA; trademarks used for some proprietary technologies; firms adopt strategies to capture rents on innovations
Know-how knowledge	All knowledge in the public domain	All knowledge in the public domain	Most knowledge protected by trade secret; firms access through contracts and collaborations
Know-who knowledge	All knowledge in the public domain	All knowledge in the public domain	Most knowledge protected by trade secret; firms access through contracts and collaborations
Developing a new variety for licensing	Cooperative field trials for yield, agronomic and quality characteristics	Cooperative field trials for yield, agronomic and quality characteristics	Dir95–01 Field testing plants with novel traits in Canada provides for confined field testing for plants with novel traits and for confidential trials Dir94–08 Assessment criteria for determining environmental safety of plants with novel traits Fertilizers Act and Regulations for novel microbial supplements

Continued

Table 10.1. *Continued*

	Pre-1985 conventional varieties	Post-1985 conventional varieties	Post-1985 varieties with novel attributes
Marketing a new variety to farmers	Seeds Act and Regulations require distinct, homogeneous stable and higher-yielding cultivars with a maximum weed content and specified germination rates SECAN seed marketing rules Recommending committee chooses best of offerings	Seeds Act and Regulations require distinct, homogeneous, stable and equivalent-yielding cultivars with a maximum weed content and specified germination rates (yield gains no longer mandatory) Plant Breeders' Rights Act, 1990. WCCRRC chooses all varieties meeting reference standards	Plant Breeders' Rights Act, 1990 Trade marks used to protect names of some varieties WCCRRC chooses all varieties meeting reference standards (yield gains not mandatory; some trade off-between yield and other input attributes allowed) Special WCCRC committee of public breeders manages contract registration system for products with output traits that will be produced under identity-preserved production arrangements (yield can be traded off for other attributes)
Growing a new variety	None	None	Contract Registration rules Company guidelines for agronomic practices (e.g. for HT crops), required crop rotations and bin-run seeding Hybrids increasingly being used to stop bin-runs
Marketing a new variety in the wholesale trade	Canada Grain Act (3 grades) Export tests for moisture, oil and fatty-acid content	Canada Grain Act (3 grades) Export tests for moisture, oil and fatty-acid content	Identity-preserved production contracts

Processing and marketing oil for human consumption	Voluntary standard of 5% maximum erucic acid, 1974 Food and Drugs Regulations set maximum of 5% erucic acid in 1985 Trademark for canola in 1978 set standards	Food and Drugs Regulations set maximum of 2% erucic acid in 1987 Trademark for canola revised in 1996 to conform with regulations Industry standards exceed regulations	Trademarks used for both canola[TM] and various canola products Health Canada uses Novel Food Guidelines of the Food and Drugs Act to inspect, monitor, test and confirm adherence to safety standards Industry standards exceed regulations No specific labelling guidelines exist
Processing for meal	Feeds Act and Regulation specify maximum of 3 mg of glucosinolates g^{-1} of air-dried meal	Feeds Act and Regulations changed in 1983 to specify maximum level of glucosinolates to be 30 μmol g^{-1} of meal	Feeds Act (Trade Memorandum T-3, 141, 142, 122, 143, 153 and 148) and Guidelines (1995) provide detailed requirements for approval

Source: Busch and Tanaka (1996) and http://strategis.ic.g.ca/SSG/bh00232e.html

Regulating discovery

Governments have been predominantly concerned about growth. They argue that private firms have not and will not do the optimal amount of research to develop new varieties for farmers. As discussed in Chapter 7, numerous studies show that research in agriculture provides high returns. This high rate of return has been explained historically by two factors. First, without any means for private investors to capture the gains from their research (e.g. patents or plant breeders' rights), they under-invest, causing higher marginal returns to research. This has been borne out by studies that show that private returns average less than half of the total benefits of research (e.g. Ulrich *et al.*, 1986). Less often stated, but perhaps as important, most studies show that farmers tended to act competitively, bidding away the gains from agronomic, yield-enhancing innovations, so that consumers ultimately gain (see Chapter 14). Hence, efforts to support farmers through yield-enhancing innovations ultimately benefit consumers through greater choice or lower prices. Although recent research by Alston *et al.* (1998b) raises some concerns about the large returns estimated by others – using a different model specification, they calculate that the internal rate of return on public research is probably below 10% and may be as low as 7% (results of our estimates for the canola industry presented in Chapter 15 are consistent with their finding) – the prevailing view in government and much of industry is that the returns remain high.

This set of arguments has been used to justify two often-conflicting types of government regulatory interventions. First, the high rates of return suggest that too little private research is occurring; the solution reached by many governments has been to introduce enhanced intellectual property rights, through the extension of patents to genetic constructs and through the introduction of plant breeders' rights. Second, governments have introduced provisions through the Patent Act, Plant Breeders' Rights Act and the Competition Act that can be used to limit monopolistic tendencies of holders of intellectual property rights, in an effort to ensure greater dissemination of research and innovation as a means of ultimately supporting consumers.

Regulating domestic markets

In the domestic research, production and marketing systems, the state attempts to mediate between public and private goals by providing a set of rules and norms that ultimately determines the extent and scope of private initiative, including research and development policy (e.g novel characteristics in Seeds Act), intellectual property rights (e.g. patent and PBR laws), competition policy (e.g. Patent Act, Plant Breeders' Protection Act and Competition Act), regulation of the seeds industry (e.g. Seeds Act), operation of the oilseeds markets (e.g. rules governing contracts and Winnipeg Commodity Exchange options) and laws relating to the environment and public safety in the production and mar-

Table 10.2. The regulatory landscape for canola in Canada and in international markets.

Objectives	R&D/discovery	Production and domestic marketing	International trade
Growth	Plant Breeders' Rights Act Patent Act	Canadian Grains Commission Trademarks	Patent Conventions UPOV Trade in Intellectual Property Rights Agreement in WTO
Equity	Compulsory licensing in Patent Act Competition Act Breeders exemption in PBR	Farmers exemption in Seeds Act Competition Act	Farmers exemptions in UPOV
Public safety (health and environment)	Seeds Acts (field trials)	Plants with novel traits regulations Environment Act Feeds Act Fertilizers Act Novel Foods Regulations Labelling	International Plant Protection Convention SPS and TBT agreements in the WTO *Codex Alimentarius* BioSafety Protocol

PBR, Plant Breeders' Rights; UPOV, International Union for the Protection of New Varieties of Plants; WTO, World Trade Organization; SPS, sanitary and phytosanitary; TBT, technical barriers to trade.

keting systems (e.g. Environment Canada's rules related to novel varieties and to the use of herbicides and the Novel Foods Regulations). In each case, the state attempts to balance the public interests of the general citizenry with the private interests of both domestic and foreign firms.

Regulating international markets

The concern to mediate between public and private goals flows over to the international marketing system, where the state works to help domestic producers to gain access to international markets on a broadly comparable basis as they have within domestic markets. With more than 140 different countries involved in these negotiations, this creates a complex set of mechanisms to ensure access, ranging from international agreements on intellectual property (e.g. International Union for the Protection of New Varieties of Plants (UPOV), Patent

Conventions and the Trade-Related Intellectual Property Agreement (TRIPS)), various standards-setting institutions (e.g. International Plant Protection Convention) and various trade and environmental arrangements (e.g. the sanitary and phytosanitary and technical barriers to trade agreements of the World Trade Organization (WTO), *Codex Alimentarius* and the Biosafety Protocol) (Table 10.2). As with both the discovery and domestic production and marketing rules, some of these institutions are designed to encourage growth, some to preserve equity and some to improve and enhance public health and safety.

The purpose of this section of the book is not to determine the appropriateness of the public objectives – that is the prerogative of voters – but to examine how these overlapping, and at times conflicting, public objectives have been translated into mechanisms that have influenced the development of the industry.

There are really three spheres of regulatory intervention: supply-side intervention, where governments intervene in order to ensure particular industrial or production outcomes, such as growth, development, equity, competition and diversification (discussed in Chapter 11); supply–demand intervention, where governments intervene for non-economic objectives, in order to ensure particular consumer, voter, citizen and societal objectives (discussed in Chapter 12); and domestic–international intervention, where the government intervenes for nationalistic reasons, either economic or non-economic. In the case of international organizations, the government confers national authority to the organization because it is pursuing congruent objectives; this is examined in Chapter 13.

11

Regulating Discovery

Peter W.B. Phillips

Introduction

With private capital and companies now dominating the research base of the canola business, governments have moved to use regulatory measures to manage the discovery and production of innovative products. In addition to direct and indirect financial support for the industry, the government has used regulations to create and allocate incentives and disincentives for private action.

The regulatory role of government can be looked at over three relatively discrete periods of time and areas of application. Before 1985 the research was predominantly driven by public agencies and universities, which reduced the need to regulate the creation, ownership and production using intellectual property. Nevertheless, as Busch and Tanaka (1996) show, the state was an active participant in establishing and managing the framework for regulating production and marketing of rapeseed and canola. Their work clearly and convincingly demonstrates that the interventions by the state created many of the conditions for rapid product and market development.

However, Busch and Tanaka miss the increasingly important role after 1985 of intellectual property rights in the canola production system, both for conventional and transgenic varieties. The introduction of IPRs fundamentally altered the canola industry, first precipitating the privatization of the seed industry and then, with the commercialization of biotechnologies, diffusing research into a significant number of private research consortia that are searching for differentiable product characteristics to exploit. Table 10.1 shows the regulatory mechanisms for those three cases.

Many authors have looked at the succession of legislative and regulatory changes in the US, Canada and elsewhere that created the incentives for private companies to enter the canola research business. What has been missing from those analyses, however, is the linkage between the intellectual property rights rules and the regulations affecting production and marketing. As discussed in Chapter 1, intellectual property rights are not enough to generate private investment. Schumpeter (1954) argued that firms will only invest if they have some expectation of earning a return on their investment. To do that, firms require some ability to price their product above marginal cost. As Romer (1995) points out, this is difficult to do, especially for knowledge-based innovations, first because they are often not easily excludable and second because these innovations often have low or no marginal costs. As a result, private investors do not have any reason to invest unless they have some means of excluding others from using their innovations. Ergo, the state grants 'monopoly' status through the patent and copyright systems to allow firms to exploit their innovations and recapture some of their investments. But the monopoly status only gives an innovator the right to try to collect their return. This return must be extracted from other actors in the production system or the final consumer markets. Hence, any discussion of regulating discovery must necessarily examine the interconnection between intellectual property rights provisions and legal and practical structures that define the operations of the production and marketing systems.

This chapter looks at the regulations and institutions that span the continuum from the production of basic, know-why research to the final consumer, in order to examine how the integrated system has evolved and to determine how its evolution has supported or impeded public and private investment in industry development.

Regulating Discovery in the Public Phase of Research

Busch and Tanaka (1996) have done a good job of documenting the regulatory and institutional developments that facilitated the development of canola as a new product. Column 1 of Table 10.1 shows the key elements of the system that existed prior to 1985. The knowledge-based research system remained fully in the public domain, eliminating the need for any regulations targeted on discovery. Instead, regulations became effective at the varietal registration stage, with a coordinated system of seed variety registration rules that ensured that the public-good benefits from the research would flow to producers and consumers. Varieties that came from the public research programmes in the universities and public agencies competed at the registration stage for the opportunity to make it to the marketplace. Varieties that outperformed existing varieties but performed poorly relative to other competing new cultivars were rejected. Given that perspective, it is clear that the varietal registration records before 1985 reflect more the net output of the research effort than the records after 1985, which tend to reflect inputs to the research effort rather than net innovations. Many of the varieties registered after 1985 were duplicates of existing varieties or simply replacement varieties with little net innovation embedded (Downey, 1998).

The rest of the production and marketing system was designed to ensure maximum adoption and dissemination of the new innovations, rather than to assist the innovators to capture returns on their investments. The absence of plant breeders' rights made it next to impossible for breeders to capture any royalties on open-pollinated canola varieties. Meanwhile, the combination of the Seeds Act, the Canada Grains Act and the operation of the Canadian Grains Commission set in place a varietal registration and product-grading system that worked to standardize the product, making the Canadian wholesale market for canola exhibit more features characteristic of commodities. In short, the firms were unable to differentiate their product by quality. At the same time, the Food and Drugs Act food market rules for quality and labelling limited firms' ability to differentiate rapeseed/canola products in the consumer market. Combined with the decision of the Rapeseed Association of Canada (now the Canola Council of Canada) to take out a trademark on the name 'canola', this made it difficult for private companies to pursue a niche market. However, as Busch and Tanaka (1996) point out, the production and marketing systems reduced the risks often inherent in new product development by standardizing the quality and managing the market.

This system was forced to change due to a number of stimuli. International pressure came through the negotiation in 1978 of the International Union for the Protection of New Varieties of Plants (usually referred to as UPOV, an abbreviation based on the initials of its name in French: *Union pour la Protection des Obtentions Végétales*) and the introduction of patent protection for plants in the US. Domestically, the increasingly tight fiscal situation in the public sector reduced its capacity to undertake all of the needed research. Furthermore, the successful development of biotechnologies opened up a wide array of new market opportunities that were more suited to private than public development. As a result, the public sector opened up to increased private involvement and the regulatory system was modified to support, and at times to manage, private-sector activity.

The Advent of IPRs

The main impetus for change came from the United States. As the single largest source of new innovations and the single largest market for those innovations, the US plays a vital role in almost all innovation-based sectors. Although the US began in 1776 with little or no patent or copyright protection, it now is viewed as the prime mover and supporter for private rights to innovations and inventions. In 1970 the US made the first substantive move to encourage increased private investment in agricultural research. Although the 1935 Plant Variety Protection Act (PVPA) had granted plant variety protection for 18 years to asexually produced plant varieties (e.g. hybrid maize), it was not until 1970 that the US renewed the PVPA and granted the same property rights to sexually reproduced varieties, such as open or self-pollinated varieties. These new rights of the innovator are limited by two exemptions. Farmers continue to have the right to use the PVPA-protected crops as seed in future years and researchers have the right to use the protected germplasm for research and development purposes.

On the technology front, in 1973 the US Patent Office granted Cohen and Boyer a utility patent on gene splicing technology, starting the race to privatize agronomic research. Over the succeeding years, virtually all of the main technologies required for genetic manipulation of a plant or animal have been patented in the US (see Table 11.2 below). To be eligible for patent, innovations must not have been known or used by others in the US, have been patented or described in a printed publication in the US or any foreign country or have been in public use or on sale in the US for more than 1 year prior to the date of application. Innovations that are novel, useful and non-obvious are then eligible for protection. Although US patent protection is not automatically accepted in other countries, most of the key technologies have been patented in prime markets. Those technologies that have not been patented in other countries still receive some effective protection, as any products or innovations based on US-patented technologies can either be blocked under the Patent Act from entering the US or the producer of those products can be sued in a US court for damages. Given that the US legal system provides for judgements that assess both real and punitive damages, there is a real disincentive for unauthorized use of patented technologies, especially if the resulting products are to be traded.

In the 1980s, a number of landmark rulings related to patenting living organisms opened the flood gates. In 1980 the US Supreme Court ruled in *Diamond* vs. *Chakrabarty* that the US patent law provides for patenting life forms. The first patent on a life form was for an oil-eating bacterium. In 1985 the first patent for a living plant was issued. Since then a number of plants have been patented. Plant patents provide additional protection over PVPA rights, in that plant patents do not provide for a researcher's exemption or farmers' rights.

These moves precipitated a response in Canada (see Santaniello *et al.*, 2000, for a broader discussion of the role of IPRs in agriculture). Canada was not a member of UPOV and did not protect private property embedded in plant varieties before 1990. Nevertheless, the debate about plant breeders' rights began as early as 1980 and ebbed and flowed until 1990, when the Plant Breeders' Rights Act was passed, providing for 18 years of protection for new varieties and allowing for both farmers' and research exemptions. This change followed shortly on the heels of a policy change in the Seeds Act in Canada. Prior to 1985 new varieties had to be superior in some characteristic(s) to those which were registered previously. The Expert Committee on Grain Crops, made up primarily of plant breeders, pathologists and quality chemists, judged each candidate cultivar from their area of expertise. If all the committee members agreed, the candidate variety was supported for registration. If more than one variety was superior to the check varieties, usually only the best variety was supported. The changes in the Act in 1985 redefined merit so that candidate varieties had only to be equal to the standard or widely grown check varieties and not cause harm to the industry. Along with the new rules came the development of the Western Canadian Canola/Rapeseed Recommending Committee (WCCRRC), made up of 30 or so public and private breeders. The new system became operational in 1988 and led to a significant rise in 'me too' varieties (Downey, 1998). Some argue that the looser rules and the larger public–private recommending committee led to a willingness by participants to be lenient with the recommendations. Few participants had any incentive to reject competitor's varieties as that might cause their varieties also to be rejected. A further refinement was added to the system to assist development of new, modified-attribute varieties. The WCCRRC evolved to provide a mechanism for contract registrations of new varieties with novel attributes. A private breeder that has created a modified oil variety can apply for a special contract registration which requires the variety be produced under an identity-preserved production system. The benefit for the breeder is that a sub-committee of the WCCRRC made up of public breeders and academics reviews the variety (ensuring confidentiality for the variety) and the merit rules allow for trade-offs between different standards. For instance, yields may be allowed to drop if the oil content can be enhanced.

Meanwhile, the Canadian patent system accommodated some of the demands for patents, extending utility patents as requested for new biotechnologies, for gene constructs and to single-celled organisms. One significant policy change that came about in 1987 was the decision by the Canadian government to cease using the 'compulsory licensing' provisions of the Patent Act for pharmaceuticals. For more than 20 years before then, Canada had issued compulsory licences for patent drugs to generic drug manufacturers in Canada, in order to lower the cost of pharmaceuticals for Canadian consumers and government drug plans. The Canadian government agreed to stop this practice after negotiating new commitments for R&D in Canada by the pharmaceutical industry. Given that many of the companies which have entered the agricultural biotechnology industry are divisions of multinational pharmaceutical firms (e.g. Zeneca, AgrEvo), the end of this contentious practice sent a signal that Canada was willing to accept and encourage private biotechnology research. The Canadian system has not, however, gone as far as US protection. To date, the Canadian Intellectual Property Office has refused to grant patents for multicellular living organisms. Furthermore, the Canadian legal system does not provide for punitive damages, which limits the deterrent factor of legal judgments against patent infringement.

Most recently, a number of companies have developed effective, commercially viable hybrid systems for canola, with the first hybrid variety released in 1989. In the succeeding decade, more than 17 new hybrid varieties were released. This science-based solution to protecting intellectual property in varieties may simply help to strengthen and reinforce the regulatory system.

Use of Intellectual Property Rights Regimes in the Canola Industry

The key to private research activity is the appropriability of the resulting gains. There are a wide range of means to ensure excludability of the results of the research, ranging from climatic or locational factors that restrict the geographical transfer of technologies, to measures that firms can undertake on their own – such as vertical integration between researchers and the unit doing the marketing, contracts and trade secrets – to legally sanctioned protection for intellectual property, as provided by patents and plant breeders' rights. All are in use in Canada (Table 11.1).

The key non-legislative approaches to ensure excludability and capture of the rents on canola research include a wide variety of measures. Selective choice of research priorities has helped to make some research results more excludable (Rosenberg *et al.*, 1992). Given that know-who and know-how tend to be found within firms or larger geographic clusters of research, there is a strong tendency for research communities to produce competitive, like-types of innovation which relate to the specific climate, soil characteristics, microbiology and industrial structure. In the canola sector, for instance, some of the varieties can only produce in the Canadian climate (certain pests or microbial phytopathogens limit or curtail production in other areas) and many of the new genetically altered varieties require a certain scale of production (e.g. total acreage or average field size) or complementary investments (e.g. mechanized seeding, spraying and harvest equipment). As a result, some of the Canadian innovation into canola cannot be transferred elsewhere, setting the base for excludability between jurisdictions.

Industrial restructuring has been, at least partly, driven by efforts to capture the returns to intellectual property. Perhaps most dramatic was the indus-

Table 11.1. IPR regimes for canola varieties, 1990–1996.

	Number of varieties	% varieties	% market share in 1997
Identity-preserving production contracts for novel traits	10	10	<5
Hybrids/synthetics	17	18	30
Plant Breeders' Rights[a]	36	[a]	70
Proprietary complementary technologies (e.g. HT varieties)	4	4	35

96 varieties registered 1990–1996.
Source: Canola Council of Canada webpage and author's calculations.
[a]Out of 186 applications as of March 1998 (CFIA communication).
The numbers add to more than 100% as almost all new varieties are protected by PBR.

trial restructuring that occurred in the chemical sector itself. As Just and Hueth (1993) point out, chemical firms had an incentive to invest in genetics to protect the value of their intellectual property rights in patented herbicides. As a result, almost all of the large chemical companies moved to partner their agrochemical divisions with genetics and seeds units. AgrEvo, in 1996, purchased 75% (subsequently raised to 100%) of Plant Genetics Systems of Belgium, an early leader in transgenics in canola and the owner of the InVigor™ hybrid technology. Dow AgroSciences bought 57% of Mycogen (the owner of several *Bt* genes) in 1996, and purchased the remainder of the company in 1998. Monsanto purchased Calgene (the owner of the patented *Agrobacterium* transformation technology for *Brassica* species) and has since acquired significant interests in Holden Seeds and Limagrain Canada. Zeneca bought Mogen, Dupont bought Pioneer Hi-Bred and BASF has bought into Svalof. As a result, the private genetics and seeds business has become almost fully vertically integrated, which has allowed the major agrochemical companies to acquire or to develop proprietary technologies that support their core agrochemical or seed businesses.

Production-input contracts (Rosenberg *et al.*, 1992) have been used by most of the companies that have developed herbicide-tolerant varieties of canola. AgrEvo's Liberty-Link™ system, for example, includes a package sold to farmers of a gluphosinate-tolerant variety protected by Plant Breeders Rights and the patented Liberty™ gluphosinate herbicide. Given that Liberty is only licensed in Canada for use on AgrEvo's varieties, it is extremely difficult for farmers to use bin-run seed for replanting in future years, as they would be unable to purchase the herbicide. Monsanto, the producer of Round-Up™ herbicide, has significant competition in the herbicide market (given that its primary patent on Round-Up™ has expired) and so has adopted another approach to marketing its herbicide-tolerant varieties. It has developed and patented a Round-Up Ready™ (*RR*) gene, which it has inserted into its own varieties but has also licensed to other breeders (as of 1998 also to Alberta Wheat Pool, Pioneer Hi-Bred, Agriculture Canada, Svalof and Limagrain). In order to acquire these new seeds, farmers are required to attend a sign-up meeting, to agree to a Technology Use Agreement (which prohibits bin-run seeding and grants Monsanto 3-year rights to inspect fields for their seed), to pay a Can\$15 acre^{-1} technology fee and to buy a package of seed and Round-Up™ herbicide. Monsanto is actively enforcing its production-input contracts – it has hired field investigators in western Canada to search for infractions and is reported to be spending approximately Can\$1 million annually (private conversations). As of 2000, Monsanto reported that it had found a few farmers using their seed without a licence. All but one was resolved through negotiation. In 2000 Monsanto prosecuted Mr Percy Schmeisser for planting 900 acres to a Round-Up Ready™ variety without first acquiring the licence and paying the TUA fee.

Identity-preserved production contracts are used increasingly to capture some of the added value resulting from canolas with novel traits. Canola has already been modified to produce a wide variety of engineered fat chains, industrial oils (e.g. laurate) and proteins, including low-value, end-of-the-scale proteins for improved nutritional value of the seeds, intermediate-value, bulk proteins such as industrial and food enzymes, and high-value proteins, mainly of interest to the pharmaceutical industry. End users, such as Procter & Gamble, Nabisco, Frito Lay, Lubrizoil, Mobil Oil, Shell Oil and Ciba-Geigy, among others, are showing significant interest. Each of these products needs to be developed and managed under an IPP production contract in order to capture from the marketplace the value inherent in the new end-use attributes. Details on these agreements are scarce as they are viewed as commercially sensitive information. As discussed in Chapter 8, these systems generally involve rules that require all seed be grown under contract, growers purchase new seed each year, the IPP crop not be produced on farms with commodity crops, routine checks of individual fields of IPP crop to estimate the yields, and separate elevators, unloading locations and storage facilities to prevent contamination of the product (Del Vecchio, 1986).

The development of effective hybrid technologies for canola has provided another technical mechanism for protecting intellectual property. It is uneconomic for farmers to replant seed from a hybrid or synthetic crop as they will lose half of the specific genetic traits with each successive planting and the resulting crops exhibit uneven growth. Firms that sell hybrid varieties are almost certainly assured that farmers will return each year to purchase new seed. The first canola hybrids were developed in the late 1980s, and during 1990–1998 17 hybrid/synthetic varieties were developed and introduced, accounting for about 18% of all the new varieties over the period. Only a few firms are actively breeding hybrids – Zeneca has about half of the varieties now while AgrEvo, with its purchase of PGS and its In-Vigor™ technology, is expanding its use of hybrids.

Few companies rely on trade secrets, except to protect their proprietary investments in germplasm. Historically, germplasm was public and both varieties and breeding lines (i.e. those lines that were not ultimately registered as varieties) were placed on deposit in public gene banks. AAFC thereby accumulated a relatively large collection of more than 650 varieties and advanced breeding lines. With the introduction of Plant Breeders' Rights, few companies now put their breeding lines into public collections. They simply meet the minimum disclosure requirements of the Act and put on deposit a sample of registered varieties. Private breeders now withhold access to virtually all of their breeding lines and use them as bargaining chips in negotiating collaborations with other private companies. This has also caused AAFC to re-evaluate its germplasm collection and has raised concerns that many of the advanced breeding lines it holds on deposit may no longer be in the public domain. Regardless of how the rights to the AAFC collection are resolved, its relative size and importance in the breeding industry has diminished. A number of foreign public collections are now larger that the AAFC collection and a number of private breeders claim to have larger collections than AAFC.

Even though the non-legal mechanisms go a long way to enabling commercial firms to control the use of their intellectual property, all additionally use one or more of the formal mechanisms, including patents, Plant Breeders' Rights or trademarks (Table 11.2).

Table 11.2. Intellectual property rights regimes related to canola breeding processes.

Technology category	Key technologies (and owner, if any)	IPR regime
Genomic information	*Arabidopsis* genome project AAFC *Brassica* genome mapping	None; public domain; on Internet
Germplasm	Public gene banks in Canada, USA, Germany, Russia, India, Pakistan, Australia, Japan and others Private gene collections	Restricted access only for private collections
Allele-specific amplification	SCARs SRSLOs Micro-satellites	100% private patents
rDNA strands/genes	HT genes (Monsanto, AgrEvo, American Cyanamid and Rhône Poulenc) Antifungal proteins (Zeneca) Antishatter (Limagrain) Fatty acids (Calgene) Pharmaceutical compounds (Ciba-Geigy)	100% private patents
Transformation technologies	*Agrobacterium* (Calgene) Whiskers (Zeneca) Biolistics (Cornell and Dupont) Mutagenesis (public domain)	100% private patents except mutagenesis

Growth promoters	Constitutive promoters (e.g. for HT-, disease-, drought- and salt-resistance, to express genes in all cells in plants:
	  35S (Monsanto)
	  Ubiquiton
	  AHAS (American Cyanamid)
		100% private patents

	Tissue specific promoters:
	  Pod/shatter control (Limagrain)
	  Floral morphology (AgrEvo and others; multiple)
	  Oil traits (AAFC and others)
		100% public and private patents

Hybrid technologies	InVigor™ (PGS) CMS System (Zeneca) Ogura CMS Systems (China?) Bolima (public domain)	All patented except Bolima, which is in the public domain
Traditional breeding technologies	Half-seed process Double-haploid process Backcrossing Gas–liquid chromatography Shuttling Computer-assisted breeding	Most in public domain; some trade secrets and potential for copyrights on computer programs

Source: Personal communications with canola researchers and patent searches.

An examination of the canola breeding system shows the dominance of private companies in key stages of the process. Virtually every step of the research process is patented, most originally by entrepreneurial start-ups which are now part of the larger agrochemical seed industry. The public sector, both in Canada and elsewhere, has been largely absent from the key areas of the know-what knowledge required to transform canola.

Both original patents (i.e. mechanical or electrical inventions) and the new extended patents (for genes or gene processes) have been actively used by the research community (Evenson, 1998). Before about 1982 all of the processes used to develop new canola varieties were in the public domain. Since then, there has been a rapid expansion of effort globally, with the result that most of the processes now are privately owned. The Canadian patent database shows that of the 634 patents issued for canola-related innovations since 1978, about 45% were issued for process inventions and 55% for products (Table 4.4). As one might expect, the public sector (universities and governments) have done significantly less patenting (16% of total patents) and 71% of their patents were for processes. In contrast, the patent data for private companies suggests that about 42% of their work is focused on processes and the rest on product development.

Looking at the data by firm, we see that among those companies patenting more frequently, the bulk of the patents have been issued to end users of canola (e.g. the big food processors such as Nabisco and Procter & Gamble), with a little interest from industrial users (e.g. Shell Oil) and the rest produced by the plant breeding and chemical companies (e.g. Pioneer Hi-Bred, Calgene and Monsanto). As might be expected from the firm-based data, the bulk of the patents issued have been to US-based companies. When one examines the data further, it becomes clear that most of the patented innovations have in fact been developed in a wide variety of locations (sometimes universities) across the US, with only a few isolated pockets of patenting of inventions by Canadian or European researchers (Table 4.12).

Trademarks have also been used by parts of the industry to distinguish 'canola'-grade rapeseed from other varieties. In 1978, the Rapeseed Association (now the Canola Council of Canada), trademarked the new low erucic acid, low glucosinolate rapeseed as 'canola.' This new type of rapeseed had 5% or less of erucic acid and less than 3 mg of glucosinolates per gram of air-dried meal. The quality standards for canola were tightened in 1996 to allow only 2% erucic acid and less than 30 μmoles of glucosinolates per gram of meal. Since then there have been seven attempts by private firms (mostly breeders or food processors) to trademark either new varieties or specific canola oil food preparations (CIPO, 1998). In addition, a number of the breeding companies have trademarked their industrial processes. For example, PGS has acquired the trademark 'In-Vigor' for its hybridization process. Finally, all of the chemicals used are trademarked (e.g. Round-Up[TM], Liberty[TM], Pursuit[TM]). In almost every case, the adoption of trademarks has been designed to supplement other intellectual property protection and to differentiate for marketing purposes the specific products.

Plant breeders' rights, finally introduced in Canada in 1990, provide some-what weaker protection than patents for new varieties. Although the period of protection is almost as long as patents (18 years from the date of registration), the holder of the plant breeder's right is required to hold on deposit and make available for research purposes a propagating sample (usually deposited in the National Seed Collection in Saskatoon) and farmers retain the right to keep back seed for their own use. Some firms have attempted to improve the protection they get against other competitors by seeking to limit access to their deposit of propagating materials; it is not clear yet whether they will be successful. Since PBRs were introduced in 1990, breeders have automatically applied for PBR on virtually all new open-pollinated varieties (186 applications by March 1998), but only 36 of the varieties were awarded a certificate as of March 1998. As noted above, some companies also seek to go beyond the protection granted by PBRs, requiring farmers to sign away their farmers' exemption in order to gain access to the seed (e.g. Monsanto's Technology Use Agreements).

Academics, public research institutions and some private companies choose not to exercise their intellectual property rights for immediate monetary gain and instead publish the results of their research in academic journals – most of this work is know-why knowledge that is vital to future direction but often has little immediate commercial application. Prior publication in the US and, except in limited cases, within a year of patent application in the European Union (EU) and Canada, effectively precludes future efforts to protect the result-ing intellectual property through patent or Plant Breeders' Rights. As such, publication effectively grants the author rights to citation by subsequent researchers (a key currency of academics) but allows the economic benefits of the innovation to become public property (i.e. non-rival, non-excluded knowl-edge). This includes the *Arabidopsis* genome project, which is being developed and so far has been put into the public domain through academic publication. The genomic information, however, has significant potential to be codified and thereby to become a commodity that is rationed based on price. (Other genome projects have been effectively privatized by firms that have taken and assembled the public information in such a way as to extract economic gain.)

Taken together, the informal mechanisms and legal rights have effectively protected the vast majority of the technologies being used and the products flowing from the canola research community.

The Regulatory System for Production and Marketing

Having the rights to exploit intellectual property is not enough to ensure actual exploitation. The operations of the regulatory systems for production and mar-keting have a major influence on whether firms can hope to gain from their innovations.

Downstream of the discovery phase, the Canadian regulatory system has evolved so that private firms are more effective in capturing the returns on their innovations. The food and feed standards continued to be tightened in the 1980s, partially in response to scientific studies of the impact of erucic acid and glucosinolates on human and animal health, and partially in an effort to consolidate Canada's leadership position in canola production. In 1983 the Feeds Act lowered the allowable level of glucosinolates to 30 μmoles per gram of air-dried meal and in 1987 the Food and Drug Act lowered the allowable levels of erucic acid to 2%. Few other canola-producing regions in the world could meet those standards because their seed stock was not as advanced as in Canada. As a result, Canadian producers and exporters were assisted in carving out a niche for canola in the global oilseeds market.

Although this would appear simply to be an extension of the regulatory system of earlier years, one notable difference is that the grading system enabled by the Canada Grain Act and administered by the Canadian Grains Commission remained static, allowing specific exporters to establish market or contract specific specifications for the trade. This allowed exporters to begin to differentiate the product. In addition, the canola trademark was not modified to conform with the new feed and food standards until 1996, so firms were able at times to sell canola branded product that differentiated from the Canadian domestic standards.

Meanwhile, the federal government made a strategic decision in 1988 to regulate the biotechnology part of the business, based on product characteristics rather than on the process used. This conformed with the approach adopted in the US but diverged fundamentally from the process-based regulatory systems developed in the EU and Japan. A second strategic decision was to use the existing regulatory system wherever possible (Doern, 1997) rather than to develop a completely new system. The EU began the process of developing a new system in the late 1980s and still did not have all the directives fully operational in 2000.

The federal government decision, or more precisely lack of a decision, to set any specific requirements for labelling of genetically modified products has also left considerable room for private companies to position their products in the marketplace. So far the federal position has been that as long as there are no scientifically verifiable health and safety risks of the genetically modified product, there is no need for specific labels. To date, no company has taken this latitude to develop a specific marketing campaign for genetically modified canola. Currently, the genetically modified canolas on the market all have herbicide-tolerant transgenes in them but their oil or meal properties are the same as those of traditional varieties. The process for creating canola oil removes all genetic material from the oil and all of the countries currently requiring labelling have said that processed canola oil will not require a label. While GM canola meal contains transgenes, it is only used as animal feed. So far no countries have passed rules requiring labelling of animal products produced using GM feeds (Phillips and Foster, 2000). When genetically modified varieties with novel oil attributes reach the market, the absence of compulsory labelling will provide private companies with latitude to position their product in the market in the most effective way to capture the returns on their investments. This will be discussed further in Chapter 13.

Although economics provides some direction and advice on how an IPR system can be constructed and how product markets can be regulated to ensure the private sector undertakes the optimal amount of innovation to generate the maximum social welfare, it is less able to explain how companies respond to differing systems in different countries. In practice, the absolute level of protection is less important than the relative level of protection for understanding commercial decisions related to investment and production.

Table 11.3 shows that the US regulatory system provides the most support for private exploitation of intellectual property, with the widest protection for innovations through the Patent Act, Plant Variety Protection Act and the civil courts, and with the greatest latitude for private exploitation in the marketplace. Australia and New Zealand have systems comparable to that of the USA, with the exception that they have more restrictive seed registration systems. Canada largely mirrors the US system, with the exception of no patents for multi-cellular organisms, weaker civil remedies for patent infringements and a more restrictive seeds registration system. Japan and the EU have comparable IPR rules for agricultural biotechnology innovations throughout the discovery system, but have much more restrictive rules for commercializing the results, which tend to diminish the value of the intellectual property protection. At the other extreme, India and China, with weak or no legal protection for intellectual property and incomplete or ineffective market rules, provide significantly less support for innovation. The area is changing somewhat as countries address their TRIPS commitment to extend property rights. The EU announced in January 2000 that it would allow patents for whole plants, while India enacted new patent provisions and China negotiated accession to the WTO.

Implications and Future Trends

There has not been a definitive exploration of the impact of intellectual property rights on the research flow or gains to research. As noted in Chapter 8, the expansion of private investment in this sector coincided with the extension of intellectual property rights in Canada and the corresponding modification of the production and marketing regulations. In Chapter 15 evidence is discussed that shows the gross benefit of this research effort has risen at a slower rate than the rise in investment. As a result, the benefit-to-cost ratio has fallen from earlier years. Chapter 16 contains analysis of changes in the distribution of benefits. In earlier times consumers gained most of the benefits, with public breeders getting no direct returns on their investments and farmers getting only a small share. Now, the innovative nature of the product has enabled farmers to capture a larger share of the benefits while the oligopolistic structure in the increasingly integrated genetics, seeds and chemicals business has enabled the varietal developers to extract a significant share of the benefits from consumers.

Table 11.3. Key elements of policy affecting commercialization of intellectual property in competing countries, 1999.

	IPR system	Production system	Biotechnology regulations	Marketing regulations
Australia and New Zealand	Process, utility and life patents; no punitive damages	Conform to UPOV 1978 as of 1989; seeds registration	Product based	Mandatory labelling
Canada	Process and utility patents; no patents on multicellular organisms; no punitive damages	PBR Act, 1990; conform to UPOV 1978; Seeds Act regulates registration	Product based; no special legislation	Voluntary labelling
China	No effective patent protections	Entered negotiations to join UPOV; not currently a member	None	None
17 member states of the EU	Process and utility patents; patents on multicellular organisms; no punitive damages; *ordre publique* and other exemptions	All conform to UPOV, 1978 or 1991, on or before 1981 or entry to EU; special seeds registration rules for GM varieties	Process based; special legislation	Mandatory GMO labelling rules
India	Weak process and utility patents; no life patents; no punitive damages	Not currently a member of UPOV	None	None
Japan	Process and utility patents; no life patents; no punitive damages	Conform to UPOV 1978 as of 1982	Process based; special legislation	Mandatory GMO labelling rules
USA	Process, utility and multi-cellular organisms; punitive damages in legal challenges	PVPA; conform to UPOV 1978 as of 1981; no seeds registration	Product based; no special legislation	Voluntary labelling

A number of other implications arise from this analysis. A key issue raised by both public and private breeders is whether their 'freedom to operate' is being reduced by the increasingly proprietary nature of fundamental biotechnology and breeding technologies, genes and gene constructs and germplasm. Many public-sector scientists have suggested that the state should invest in the development of duplicate technologies or germplasm development in order to provide an accessible base of technologies for smaller private and public breeders. The fear is that multinational ownership of all of the technologies could shut down public or new private competing breeding programmes. From an economic perspective, investing to develop duplicate technologies does not appear to be cost effective. Instead, the state would be wiser to use the powers vested in its intellectual property rights regime or competition laws to encourage greater dissemination of non-rival, patented innovations in order to generate more access and hence greater spillover effects (Lesser, 1994). Both the Canadian Patent Act and the Plant Breeders' Protection Act provide for compulsory licences to remedy what is called 'abuse of patent rights.' The CIPO (1998) says that:

> if firms use their patents to 'hinder' trade and industry – i.e. not meeting demand in Canada, hindering trade or industry in Canada by refusing to grant a license (if such a license is in the public interest), attaching unreasonable conditions to such a license, using a process patent to unfairly prejudice production of a non-patented product or allowing the patent on such a product to unfairly prejudice its manufacture, use or sale – another company or the state can challenge them as soon as three years after the patent grant.

Meanwhile the anti-combines provisions of the Competition Act allow for the state to curb anti-competitive behaviour through investigation and prosecution. Neither provision has yet been used in Canada in the area of agricultural biotechnology.

Another issue facing Canada is a continually rising level of protection for intellectual property rights in other countries. To some extent one could view Canada as in competition with the other major canola research or producing regions for the role as seed developer to the industry. As Table 11.3 shows, Canada currently has a set of IPR and regulatory measures that compares favourably with the EU, Japan and many other countries. Only the US has a more innovation-supportive system. There will, however, be pressure to change over coming years. The US has raised concerns that the absence of patents for living organisms gives less than full protection to breeders. In countries without patents for living organisms, plant variety protection rules do not protect patented genes that can be transferred through traditional backcrossing methods (Lesser, 1994). This concern led the US, through the last round of multilateral trade negotiations, to press for a stronger agreement for trade related to intellectual property. The resulting TRIPS agreement requires WTO countries (Canada included) to either grant plant patents or to effect some form of *sui generis* protection (e.g. separate law such as PBR). If Canada declines to do so, some of the research effort located in Canada or directed at the Canadian market might shift to other areas that are more adequately protected.

Regulating Domestic Markets

Grant E. Isaac and Peter W.B. Phillips

Introduction

The purpose of this chapter is to address three questions. First, what motivated domestic government intervention, in the form of regulations, in agricultural biotechnology market transactions? Second, how do domestic governments regulate the market transactions? Third, what is the impact of domestic regulations upon the development and commercialization of agricultural biotechnology?

With respect to the first question, domestic regulations are imposed to modify the relationship between the supply and demand for agri-food products in order to protect both consumers and producers. A large information gap that exists between producers and consumers, especially in new product areas, creates a market failure that opens the way for government regulatory action. Government regulatory intervention for producer protection is driven by the economic pressures to develop an internationally competitive agricultural biotechnology sector. In order to facilitate the development of the domestic industry, regulatory intervention may be used to protect the industry from more competitive import products. Given the significant 'social dimensions' of domestic agricultural biotechnology regulations, there is significant potential that even the most objective regulator might interpret fairly basic non-economic goals in ways that could support mercantilist economic and trade objectives.

With respect to the second question, there are several key characteristics of government regulatory intervention to identify. Generally, domestic regulations are a function of the traditional role of the state and must be understood in this traditional context. More specifically, domestic biotechnology regulations differ according to four fundamental principles. First, domestic regulations can focus on the *technology* used or on the *products* that result from agricultural biotechnology. Second, regulations can either be an extension of existing regulatory frameworks or completely new. Third, regulatory decisions can be taken according to the *certain* or *reasonably certain* interpretation of the precautionary principle. Fourth, different systems permit different actors to participate directly in the regulatory decision-making process. In order to answer how domestic governments regulate the domestic market, it is necessary to identify the regulatory tradition of the state, as well as how the regulatory framework relates to the four principles. Using the four principles, it is possible to identify three types of biotechnology regulations currently in the world – the North American regulatory model, the European regulatory model and the Japanese regulatory model. Beyond a handful of other countries, there are no effective regulatory systems.

With respect to the third question, domestic regulations can have several significant implications upon the development and commercialization of agricultural biotechnology products. Consumer concerns that result from the information gap may adversely impact consumer acceptance, which is a crucial factor in the potential growth and development of new agri-food products. As shown in Europe, formal government approval for a new product does not always assuage consumer concern and result in consumer acceptance. In practice, it is the latter that is critical to the long-term sustainability of research and product development. Moreover, the lack of consumer choice and its ensuing adverse impact upon consumer acceptance affects all firms producing new products, both those using modern biotechnology and those using conventional technologies, because of the inability of consumers to differentiate their concerns among firms. A recent example of this phenomenon is the marketing policies of the American agricultural biotechnology company Monsanto. Although Monsanto secured EU regulatory approval for a GM soybean variety in 1996, it made no effort to separate GM soybeans from non-GM soybeans, effectively eliminating consumer choice. Other firms producing and selling soybeans in the EU complained that the adverse consumer response to the presence of GMOs affects them all, not just Monsanto (Palast, 1999).

Therefore, it is important to distinguish between product approval and product acceptance. Product approval is a regulatory decision, which is most often the result of an exchange of information between the producer and a regulator. Acceptance of a new product, on the other hand, is a consumer decision associated with access to information (provided by both government regulators and private companies) and the resulting choice. Consumers will sacrifice perfect information as long as they retain choice. Information that addresses consumer concerns improves consumer sovereignty and rationality, and ultimately creates choice and improves consumer acceptance.

Further, this chapter demonstrates that there is a symbiotic relationship between consumer acceptance and formal regulatory approval – regulators are beginning to respond to consumer pressures to delay or refuse regulatory approval for new products if consumers will not accept them, while consumers will not accept new products without approval. This creates a commercially adverse circularity, posing a conundrum for companies wishing to place new products in the market, as acceptance requires approval, which in turn requires acceptance.

In this chapter there is a brief discussion of how products of agricultural biotechnology raise crucial consumer concerns that may adversely impact consumer acceptance and which motivate government regulatory intervention. Specifically, this involves a discussion of how new biotechnology-based canola products differ from the perfect information paradigm of consumer theory. Next in the chapter is a discussion of alternative theoretical specifications that may be more congruent with both the evolving nature of knowledge-based agri-food products and the asymmetry of consumer concerns across both products and regions. There follows an examination of the evolving domestic regulatory and marketing systems in Canada, the US, Japan and the EU and a cross-section of developing countries. These comparisons illustrate the critical role of both governments and producers in managing the marketing of a knowledge-based product. This is followed by an examination of some new and evolving consumer concerns about modern agricultural biotechnology. Finally, the chapter examines the implications of government regulatory intervention upon the development and commercialization of agricultural biotechnology products, in an attempt to identify those regulatory interventions most conducive to development and commercialization.

Motivations for Government Regulatory Intervention

Using a strict economic definition, we would consider the consumer to be 'any economic agent responsible for consuming final goods and services, including individuals, groups of individuals or more formal organizations' (Pearce, 1996). Tirole (1988) identifies three types of goods: search goods, where consumers can visually identify attributes before consumption; experience goods, which require consumption to determine the attributes; and credence goods, where the full attributes of consuming a good may only be known after a long period, if ever.

Neoclassical consumer theory assumes that consumers, driven by the principle of non-satiation, consume normal goods (i.e. goods with an income elasticity of demand between zero and one) based on: the 'attributes' of the product, including price, quality, safety, etc.; and the consumer's budget constraint, where more income would entail more consumption of the good in order to maximize consumer utility. Under this paradigm the consumer has access to perfect information about all of the attributes of the product being consumed, including information on the inputs, the processing and the production techniques as well as the cost per unit to produce the good. Also, the consumer has perfect information on the impact of consumption on broader concerns such as human health, the environment, morality, ethics and religious beliefs. In this sense, there are four broad types of consumer concern to consider: traditional economic concerns, such as the price of the good; credence factors, such as human health and safety, which may or may not be properly reflected in the price of the good; externalities on the environment or biodiversity which often are not reflected in the price; and morality, ethics or religious beliefs, which transcend the price system. Perfect access to all of this information allows the consumer to make *rational* consumption choices. These assumptions entail *consumer sovereignty* – that is the consumer is the best judge of the implications of consumption on his or her own welfare and does not require market interventions to enhance that judgement. The government cannot make the consumer or society better off than can the self-interested market exchange.

It should be noted that a consumer might still be rational and sovereign even without actually acquiring perfect information. This is because it is access to information and choice that is crucial. A consumer may choose not to be fully informed because, for instance, the costs of being fully informed are too high or the time required is not justified by the perceived risk of consumption, or the consumer trusts that the partial information received is accurate. In this case, the consumer is said to be boundedly rational (Williamson, 1985). The consumer is still sovereign and rational because it is the consumer's choice to have partial information. Ultimately, it is rational for the consumer to sacrifice information as long as the consumer retains trust and choice.

Although some might argue that the current marketing of agricultural products provides consumers with the information, trust and choice that are necessary to ensure consumer rationality and choice, the history of the introduction of rapeseed and canola into the edible oil market in Canada demonstrates that it does not now and likely never did do so. As early as 1951 a study on possible health hazards of high erucic acid rapeseed was in the public domain (NRC, 1992). This information, in the form of a highly technical study, was not effectively available to consumers, limiting their access to perfect or complete information. So even though the information was 'technically' available, many consumers were unable to use it when choosing those margarines and salad dressings that contained rapeseed oil. In 1956, the Food and Drug Directorate of the Department of National Health (DNH) responded and ruled that rapeseed 'was not approved for human consumption' and hence, initiated new technical studies into the potential health hazards of the consumption of rapeseed oil. Within days of the ruling, pressure by rapeseed promoters caused the DNH to rescind its objections to the use of rapeseed as an edible oil. The Food and Drug Directorate, however, took the opportunity to begin studies on the health impacts. A state of unproved concerns with the erucic acid in rapeseed remained until the 1970 St Adele rapeseed conference, when two European papers posed new concerns about the health hazards of erucic acid. Even then, although the futures markets for rapeseed in Winnipeg reacted strongly negatively, the regulatory system did not ban the use of rapeseed. Instead, the scientists, the regulators and the industry worked together to manage a changeover from traditional rapeseed varieties to low erucic varieties then available. The changeover took more than 3 years to complete (NRC, 1992). The technical studies did little to provide information necessary for consumers to make rational choices about the consumption of high-erucic acid rapeseed oil products. Further, the regulatory framework did little to address this consumer–producer information gap. In essence, the regulatory system was used by the industry to manage its market needs.

Rather than improving, it seems that more recently the picture has become more complicated and the information gap is widening, not narrowing, forcing the regulatory system to take a more proactive stance in managing the introduction of new products and the protection of consumers and the public interest. There are two difficulties. First, due to the degree of scientific sophistication associated with products of modern biotechnology, there tends to exist a natural and insurmountable information gap between producers (who possess extensive scientific knowledge, some of which is proprietary and not generally available) and consumers (who are generally less scientifically knowledgeable and do not have access to proprietary knowledge). Modern biotechnology represents sophisticated technological processes or production techniques, which involve the manipulation of genetic material beyond traditional plant breeding or animal husbandry. Instead, modern techniques involve recombinant DNA (rDNA) engineering (Wiegele, 1991). A simple metaphor used in *The Economist* news magazine (13 June 1998: 79) is that traditional plant breeding is like shuffling a given card hand, while rDNA engineering is like adding new cards to the hand. These techniques are applied to living organisms in order to enhance valuable commercial attributes that are either naturally occurring traits or to induce new or novel traits. As illustrated in the previous chapters, it is possible to group the genetic modifications being performed in the canola sector in the 1990s into two types: production- or input-attribute modifications; and output-, marketing- or commercial-attribute modifications. The objective of production-attribute modifications is to ensure supply-side gains of higher yield, enhanced productivity, increased viability and reduced risk of failure. The attributes pursued to achieve these gains include decreased need for fertilizer, increased resistance to pests and to fungi and increased tolerance to herbicides required to control weeds that normally compete with the agricultural crop. On the other hand, the objective of marketing- or commercial-attribute modifications is to improve crop characteristics such as delayed ripening and rotting and improved colour, taste, texture and dietary properties. The array of scientific principles and methods used to achieve these attributes exceeds the comprehension of most consumers and all but a few specially trained scientists. For the most part, consumers are forced to accept the judgement of others about the benefits or risks of these new products. In short, with the research and development which underlies the application of modern biotechnology advancing rapidly in many nations, products of modern biotechnology are on a dynamic trajectory where science greatly exceeds the understanding of most consumers, and hence the information gap is widening. Baltimore (1982) argues that it could take three decades (or a generation) for consumers to gain the scientific sophistication necessary to fully understand modern biotechnology.

As a result of this information gap, consumers also face the challenge that they cannot completely know, or understand, all the attributes and the efficacy of the good, either before or immediately after consuming the product; in many cases the ultimate effect of the consumption is not known for an extended period of time. As such, rapeseed/canola is a credence good, with all its attendant challenges (Bureau *et al.*, 1997). Essentially, the consumer lacks access to the perfect information that must be present for neoclassical consumer theory to deliver an optimal supply and demand of the product. That is, even if the consumer wanted to have perfect information, the degree of scientific sophistication necessary is prohibitive and many of the impacts are unknowable at the point of consumption.

A further problem with the introduction of the new technology is that consumers usually do not have the ability to distinguish between products produced with traditional technologies and those products that used modern biotechnology. This is particularly relevant to the agri-food system. The nature of the current global agri-food handling and distribution system for bulk commodities makes it virtually impossible to ensure that GM production is fully segregated from non-GM production (Isaac and Phillips, 1999). With the possibility of co-mingling, consumer choice is effectively eliminated. Given that the first agri-food biotechnology-based products are edible oils (maize, soybeans and canola), which are constituent ingredients in more than 6000 processed foods, there is a high probability that virtually any product you consider consuming could incorporate some biotechnology component. Therefore, with respect to products of modern biotechnology, the consumer facing a significant, and growing, information gap is unable to make rational consumption choices regarding the biotech products because of a lack of access to perfect information. And the consumer does not really have any choice once the products are approved for production, importation and sale.

Instead, the consumer must have confidence that the product does not contravene their concerns in order to accept the innovations and choose to consume the product. In such a circumstance, the consumer must trust that the information being received is accurate, benevolent and primarily concerned with consumer protection. However, there is a danger with the consumption of credence goods and it is that the consumer may be 'misinformed' by those who are entrusted with providing accurate information. In the case of the safety of the food products, the misinformation may lead to illness or perhaps death.

In consumer theory, the presence of market failure is used to justify government intervention in the market. Governments have traditional jurisdiction over issues such as health, safety and the environment, as well as preservation of social norms. The imperfect information gap that characterizes the relationship between producers and consumers with respect to products of modern biotechnology is often intertwined with these non-economic concerns under the jurisdiction of government. Therefore, the information gap is a driver for government regulatory intervention and regulation over the application of the techniques and procedures of modern biotechnology. The goal of the intervention is to provide the consumer with enhanced market knowledge in order to reduce the information gap and enhance consumer rationality, choice and sovereignty.

It is illustrative to discuss some common consumer concerns associated with agricultural biotechnology in order to identify just how the information gap can adversely impact consumer acceptance. One particular concern is that modern biotechnology involves inserting all sorts of genes into an organism, possibly creating uncontrollable, alien or unnatural organisms that may result in aggressive, toxigenic, pathogenic, infective or invasive changes to the plant which are unstable over time. This concern captures both human health and environmental safety concerns. However, it must be recognized that there is both an economic and a technological constraint that necessarily limits the development of novel super-organisms. With respect to the economic constraint, there is very little incentive for researchers to develop entirely novel genetically modified agricultural plants, unlike anything produced before. The current state of capital equipment at the farm level and the bulk handling and distribution nature of the global crop marketing system in reality limit any modifications to agricultural crops. The agricultural industry would be unwilling, and indeed may be unable, to undertake the investment necessary to completely adopt new agronomic systems to accommodate novel plants. Instead, the incentives remain for researchers to develop new varieties of conventional crops with enhanced yield and value attributes, but which remain compatible with existing agronomic practices and crop marketing systems. Further, with respect to the technological constraint, it has been argued that plants are the result of an evolutionary process which has resulted in finely tuned organisms particularly suited to survive in particular agronomic conditions. Disrupting the genetic make-up even a little bit is likely to create an organism that is incapable of surviving rather than a super-organism capable of destroying biodiversity. Therefore, 'the greatest task for agricultural researchers is not how to keep the organism from getting out of control, but devising modifications that allow it to survive long enough to perform its beneficial activity' (Brill, 1988).

A second common concern is that planting herbicide-resistant crops, in particular, will be detrimental to both human and environmental health because farmers will apply the herbicide widely in a reckless, irresponsible fashion. Again, there is an important economic constraint, which limits this concern. In reality, it is very unlikely that excessive use of herbicide will ever occur as long as herbicide purchases remain such a significant portion of a producer's input costs. The cost-minimizing producer will always use herbicides economically, not waste them. Further, it may be added that in most cases the producer's wealth is embedded in the land. There is a strong long-run incentive for producers to ensure the health of their land in order to protect its value.

A third common concern is that private interests from the developed world will circumvent their own domestic regulations on testing and release by conducting these activities in the developing world where little or no regulation exists (Mulongoy, 1997). There are two challenges to this concern. The first challenge relates to technological capacity. The application of modern biotechnology requires a significant research capacity, including a skilled workforce, capital-intensive research equipment and sources of funding. Those countries that have the requisite research capacity also have the domestic regulations to monitor the development and application of modern biotechnology. This capacity is not easily duplicated; it takes years to establish and requires heavy investment in training and equipment. Therefore, firms have very little incentive to move their research activities to another country because of the regulatory climate – the research capacity is far more important than the regulatory environment. The second challenge has to do with the economics of potential adopters of GM varieties. These varieties, with their enhanced attributes are being targeted to producers in the developed world, and therefore it is crucial to ensure that the varieties are viable in their region. Field testing in a developing country is not an effective substitute.

In addition to the concerns noted above, a further complication for government regulatory intervention is that consumer concerns vary widely across both products of modern biotechnology and across regions. This non-homogeneity ensures that domestic regulatory intervention will be a function of divergent domestic regulatory traditions and current social dimensions. The degree of consumer acceptance of products of biotechnology is most directly associated with consumer perceptions about the primary beneficiary of the innovation. Thus, in general, biotechnology applied in the medical and pharmaceutical industries has been met with greater consumer acceptance than biotechnology applied to the agricultural sectors. This differential rate of acceptance is also associated with the manner in which modern biotechnology is applied (*The Economist*, 13 June 1998, p. 79). Medical research is often seen as less risky because it is done in closed laboratories under 'controlled' conditions, while consumers perceive that agricultural biotechnology is done in an 'uncontrolled' manner because the products are tested in open fields and the seeds or animals are ultimately released into the environment. The more open production system for agri-food products is perceived to increase the risk that unanticipated risks will jeopardize consumer health and environmental biodiversity.

As was discussed above, the application of the techniques and procedures of modern biotechnology to agricultural production is generally perceived to yield only supply-side or production gains. GM varieties of canola, most of which involve herbicide tolerance, have proven extremely popular with western Canadian farmers, with adoption rates in excess of 75% after only 3 years of unconfined release and unrestrained production. At the same time, biotechnology-based growth hormones are used extensively in the North American livestock and dairy industries because of the real and obvious productivity gains which they produce for ranchers, feedlot operators and dairy farmers. Yet, until the productivity gains result in significant price decreases, consumers do not see or perceive that they receive any benefits from these technological adaptations.

From the point of view of canola, a crucial factor contributing to the information gap is consumers' perception of the distribution of benefits. Consumer acceptance may rise at least in part when consumers begin to see and receive some of the benefits from the application of modern biotechnology to agricultural products.

So far we have discussed the information gap between producers and consumers in the context of individual markets. There is also an international dimension to the information gap. Because there are no universally accepted norms related to human and environmental safety and product quality, people with different experiences and interests have widely different perspectives. As a result, different states have different notions of these norms, and these differences are reflected in a wide range of government intervention. In this context, it is possible to talk about North American and European consumer acceptance as separate and distinct.

With respect to agricultural biotechnology, the industry is indeed aware of the asymmetry in consumer concerns that exists between North American and European consumers. In 1998, both Monsanto and EuropaBio launched consumer-oriented campaigns in Europe to address the relatively low level of acceptance among European consumers. Monsanto alone spent an estimated US$5 million in the EU in 1998/99 on a high-profile advertising campaign (*The Financial Times*, March 1999). This approach was not overly successful. For the most part, greater awareness did not translate into greater confidence. Looking at cross-national survey results (see Table 12.1), we can see that countries with greater consumer awareness tend to have greater concerns with safety (there is a positive 0.335 correlation between awareness and the perception of biotechnology as a serious health hazard). As a result, greater awareness tends to be negatively correlated (−0.377) with a willingness to purchase a biotechnology-based food product. More recent surveys (e.g. Angus Reid Group, 1999) show that the gap remained.

In Europe, resistance to GM products was growing during the 1998–2000 period. Seven large supermarket chains joined forces to eliminate GM ingredients in their own-label products: J. Sainsbury and Marks and Spencer in the UK; Carrefour in France; Migros in Switzerland; Delhaize in Belgium; Superquinn in Ireland; and Effelunga in Italy (*The Financial Times*, March 1999). As an example, Carrefour, France's largest supermarket and the world's third largest retailer, carries 1783 own-label products. They report that 516 products contained GMOs. Carrefour replaced GM ingredients with substitutes not subjected to genetic manipulation in 286 of the products; for 221 products where alternate ingredients were not available, Carrefour offered a guarantee of origin and demanded that its suppliers guarantee and prove their products do not contain GMOs. Nine product lines were discontinued because it was impossible to guarantee their GMO-free status (*Ram's Horn*, 1999). The UK response has been greater, as Tesco and Sainsbury's, Britain's two largest chains, have been joined

Table 12.1. Consumer attitudes to biotechnology-based foods, 1995 and 1996. (*Source:* Hoban, 1997.)

Country	Awareness (%)	Willing to buy (%)	Perceive as health risk (%)
Austria	90	22	60
Belgium	57	62	44
Canada	67	74	na
Denmark	89	55	44
Finland	79	55	41
France	55	60	38
Germany	91	30	57
Greece	39	60	33
Ireland	57	50	48
Italy	47	53	30
Japan	89	69	na
Luxembourg	56	43	38
Netherlands	70	64	48
Norway	68	49	28
Portugal	51	71	62
Spain	35	59	49
Sweden	75	51	65
UK	57	63	39
USA	65	73	21

Canada, the USA and Japan are 1996; the European results are 1995.
Awareness: amount heard or read about biotechnology (a lot, some, a little).
Willingness to buy: percentage likely to purchase produce developed through biotechnology to resist insect damage.
Perception of risk: percentage rating genetic engineering as a 'serious' health hazard.
na, not available.

by Marks and Spencer, Asda and Iceland in adopting similar or more stringent policies. Meanwhile, Unilever and Nestlé in the UK, both large food processors, announced in 1999 that they will remove GMOs from their products (Bowditch Group, 1999).

Consumer acceptance would appear to require more than awareness – it may require real and visible transparency. For example, one can look at the Swiss 1998 referendum on a proposal to ban outright all domestic biotechnology R&D. The Swiss industry, in an effort to shore-up support for continued biotechnology R&D opened up their laboratories to greater public scrutiny. This increased flow of information about what was actually being done in the laboratories, rather than complex advocacy messages, helped to close the information gap. The result was increased consumer acceptance, with the result that 66% of Swiss citizens voted to reject the proposed ban on domestic biotechnology R&D (Braun, 1999).

Comparative Assessment of Domestic Regulatory Systems

Given that governments are motivated to regulate biotechnology products, it is crucial to examine how regulatory intervention is pursued. There are several key characteristics of regulatory intervention to identify. Generally, domestic regulations are a function of the traditional role of the state and must be understood in this context. More specifically, domestic biotechnology regulations differ according to four fundamental principles. First, domestic regulations can either focus on the *technology* used or on the *products* that result from agricultural biotechnology. Second, domestic regulation can be based either on an extension of existing regulatory frameworks or on new regulations. Third, regulatory decisions can be taken according to the *certain* or *reasonably certain* interpretation of the precautionary principle. Fourth, a wide variety of actors can be permitted to participate directly in the regulatory decision-making process. In order to consider how governments regulate the domestic market, it is necessary to identify the regulatory tradition of the state, as well as the regulatory framework according to these four principles.

Currently there appear to be two distinct regulatory traditions, with a third, intermediate model evolving. North America and Europe tend to represent the two extremes: North America pursues a *laissez-faire* market model and Europe has developed a social market model (Table 12.2). Other countries span the spectrum between the two models.

It has been argued that traditional government regulatory intervention in North America derived from a legalistic approach, while in Europe it derived from a political control approach (Woolcock, 1998). The North American legalistic approach prescribes government regulatory intervention only in reaction to market failure, but the regulatory intervention still preserves the fundamental focus on market forces. Essentially, this means that regulations face the objective of ensuring market efficiency or effectiveness in correcting apparent market failure (Majone, 1990). The discretionary decision-making power

Table 12.2. Competing regulatory approaches for agricultural biotechnology.

	North American approach	EU approach
Orientation	Supply push	Demand pull
Tradition	Legalistic	Accountability
Features of regulatory system		
Trigger	Novel product attributes	Use of biotechnology processes
Regulatory base	Vertical through existing regulations and agencies	Horizontal through new regulations and agencies
Precautionary principle	Reasonably certain	Certain
Access to system	Closed to interest groups	Open to interest groups
Labelling	Discretionary	Mandatory

wielded by regulators is kept in check through the 'transparency and openness of the decision-making process' (Woolcock, 1998). This approach provides both public scrutiny and limits the influence of populist politics of day-to-day public interest on the regulatory pursuit of efficiency and effectiveness. Indeed, with respect to modern biotechnology, it is quite appropriate to suggest that North American regulations have been pro-competitive, and focused on removing market failure to enhance the efficiency and effectiveness of market operations.

The European political control approach to government regulatory intervention is traditionally dominated by concerns over the democratic accountability of the discretionary decision-making power of regulators (Majone, 1994). In achieving accountability, the objectives of market efficiency and effectiveness tend to be subordinated. Regulatory decision-making resides with elected public officials to ensure accountability. As a result, any regulatory intervention is very much subject to the day-to-day public interests which dominate the concerns of elected officials. On one hand, it is positive that market failure is corrected with respect to the 'social dimensions' (European Commission, 1983) of the public interest. On the other hand, public concern, which is ever fluid and based on perceptions rather than objective assessments of risks, can result in regulatory intervention adversely affecting market efficiency and effectiveness.

The divergent traditions in government regulatory intervention lead to different results. In Europe, regulations must be very sensitive to the public perceptions of risk and must consider the broader social dimensions of agricultural biotechnology, while in North America, although public interest is indeed crucial, the regulatory tradition supports a more arm's-length approach to regulatory development and the focus of regulatory intervention is on efficiency and effectiveness of market operations. Other countries come somewhere in the middle.

With respect to the current regulatory frameworks, there would appear to be a number of alternative approaches that governments have adopted to handle the information gap between producers and consumers. Currently, a handful of key countries – Canada, the USA, Australia, New Zealand, Mexico, Japan and the EU – have adopted extensive specific regulations to manage biotechnology-based agri-food development. The approaches range widely, although it is possible to categorize the countries into three distinct groups: Canada, USA, Mexico, New Zealand and Australia generally follow the North American model; the EU follows the European model; and the Japanese have developed their own, intermediate model. It is possible to identify these three general models by examining the regulatory frameworks according to the four fundamental principles identified above.

First, domestic regulations associated with biotechnology appear to follow one of two approaches. They either focus on the *products* created through the use of biotechnology or they focus on the *technology*, that is, how the products are made. There are major implications of the choice of focus because biotechnology as a production and processing method cuts horizontally across many industries, such as agriculture, forestry, fisheries, pharmaceutical, medical and environmental applications. These industries are traditionally regulated independently and pursuant to specific, and often quite divergent, mandates. Choosing to regulate *products* of biotechnology allows for preservation of the independent and vertical regulatory jurisdictions. Choosing to regulate the technology *per se*, requires regulatory intervention that cuts across the many applications and divergent mandates.

Second, the decision to regulate biotechnology according to existing legislation or according to new legislation is closely associated with the technology vs. products debate. Because in all the countries under study vertical regulatory jurisdictions are the tradition, regulating the technology essentially requires the development of new horizontal regulations, while regulating the products of biotechnology is congruent with the existing, vertical regulatory jurisdictions. Regulating according to existing regulations at its root assumes 'substantial equivalence' whereby the risk of a biotechnology-based product will be compared to the prevailing risks of the conventional product. The key benefit of regulating products is that existing vertical regulatory jurisdictions may be employed, building on the expertise and capacity in existing vertical agencies. This allows immediate regulatory oversight, without the inevitable delay and difficulty associated with establishing new regulations that cut horizontally across divergent, often competing, departments and agencies. Nevertheless, there are potential costs to regulating products only. Biotechnology is a nascent technology still associated with much uncertainty about risk. As it remains virtually the same regardless of the application, regulating the technology provides an integrated approach to dealing with risk, rather than having divergent regulations in different agencies. Further, the mandates of traditional vertical regulatory agencies in most cases were developed prior to the biotechnological revolution and hence, were not designed to address the type of risk associated with this new technology. A new, horizontal regulatory approach focused on the technology may provide a more appropriate level of oversight, regardless of the product application.

Third, when developing regulations, it is essential to determine tolerance levels for potential risks. In the countries under study, the regulations are established according to the precautionary principle. However, there are two interpretations of the precautionary principle. Some countries accept that regulators may be only *reasonably certain* that no adverse effects will occur. This approach explicitly accepts that some level of risk is inevitable, yet tolerable. It supports research, development and commercialization efforts because they further the knowledge base and lead to greater understanding of the risks. Other countries direct their regulators to be *certain* that no adverse effects will occur. This approach explicitly pursues zero risk. It assumes the worst-case risk scenario, heavily scrutinizes research and development, and does not permit commercialization until the assumption of risk can be proven wrong. Clearly, the interpretation of the precautionary principle employed can lead to significantly different regulations.

Fourth, the actors allowed to participate in regulation development and decision making heavily influence the extent of government regulatory intervention in biotechnology. In a traditional, independent regulatory jurisdiction such as agriculture, regulations have a clear set of involved actors. Modern biotechnology tends to dramatically increase the number of interested actors because it deals with all four types of consumer concerns (i.e. economics, human health and safety, biodiversity and religion, morality and ethics). The greater the number of actors, the greater will be the number of competing interests and, by implication, regulatory decision-making will become more complex.

Together, the regulatory tradition and the four fundamental principles of biotechnology regulations characterize government regulatory intervention. Three divergent regulatory models will be assessed: the North American model, the Japanese model and the European model. These three models, in general, characterize the current array of biotechnology regulations. The remaining countries in the world for the most part have not adopted any specific regulatory systems to evaluate and to manage the specific risks of biotechnology-based agri-food products. As shown in the next chapter, these countries are hoping that systems such as the *Codex Alimentarius* and BioSafety Protocol will be able to provide them with a regulatory system that will work for them. It is worthwhile to examine the three divergent regulatory models in practice to show how the differing systems influence the market access and market placement of knowledge-based agri-food products.

The North American regulatory model

The North American regulatory model for biotechnology is employed in Canada, the US, Mexico and Australia. The regulatory tradition in this model is one of *laissez-faire* market intervention that is pro-competitive and tends to be focused on efficient and effective regulatory intervention. According to the four principles, this model employs product-based regulations within the existing vertical regulatory jurisdictions, where supplemental regulations or guidelines have been developed to deal with new concerns and risks associated with novel organisms and products. Further, regulatory decision-making employs the reasonably certain interpretation of the precautionary principle and there are only a limited number of actors that directly influence regulatory decision-making. Within this general North American regulatory model, there are some differences to identify, therefore the Canadian and US systems will be discussed in greater detail.

The Canadian regulatory system has evolved significantly in recent years. In earlier periods the health and safety regulatory system supported and assisted the canola industry to develop, and tended to take its lead from the industry. Although the same regulatory base is still in use, the system has been expanded to incorporate more intensive examinations of new biotechnology-based products (see Table 12.3). Following the rulings that canola was acceptable as a food and feed, the regulatory system operated smoothly, with the single largest hurdle being the provision in the Seeds Act that required new varieties be better than a reference variety. There was little or no further review of the varieties for their health or safety risks. In the mid 1980s, two major changes took place. First, the Seeds Act was revised, allowing any new varieties that were equivalent with the reference variety to be registered; shortly thereafter rules were enacted to allow for 'confined' releases to assist with breeding and regulatory compliance. By 1988, the federal government had reviewed and worked to coordinate the existing system of regulations and added the Novel Foods Guidelines to complete the system that would review biotechnology-based products. The first test of the system came with the herbicide-tolerant canola varieties produced by Monsanto and AgrEvo. By 1989, after more than 6 years of research in the laboratories and greenhouses, Monsanto had identified specific genes and expressed them into superior breeding lines that were candidates for commercialization. AgrEvo was about 1 year behind but caught up in 1994. They successfully applied for approval for confined field trials and began the fieldwork in earnest. By 1992 each of the companies had

Table 12.3. Canadian regulatory system.

Agency	Product	Act
CFIA, Science and Technology Services	Agri-food products (meat, dairy, eggs, fruits, vegetables, honey, maple products)	Meat Inspection Act Canada Agricultural Products Act
CFIA, Feeds Section, Plant Products Division	Livestock feeds, additives (e.g. novel feeds)	Feeds Act
CFIA, Fertilizer Section, Plant Products Division	Fertilizers, supplements (e.g. biofertilizers)	Fertilizers Act
CFIA, Plant Biotech Office, Plant Products Division	Plants (including plants with novel traits and with genetically engineered microorganisms)	Seeds Act (field trials) Plant Protection Act
Health Canada, Food Directorate, Food Inspection Directorate	Genetically engineered (novel) foods	Novel Food Guidelines, Food and Drugs Act: regulates GE foods in same manner as if produced conventionally; reviewed for safety prior to reaching the food system; no labelling required

Source: CFIA at http://www.cfia-acia.agr.ca/english/

identified the cultivars for which they would seek regulatory approval to commercialize. Over the 1989–1994 period, the two companies each conducted more than 400 confined field trials, first to select their commercial lines and then to provide the scientific evidence to satisfy the regulatory system. Over the 1992–1995 period, the companies provided data and information to Health Canada to meet the Novel Foods Guidelines, to Agriculture Canada for animal feed approval and variety registration and to Environment Canada for environmental approval. In each case, the approvals were bunched into a short time period: all approvals to proceed came within 6 months for each variety (CFIA). Since then, the regulatory system has been streamlined, with the Canadian Food Inspection Agency assuming responsibility in 1997 for all the regulatory functions except the Novel Foods Guidelines, which continue to be managed by Health Canada.

The Canadian system has operated relatively efficiently, approving by 1997 more than 3800 field trials, 29 plants with novel traits for feed, 34 trials for veterinary biologistics, and three novel foods. It has been able to achieve this flow of regulatory decisions partly by assuming 'substantial equivalency' in the rules and guidelines for action, and partly by limiting 'standing' in the process to the proponent and the regulators. Citizens, consumers, environmental groups and provinces, while allowed relatively free access to applications and specific decision documents through the CFIA website, are not allowed to have any say on individual product approvals in the science-based system. So far, in Canada, that has not created any significant backlash by either the major lobby groups or the general public. While respondents to surveys indicate concern with biotechnology, there has been little general public debate.

The US, Australian and Mexican regulatory systems have many similar attributes. Given that Australia and Mexico have largely patterned their process and systems off the US or Canadian systems, it is worthwhile examining the US system in some depth. The US regulatory system, in particular, has led the rest of the world in developing the regulatory base for biotechnology products. In 1976 the National Institutes of Health (NIH) issued the *Guidelines for Research Involving Recombinant DNA Molecules*, which focused on controlled research not on experimental or commercial environmental release. The key feature of the guidelines was the linking of the degree of containment with the amount of scientifically determined hypothetical risk (Chen and McDermot, 1997). This provided the base for the US regulatory structure. With the growth of research and development in agricultural biotechnology, pressure began to mount for the NIH to update their guidelines to deal with environmental release. This issue was addressed in the 1986 Co-ordinated Framework for Regulation of Biotechnology. In both releases, the NIH advocated that there was no significant risk inherent in the use of biotechnology which required technology-based regulations.

Although in earlier years there was a coordinating framework and committee for regulating biotechnology, three key US agencies currently share responsibility for biotechnology-based agri-food product regulation. The regulations are based on a sectoral or vertical approach so that biotechnology is dealt with when it is employed in the production or processing methods of products according to traditional sectoral jurisdictions. The Animal and Plant Health Inspection Service (APHIS) of the USDA is responsible for environmental assessments of plant risk, issuing permits for field testing, and for regulating the importation and interstate movement of genetically modified plants (Table 12.4). USDA's APHIS examines all organisms and products altered or produced through genetic engineering that are or have the potential to be plant pests. To speed the process, APHIS regulations provide for a petition process, where the proponent can request their product be granted a non-regulated status. The petitioning company submits the necessary evidence to prove that the genetically modified plant does not pose a plant pest risk, APHIS reviews the evidence and, if approved, the plant product (and all its offspring) no longer requires APHIS review for movement or release in the US. The US Environmental Protection Agency (EPA) is responsible for the environmental release of both bio-engineered pesticides and bio-engineered plants with pesticidal characteristics, such as *Bt* varieties. EPA's environmental assessments consider adverse impacts upon humans, non-target organisms and biodiversity. The main impact on new canola varieties is the requirement that the EPA establish tolerances for residues of herbicides used on novel herbicide-tolerant crops. The third key institution in the US system is the Food and Drug Administration (FDA) of the Department of Health and Human Services. The FDA has traditional responsibility over ensuring the safety of food and feed use of plants. It is important to note that FDA consultation is not mandatory but recommended prior to the market release of genetically modified food and feed. Biotechnology-based products only come under direct FDA jurisdiction if they are determined to be a food additive. This is because the FDA regulates foods and feed derived from new plant varieties under the authority of the Federal Food, Drug, and Cosmetic Act. FDA policy is based on existing food law, and requires that genetically engineered food additives meet the same rigorous safety standards as is required of all other food additives. FDA's biotechnology policy treats substances intentionally added to food through genetic engineering as food additives if they are significantly different in structure, function or amount than substances currently found in food. Many of the food crops currently being developed using biotechnology do not contain substances that are significantly different from those already in the diet and thus do not require FDA's pre-market approval. As a result, in the US canola varieties produced elsewhere have been imported without great difficulty. Even for those canola varieties that qualify as novel foods, such as Calgene's Laurate canola, the regulatory process has moved quickly, taking no more than 2–3 years to complete. Before commercialization, genetically engineered plants must also conform with standards set by state and federal marketing statutes, such as state seed certification laws, but there are no national requirements for varietal registration of new crops. One issue that has arisen at various times and in certain places is the potential for states and local governments to enact their own laws to address environmental concerns

Table 12.4. US biotechnology review system.

Agency	Products regulated	Authority
US Department of Agriculture (APHIS)	Plant pests, plants, veterinary biologics	Federal Plant Pest Act
Environmental Protection Agency	Microbial/plant pesticides, new uses of existing pesticides, novel microorganisms	Federal Insecticide, Fungicide, and Rodenticide Act (FIFRA) – 7 USC 136 Federal Food, Drug, and Cosmetic Act Toxic Substances Control Act
Food and Drug Administration	Food (except meat, poultry and egg products), feed, food additives, veterinary drugs	GRAS status Federal Food, Drug, and Cosmetic Act Statement of Policy: Foods Derived from New Plant Varieties

(Coombs and Campbell, 1991); there has over time been a tendency in the US for *subsidiarity*. Several states have enacted legislation regulating field trials requiring either notification of the release of GM varieties (Hawaii, Illinois, Wisconsin) or requiring formal permits for trials (Minnesota and North Carolina) (OTA, 1991). Nevertheless, once the product has been approved for unconfined release, the states lose any say in control of the system.

The US regulatory system provides industry with predictable, timely decisions. APHIS acknowledged, between 1988 and 1998, more than 3930 notifications of movements of genetically engineered materials and issued 886 permits for release. Only about 4% of the requests were denied. Between 1992 and 1999, APHIS received 65 petitions for deregulation: 49 were approved (the average time for the decision was about 4 months, with the range between 1 and 10 months), 11 were withdrawn/voided and the rest are pending. EPA reviewed 34 proposals for biopesticides in the 1987–1996 period, while the FDA reviewed and approved 43 food products between 1994 and 1998. From the consumer's perspective, relying on the National Institutes for Health (NIH) rules and the FDA processes provides confidence in the system, as they view both institutions as credible information sources on biotechnology (Hoban, 1998b, 2001). If the USA had developed a new set of rules or created a new regulatory agency, that confidence would have been lost and would have had to be rebuilt.

Given the significant linkages between Canadian and US agriculture, and especially the respective research efforts, the two lead regulatory agencies (APHIS of the USDA and CFIA) have worked to streamline the approvals of products in the two countries. In the past the two agencies have undertaken simultaneous reviews of transgenic plants prior to their commercialization, and have informally shared data and observations. In 1998, USDA, CFIA and Health

Canada regulatory officials met to compare, where possible, and to harmonize the molecular genetic characterization components of the regulatory review process for transgenic plants. Ultimately, the two agencies believe the meeting, and other activities, may lead to mutual acceptance of assessments in the future. The Canadian–US coordination of agricultural biotechnology regulations is congruent with the agriculture agreement under the North American Free Trade Agreement (NAFTA), which calls for a long-term regulatory coordination under a strategy based on equivalence. In the interim, the exchange of information is believed to expedite the review process. This level of coordination between the two countries, and the similar features in the other countries with product-based review systems, ensures that there are few insurmountable regulatory hold-ups.

The European regulatory model

The current European regulatory model is built on the tradition of political control over regulatory decision-making in order to preserve democratic accountability. As a result, the influence of public interests and the dominance of 'social dimensions' over market efficiency and effectiveness objectives are significant. In general, the European regulatory model is a technology-based approach, employing new, horizontal regulations governing biotechnology as a technology, rather than biotechnology-based products. As well, the interpretation of the precautionary principle for regulatory decision-making is based on certainty. Due to the dominance of 'social dimensions', many actors influence regulatory decision-making. An important aspect of the EU regulations is the fact they represent minimum essential requirements, rather than a harmonized approach for all member states. There are two reasons for this: the European Commission recommended that biotechnology regulations be based in part on the social dimensions of biotechnology, without accurately defining what was meant by social dimensions (European Commission, 1983); and DG-XI insisted that the horizontal biotechnology directives be primarily about community environmental safety, not enhancing the internal market. As such, member states can unilaterally impose regulations more stringent than the EU regulations in order to protect health and safety.

Prior to the 'European Unionization' of the biotechnology issues, there were divergent views among member states on the need to regulate biotechnology and on how to regulate biotechnology. At one extreme, France allowed the industry to control biotechnology, with little regulatory intervention; in contrast, Denmark established a system of significant government regulatory intervention (Cantley, 1995). Even at the EU level there was, and remains, a great deal of intra-commission disagreement on an appropriate regulatory framework. However, despite the internal European conflicts associated with regulation building, the goal here is to examine regulatory outcomes, and hence, the focus here is on the 'European' regulatory model.

Complicating the European regulatory model is both the recent coverage of genetically modified products in the UK press and the BSE crisis in the UK, which together have brought all food regulators in the EU into disrepute. Consumers have become increasingly aware of the presence of biotechnology-based food products and are uneasy about their role in the food chain. In part, the EU and its member states are attempting to redesign the regulatory system, which is logically more able to bridge the information gap, to handle highly politicized issues associated with a significant loss of trust and faith in public regulation. The result has been more stringent and less responsive regulation, without any noticeable increase in confidence in either the food safety or regulatory systems.

Operationally, the European system has been likened to a 'gigantic maze' (Hedley, 1998). Given the general EU policy of subsidiarity, much of the administration of the European policy on food and environmental safety is administered at the nation-state level. Hence, the European regulatory model, while creating new, horizontal regulatory directives, can be exceeded by unilateral action by any of the 15 national member governments, as well as by the European Commission (three separate Directorates General). As submissions advance through the regulatory system, bureaucrats and politicians, at both the national and EU level, evaluate them. In addition, the EU has interpreted the precautionary principle in such a way that regulatory decision-making must be certain of no adverse affects, as opposed to reasonably certain. This interpretation very much reflects the influence of DG-XI in the development of the two Council Directives, 90/219 and 90/220, as this use of the precautionary principle has its roots in environmental regulations (Tait and Levidow, 1992) (Table 12.5).

Since 1990, deliberate release of GMOs into the environment for research or commercialization has been regulated by Council Directive 90/220/EEC. The entry point for seeking EU approval is through the competent regulatory body of a member state or *rapporteur* chosen by the company making the submission. Most companies have made their submissions to the UK, France or Germany. Traditionally France has provided the simplest access, with largely voluntary notification and disclosure rules. The UK was next, with a fairly extensive set of compulsory rules, but a generally neutral administration. Germany had the most imposing set of restrictions and further complicated its system by delegating the enforcement to the Länder. The *rapporteur* member state is required to review the request against Regulation 90/220/EEC and provide a specific recommendation for approval or rejection to the Commission. If the product is recommended for approval, the Commission forwards the dossier to all other member states. If there are no objections within 60 days, the Commission will inform the originating member state to proceed with written consent to place the product on the market. If another member state objects, the Commission will convene and chair an Article 21 committee of member states to resolve the issue by qualified majority vote. Once a decision is made, it is binding on all member states. In this process bureaucrats make many of the decisions, but political input is required both at the Commission level (commissioners are

Table 12.5. The EU regulatory system for agri-food products.

Agency	Authority	Application	Status at end of 1999
Horizontal legislation			
DG-III and DG-XI	Council Directive 90/219/EEC of 23 April 1990	Contained Use of Genetically Modified Micro-Organisms: covers contained use of genetically modified microorganisms (GMMs), both for research and commercial purposes	Implemented
DG-XI	Council Directive 90/220/EEC of 23 April 1990 and Directive 94/15/EC	Deliberate Release of Genetically Modified Organisms into the Environment: covers experimental and marketing-related aspects of genetically modified organisms (GMOs), which covers any R&D release of organisms into the environment and contains a specific environmental risk assessment for the placing of any product containing or consisting of such organisms on to the market	Implemented
DG-III	Reg 258/97/EC, 15 May 1997	Regulation on Novel Foods regulates the placing on the market of foods and food ingredients for human consumption containing, consisting of, or derived from GMOs. Novel Foods Directive still granted 'essential equivalence'	Implemented
DG-XI	Annex III of the 90/220 as amended 18 June 1997	Sets labelling and information notification requirements for all GMO approvals for putting products on the market in the EU. This annex supersedes the Novel Foods directive by eliminating the essential equivalence and requiring labelling for all GMOs	Implemented
DG-XI	Revisions to Directive 90/219/EEC, November 1997	Proposed that authorizations to place GMOs on the market, issued under 90/220, be valid for a period of 7 years only; if the authorization is not renewed after the 7-year period, the product must be withdrawn from the market	Pending, adopted by the Council of Ministers

Continued

Table 12.5. *Continued*

Agency	Authority	Application	Status at end of 1999
Product legislation			
DG-VI	Directive 93/114/EEC, amending Directive 70/524/EEC	Feeding stuffs. This amendment introduced new categories of additives, including, among others, additives containing or consisting of GMOs into the existing legislation: the amendment entered into effect as of 1 October 1994	Implemented
DG-VI	Decision 94/730/EEC	Establishing simplified procedures for the release of genetically modified crop plants (first simplified procedure)	Implemented
DG-VI	Directive 98/95/CE, 14 December 1998	Establishes terms and conditions for the registration of GMO varieties in official catalogues; specifies that GMO varieties must be indicated in catalogues	Pending

inherently politicians) and in the originating countries (ministries most often make decisions in the name of the ministers).

Although the system has time lines and a formal process, it has not been working as planned in Directive 90/220/EEC. Both the member states and the European Commission often ignore the time lines and in some cases decisions under 90/220/EEC are not being adhered to by member states. The tortuous regulatory road travelled by canola illustrates the problem. A good example is AgrEvo's attempt to get approval for its Liberty-Link™ canola. In 1994 AgrEvo made a submission to the government of France. The French government subsequently recommended to the European Commission a favourable decision. The Commission then undertook its consultative process and, albeit with delays, ultimately informed the French government that it could provide written approval for the variety. In the interim, the French government had changed and the new Ministers were less supportive of biotechnology. The French government has yet to implement the Commission decision and the Commission has launched a legal case against France for its delay.

The European Commission is currently considering revision or modification to Directive 90/220/EEC. Potential revisions include a limited term for market approvals for genetically modified products, where, upon expiration, market approval would have to be renewed. The current proposal is that the limit could be 7 years and that during that period there would be mandatory market monitoring of all approved products. As well, market approval would be subject to consultation with a Scientific Committee and final approval decisions by the Commission could be overturned by the Council of Ministers by a simple majority vote. These proposed changes reflect the politicization of the biotechnology issue and the desire to regulate the 'social dimensions' of biotechnology.

The EU regulatory system has not met the development and commercialization needs of producers. Some products have faced a near impenetrable regulatory barrier to commercialization. Some 18 genetically modified products had been provisionally approved for use as of April 1999, but the four most recent applications were rejected. In addition, Denmark, Britain and France have called a partial halt to GMO approvals in their countries while Austria, Luxembourg and France have all imposed unilateral bans on certain new crops.

The Japanese regulatory model

The Japanese Regulatory Model operates somewhere between those of the US and the EU. The key difference is that the Japanese system regulates plants based on how a plant has been manipulated rather than whether a novel trait has been introduced into the plant. That is, the Japanese approach is technology based. As an illustration, Smart™ canola, in which the resistance to Pursuit™ (a novel trait) was introduced by mutagenesis technology (a non-transgenic technology), was subject to full environmental, food and feed safety assessments in Canada and the USA but did not require any safety assessments in Japan before it could be placed on the market.

Before 1996, Japan had guidelines to assess the environmental risks of GMOs but did not have any guidelines for the assessment of food safety. In 1996 a set of guidelines for food and feed safety were adopted. As in Canada, companies wishing to obtain approval of a genetically modified crop in Japan must deal with two government departments. The Ministry of Agriculture, Forestry and Fisheries is responsible for assessment of environmental and feed safety while the Ministry of Health and Welfare is responsible for the assessment of food safety. Apart from the focus on process rather than product, the Japanese system is similar to the US system, except that it operates on a batch approach, under which submissions are received only until a specified date for consideration by advisory committees at their quarterly meetings. Another constraint has been the need for companies making submissions to have a representative based in Tokyo, to deal with the Japanese government officials. In short, the Japanese regulatory model blends the two systems, being technology based and employing new horizontal regulations, yet administration is through existing regulatory channels, with limited actors participating in the decision-making process and using the reasonably certain interpretation of the precautionary principle.

Biotechnology Regulation and Evolving Consumer Concerns

The Economist (1998) argues that the information gap between consumer and producers continues to widen, at least partly because of a dual lack of transparency which the regulatory systems have not yet addressed. The producers and supporters of products of modern biotechnology 'have not always been honest about their possible costs' while the detractors 'have not always been honest about their benefits'. As a result, consumer confusion grows and industry frustration mounts. The problem with this information gap for research, development, application and adoption of modern biotechnology is that it tends to create a 'fear of the unknown' reaction in consumers that is generally negative towards biotechnology and which decreases the likelihood of consumer acceptance.

Each of the countries that regulate GMOs has considered using labels to help to bridge the information gap by signalling the presence of these new production methods to consumers. The rationale is that because biotechnology products are credence goods, there exists some degree of consumer uncertainty that cannot be factored into purchasing decisions (Bureau *et al.*, 1997). There are four types of risk and uncertainty that the market needs to manage (Fig. 12.1). First, there are quantifiable 'risks' (area R), such as introduction of a new allergen into a product (e.g. peanut genes into maize) or the risk of genes in the plant mutating and it becoming a weed. Most scientists assert that these risks are very small. Given that these risks are quantifiable (i.e. their probability and economic impacts are definable), any marketplace with appropriate liability laws and procedures would lead producers to identify specific risks and factor them into the price, enabling consumers to make informed choices. Second, there is some uncertainty (true 'credence' factors) related to all products (area U1) and possibly some new ones for genetically modified products. For non-GM foods, this includes the presence of trace quantities of harmful substances, such as carcinogens; as science progresses and we understand more about disease, we often discover unanticipated interactions that lead to harmful (sometimes, beneficial) effects. Given the recent introduction of GM products, there is no way to quantify, either in terms of probability or impact, the potential of new carcinogens or toxins resulting from consumption of GMOs. Third, there are rising concerns and uncertainties about the ability of our regulatory systems to deliver safe and nutritious food (area U2). Recent evidence – including endemic *Salmonella* and episodic *E. coli* poisonings in most countries and the BSE contamination of British beef – has led many consumers to distrust both governments and scientists. Many consumer and environmental groups argue that if the food safety system cannot deliver safe conventionally produced food, it cannot be relied upon to deliver novel foods produced using new, partly unknown or poorly understood processes. This fear varies by country and product, but it is currently quite large and possibly unmanageable as many consumers seek ways to send a message of dissatisfaction to food-safety regulators. Fourth, there are significant amounts of both honest ignorance and perfidy that affect peoples'

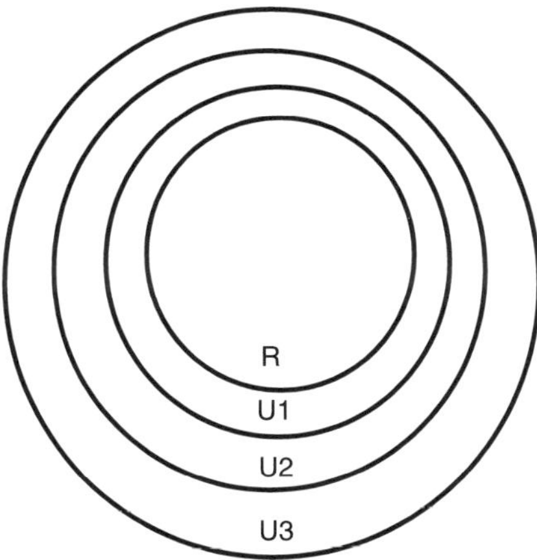

Fig. 12.1. Risk and uncertainty for normal and credence goods. R, normal good; R + U1, regular credence good; R + U1 + U2 + U3, GMO good.

perception of risks of GM products (area U3). Because the technology is highly complex, many people instinctively reject it as unwise or unacceptable; others wilfully muddy the waters with outlandish claims about the impact of biotechnology (e.g. some attribute both human immune deficiency virus (HIV) and Ebola virus to uncontrolled biotechnology experiments).

This presence of imperfect or asymmetric information causes market failure – the market fails to provide all the information required by consumers in order to make rational consumption decisions. Labelling is proposed by many as one way to remedy this market failure, because it involves a transfer of knowledge from the supply-side to the demand-side of the market. Through the use of labels, the argument goes, the information gap between the industry and the consumers may be minimized. This can be a challenge, however, as GM-labelled goods would bear the full cost of incomplete information (area R+U1+U2+U3 in Fig. 12.1), which could more than offset any benefits and thereby impede further biotechnology development.

Whether labelling is a threat or opportunity depends on the type of products being pursued and the degree of uncertainty for that product. Labelling may be either a private or public good, depending on the type of good involved. 'Private-good' labelling policies are based on the assumption that industry wants to segment the market by identifying GMO-based products and their benefits. This may be done through voluntary labelling, where consumers are provided with the information necessary to distinguish those products which use agricultural biotechnology from those that do not. 'Public-good' labelling policies are based on the assumption that industry is unable or unwilling to identify the risks inherent in their GM products. Therefore, the government intervenes in the market with mandatory labelling policies designed to identify the use of GMOs in products. In this capacity, the government is acting to ensure consumer protection from potential human health and safety risks associated with the consumption of GMOs.

Mandatory public-good labelling would force producers to assume the costs of all the risks and uncertainties, with the result that they would likely suffer a discount for their good in the market; this would dampen both production and consumption of this product. This is not socially desirable as firms are required to bear, through government action, uncertainties related to the food safety system (U2) and misinformed judgement (U3). This sort of labelling could seriously disrupt domestic and international food markets and would constitute a threat to the entire industry. In contrast, voluntary private-good labelling would allow producers to market their products in a way that both narrows the information gap and enables them to seek the higher returns from those consumers most likely to buy their product.

Currently the biotechnology industry is relatively unsophisticated in how it deals with this issue. They perceive labelling as a threat, regardless of whether it is mandatory or voluntary. That orientation in the agri-food sector most certainly is because most of the biotechnology innovations have only changed the agronomics of the products and have not fundamentally changed the attributes of the good for the consumer. As noted in Chapter 16, the benefits from herbicide-tolerant canolas have been mostly captured in the supply chain, initially shared by farmers and the agrochemical companies and over time increasingly realized by the oligopolistic agrochemical ventures. As a result, consumers have not gained any benefits. The potential public-good benefits of lower pesticide residues have not been marketed in any proactive way by the industry. This may turn out to be a short-term problem because, as noted elsewhere, second- and third-generation biotechnology products, just beginning to enter the market in the late 1990s, have engineered attributes that consumers are likely to value more highly. For these products, private-good labelling will be both desirable and necessary to extract the return from consumers, and will go a long way to closing the information gap.

Even so, given that awareness and acceptance levels vary greatly between countries, as does confidence in the food and environmental safety systems, labelling will have different impacts in different countries. It is difficult to anticipate exactly what those impacts might be. One related example that sheds some light was the introduction of irradiated foods. This technology, introduced in the 1970s, enables food processors to kill any microorganisms on processed food products, significantly lowering the risk of food-borne ailments. The scientific community asserts that the resulting product has no tangible differences, except for the lower risk of microorganisms. When the technology was introduced, there was significant debate about how to signal to consumers its presence. Most countries, supported by an agreement in *Codex Alimentarius* (5.2.1), decided to require positive labelling of the technology. As seen for biotechnology, awareness led to greater concern. Hoban (1997) shows that in the US, food irradiation, which is positively labelled, is regarded by 65% of consumers as a serious risk. As a result, that technology has never been effectively commercialized, even though more than 85% of US consumers believe that bacterial contamination is the most serious risk in the food system. Biotechnology (genetic engineering), in contrast, does not have any specific labelling requirements and is viewed by only 44% of US consumers as a serious risk. Compulsory labelling of biotechnology-based agri-food products, therefore, might have large impacts not only in the EU, where there is already significant awareness and concerns, but also on the marketplace in the US, Canada, Australia, Japan and Mexico, where lower levels of awareness seem to contribute to lower levels of concern. Furthermore, any decision to require labels in a developed country could have a knock-on effect in many other markets, especially in the developing world, where consumers have little awareness of the technology but also have relatively little confidence in their regulators and their regulatory systems. In either case, voluntary labelling is more likely to lead to both a socially and commercially optimal solution.

Conclusions

The evidence offered in this chapter has shown that government regulatory intervention is driven by consumer concerns about biotechnology associated with consumer safety and health, environmental or biodiversity protection, as well as moral, ethical and religious beliefs. Consumer concerns are exacerbated by an information gap, which drives a wedge between producers and consumers with respect to credence goods such as agricultural biotechnology products. Consumers cannot completely know or understand all the implications of consuming the product. In essence, the market fails to provide consumers with the relevant information to maintain rationality, sovereignty and choice.

Governments have traditional jurisdiction over consumer and environmental protection issues as well as over the preservation of social norms. Hence, government regulatory intervention is imposed in the market in order to address consumer concerns in the face of the apparent market failure. However, the type of government regulatory intervention employed to bridge the information gap and address consumer concerns significantly impacts the development and commercialization of agricultural biotechnology.

It appears that the North American regulatory model is the most conducive to the development and commercialization of agricultural biotechnology. Inversely, the European regulatory model tends to create a regulatory environment that significantly challenges development and commercialization. Technology-based, horizontal regulations can be detrimental to commercialization because they lack the specificity obtained from vertical or sectoral regulations based on products. For example, although the application of biotechnology may be fundamentally the same in the agri-food and pharmaceutical industries, the resultant biotechnology-based products may share little similarity and risk in the two sectors. Applying a 'one-size fits all' regulation to these diverse products may fail to address opportunities and challenges particular to each sector. Also, horizontal regulations, like those imposed at the European level, tend to create a regulatory floor rather than a regulatory ceiling. From the minimum essential requirements, member states can set more stringent regulations not necessarily coordinated among other member states. The result is a fragmented European market with different rules in different national jurisdictions. From a commercial perspective, this market fragmentation limits the achievement of economies of scale and increases commercial uncertainty.

Establishing new regulations may also be detrimental to commercialization. Existing regulations tend to be more effective because they have a history of practice and precedence. Revising existing regulations to deal with new biotechnology products has tended to provide more timely regulatory oversight, contributing to commercial certainty. In contrast, new horizontal regulations face the inevitable political challenge of appeasing all actors, which at times leads to over-regulation and anticompetitive restrictions on the market which limit commercial development (Cantley, 1998).

Basing regulatory decision-making on the *certain* interpretation of the precautionary principle also significantly challenges the development and commercialization of agricultural biotechnology. This approach assumes the worst-case risk scenario. Based on this assumption, all research activity is highly scrutinized and commercialization and environmental release is not permitted until the assumption is proven wrong. This approach can, in fact, seriously damage an industry's competitive position because agricultural biotechnology is a rapidly developing field, and any disruption to the technological trajectory may result in a permanent disadvantage.

Although the North American regulatory model is the most congruent for the commercialization of agricultural biotechnology, it may not be able to bridge the information gap and address consumer concerns that now impede market acceptance. There is a growing circularity to regulatory decision-making, where consumers will not accept products without government approval, yet governments, ever-sensitive to public interest, will not approve products unless consumers accept them. In fact, it may be too optimistic to assume that government regulation will focus on actual risk and not be captured by the public interest of the day.

In order to break this destructive cycle, producers may be forced to take a more proactive role in addressing consumer concerns and increasing consumer acceptance. Given that consumer acceptance is a function of consumer understanding of the risks and perception of primary beneficiary, producers may be advised to provide more transparent information to enhance consumer understanding of both the risks and the benefits. For instance, although many of the monetary benefits of the current crop of genetically modified plants are captured in the supply chain, there still are positive results for consumers, through reduced use of herbicide and pesticide treatments. In the longer run, consumers would likely welcome the potential for fully renewable and biodegradable bio-engineered commodities to replace non-renewable and non-biodegradable chemical and petroleum based processes. Focusing both research and marketing on attributes consumers value, could circumvent the commercially adverse circularity of politically sensitive government regulatory intervention.

Regulating International Trade in Knowledge-based Products

Peter W.B. Phillips and Grant E. Isaac

Introduction

International market access is becoming the most important issue in developing new knowledge-based products. In the past, international market access was limited or denied by governments in an effort to achieve a variety of economic and social objectives, such as domestic income support, greater production and more exports. In the knowledge-based world, time is money. Research-intensive production, short product life cycles and niche markets combine to make it imperative that products entering the marketplace get the widest possible access at the earliest opportunity. Even delays of one growing season have the potential to halve the expected return on the product.

Trade negotiators have been working over the past few decades to make market access and intellectual property protection more certain and timely, in an effort to ensure optimal growth in knowledge-based production. This effort has involved international negotiations at the General Agreement on Tariffs and Trade (GATT), resulting in the World Trade Organization Agreement of 1995, which extended trade disciplines in the agri-food sector through binding commitments by member states. Specifically, member countries agreed to reduce competitive domestic subsidies to their farm sectors, to lower their use of export subsidies, to convert quantitative import measures to tariffs and then to lower them, and to abide by new limits on the use of technical and sanitary standards. Meanwhile negotiations and agreements through UPOV and the Agreement on Trade Related to Intellectual Property have extended protection for intellectual property in the agri-food sector.

During the 1990s, however, there has been countervailing pressure from a variety of sources to slow or reverse the market liberalization that has already been achieved. In particular, there is a growing conflict between globalization and trade liberalization, on the one hand, and rising consumer and citizen demands for greater accountability and certainty in the regulatory system. In essence, the absence of public and consumer faith in domestic systems that work to protect consumers, the environment and society (discussed in Chapter 12) has created pressures to slow international trade flows. The public and consumers have pressed for their governments to use the full discretion available in both domestic and international rules to slow the distribution of biotechnology-based products. This tension is beginning to be felt in the marketplace, as trade rules work to open markets while domestic regulations and international undertakings designed to protect or promote development, health, culture and the environment have become new barriers to trade. During the late 1990s, the international effort to manage trade got a boost from the Convention on Biological Diversity's negotiations to develop a BioSafety Protocol.

This chapter will examine the tension created in the marketplace by efforts to create free access to international markets and the opposing efforts of consumers and others to slow the adoption of new technologies. The analysis confirms that market access is increasingly dependent both on meeting objective regulatory requirements and on satisfying subjective consumer and citizen concerns. In practice, this is a simultaneous problem, as the regulatory systems in a number of countries tend to key off public concerns, which in turn are being fuelled by a lack of faith in the regulatory systems themselves. This 'catch-22' makes it difficult for producers and researchers to sort out how or where to invest their research funds.

International Trade in Canola and its Products

The rapeseed/canola industry has undergone a transformation in the past 30 years, changing from a small, niche product produced in a few developing countries to the third largest source of oils for human consumption, produced and used in many developed agricultural nations.

As the research effort into canola has increased, rapeseed/canola production has shifted towards countries that are managing canola intensively as a knowledge-product, and away from countries that cannot or are not competing on the knowledge front (excluding China, which has not generally allowed market forces to determine production). The EU and Canada, both leaders in the research and development of canola, have more than doubled to 45% their share of global production over the past 30 years. India, Pakistan, Poland and Japan, meanwhile, invested little in R&D for new varieties and have seen their share of global production drop to about 21% in 1992–1999, compared with 50% in 1961–1965 (Table 13.1). In addition, a number of new producers are on the horizon with commercial quantities of canola. The UK, the USA and Australia, each significant investors in canola-related research, are notable for entering and significantly expanding their market shares in the 1980s (equal

Table 13.1. Distribution of global rapeseed/canola production (%).

	1961–1965	1976–1981	1992–1999
Australia	0	0	2
Canada	7	25	19
China	18	23	26
EU	13	16	25
Denmark	1	1	1
France	5	7	9
Germany	7	6	9
UK	0	2	4
India	32	17	17
Japan	4	0	0
Pakistan	6	2	1
Poland	8	5	3
Sweden	4	3	1
USA	0	0	1
Rest of world	5	6	5
World	100	100	100

Source: FAO statistics.

to 8% of global production in 1992–1999). Although none of these countries is producing enough to challenge the key producers of EU or Canada yet, their future role cannot be ignored, especially in the product end of the business.

After research changed rapeseed into canola, reducing erucic acid and glucosinolates to trace amounts, the consumption patterns of the crop shifted dramatically. Global consumption rose 1175% between 1961 and 1999, faster than most other food products. The growth has not been equally distributed. There was much more rapid uptake of the new 'healthy' canola oil by consumers in developed countries than in developing countries (Table 13.2). As a result, the developed countries consumed about 43% of all the canola produced globally in the 1990s, up from only 31% in the 1960s. In the developing world, only Mexico and China recorded consumption growth faster than the average. As a result, the developing-country share of total consumption dropped from 69% in the 1960s to about 57% in the first half of the 1990s.

Canola would appear to exhibit the attributes of a superior good, with rising per capita consumption strongly positively correlated with per capita incomes (Fig. 13.1). In per capita terms, Canada, Sweden, the UK and Hong Kong have the highest per capita consumption of canola oil. Per capita annual consumption in the developed world ranges from less than 1 kg per person in the USA to over 14 kg per person in Canada. The average annual per capita consumption is 3.6 kg per person in developed countries. In contrast, the average annual per capita consumption of canola and rapeseed in developing countries is less than 1.5 kg per person.

Table 13.2. Apparent global canola consumption (%).

World	1961–1967	1991–1997
Developed	**30.9**	**43.1**
Australia	0.1	0.9
Canada	3.8	5.6
EU (15)	20.5	24.9
Belgium–Luxemburg	0.6	0.8
Denmark	0.4	1.9
France	3.1	4.7
Germany	7.9	6.8
Sweden	3.0	1.0
UK	1.8	4.8
Japan	5.9	7.0
USA	0.6	4.7
Developing	**69.1**	**56.9**
Bangladesh	4.3	1.2
China	17.0	26.9
India	28.5	16.3
Mexico	0.1	1.8
Pakistan	2.3	0.8
Poland	7.6	1.7
Rest of world	9.3	8.1

Source: Author's calculations using FAOSTAT data on production, exports and imports for oilseed/rapeseed seed, oil and cake.

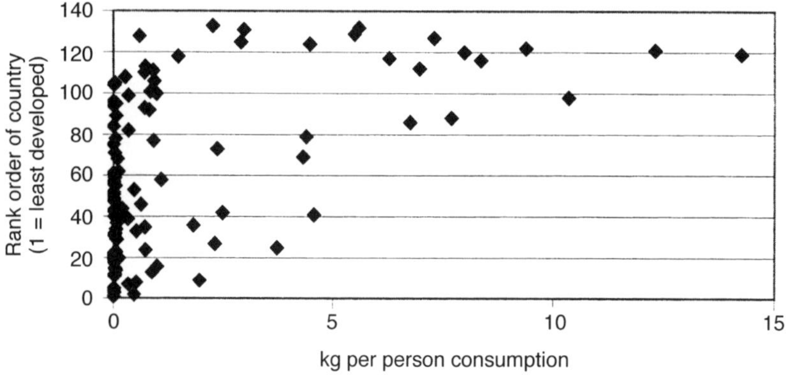

Fig. 13.1. Per capita canola consumption by level of development.

The combination of production beginning to concentrate in fewer countries and the widely dispersed consumption has led to a greater reliance on trade. Between 1961 and 1996 the volumes of trade expanded by a multiple of 18 (production rose only by 8.5 times), so that the internationally traded share of production rose from under 10% in 1961 to more than 18% by 1996. The trade dependence of canola almost doubled over that period, to a level much higher than many other agrigoods (Table 13.3).

The structure of international trade in canola products has also changed. In the 1960s, approximately 68% of the exports were in the form of seed, while 32% were in the form of oil or meal. During the 1990s, the seed trade only made up 46% of the canola-related trade; oil and meal exports accounted for 54% of the global trade. Canada is the most export-dependent of all the canola producers, exporting on average 80% of its production in the 1990s, either as seed, oil or meal. As one might expect, oil exports, which are perishable, go primarily to the US and Mexico, while seed makes up most of Canada's exports outside North America (Table 13.4).

Canada had a commanding control over the seed export market throughout the period. Even in 1961, with only 7% of world production, Canada accounted for 40% of total exports. As Canada's share of production rose, so did its share of trade, reaching a high of 60% in 1996 (Table 13.5). Europe, the only other large exporter, saw its share of trade rise from 11% to 26% over the same period. Meanwhile, as production stagnated relatively in most of the rest of the countries in the world, the export market share shared by those countries dropped, to 14% in 1996 from 49% in 1961.

The global statistics mask a potentially more important trend. As canola has become a knowledge good, it is beginning to exhibit 'product' attributes, with two-way trade between producing and consuming countries (Table 13.6). There are at least three reasons that intrasectoral trade has risen. The least interesting but largest increase in two-way trade in canola has occurred in the EU, at least partly due to the system of price supports and the operations of the EU agri-monetary system, which creates opportunities for trading profits by arbitraging between member states.

Table 13.3. Trade dependence of global rapeseed/canola production relative to other agri-food products.

	Canola	Wheat	Pigs
1961–1970	21.6%	21.1%	0.7%
1971–1980	30.5%	29.2%	1.0%
1981–1990	36.1%	33.2%	1.4%
1991–1995	40.4% [a]	26.2%	1.5%

Trade dependence = exports/total production; [a]data for canola covers 1991–1997.
Source: FAO statistics.

Table 13.4. Canadian canola production and export destinations.

	Production (000t)	% consumed domestically	% of production flowing to key export markets				
			Total	Japan	EU	USA	ROW
1992–1993	3872	22	78	42	8	21	7
1993–1994	5480	16	84	33	17	24	10
1994–1995	7233	26	74	25	17	19	13
1995–1996	6436	30	70	28	6	24	12
1996–1997	5037	15	85	37	4	31	13
1997–1998	6187	16	84	30	1	33	20
1998–1999	7288	22	78	25	0	25	28
			% distribution of sales (1992–1999)				
Seed			65	94	68	17	73
Oil			12	1	1	26	16
Meal			24	6	31	58	11

Source: Exports are the sum seed, oil and meal trade; retrieved from the worldwide web, 24 November 1999 from http://www.canola-council.org/stats

Table 13.5. Global canola production and seed trade, 1961 and 1998.

	Share of global production (%)		Export market share (%)	
	1961	1998	1961	1998
Australia	0	4	0	8
Canada	7	21	40	47
China	11	23	0	0
European Union (15)	15	27	28	37
France	3	10	10	28
Germany	7	9	1	4
UK	0	4	0	3
India	37	14	0	0
USA	0	2	0	3
World totals (million tonnes)	3,596	35,869	310	8,717

Source: FAOSTAT.

Table 13.6. Intrasectoral trade in rapeseed and canola seeds.

	1961–1970	1971–1980	1981–1990	1991–1995
Net exporters in 1995				
Australia	0.00	0.18	0.16	0.08
Canada	0.00	0.00	0.00	0.05
France	0.29	0.58	0.20	0.32
Poland	0.08	0.00	0.00	0.27
Net importers in 1995				
Belgium–Luxemburg	0.09	0.45	0.06	0.14
China	0.09	0.02	0.01	0.23
Denmark	0.07	0.04	0.02	0.53
Germany	0.35	0.37	0.32	0.82
Netherlands	0.60	0.34	0.15	0.21
Sweden	0.01	0.00	0.10	0.23
UK	0.01	0.04	0.65	0.50
Importers in 1995				
Bangladesh	0.90	0.36	0.16	0.00
India	0.05	0.02	0.01	0.00
Japan	0.00	0.00	0.00	0.00
Mexico	0.00	0.00	0.00	0.00

0 = no intrasectoral trade; 1 = complete intrasectoral trade.
Methodology: Grubel and Lloyd index of intra-industry trade: Index =
$1 - [ABS(x-m)/(x+m)]$
Source: FAO data.

A second, and more interesting, reason is that among the net importers, two-way trade is at least partly due to importing countries continuing to produce rapeseed-quality output but seeking to purchase canola-quality seeds for human consumption. Third, and most interesting, is the two-way trade that is being observed in Canada and Australia, where canola varieties with different agronomic and oil properties are proliferating. Given that each of these varieties is effectively a niche product with a limited global market, there is increasing potential that traditional net exporters may both export commodity canola and then import some of those novel varieties as feedstock for industrial processing.

The end result of these market changes is a global industry that depends heavily on international trade to commercialize the product. Production of the commodity product, which should theoretically only flow from producing to consuming countries, is now flowing to a larger number of countries as domestic demand tends to outstrip production capacity. This is exacerbated in low-technology countries where the technology and productivity gap with research-intensive countries is widening. Moreover, the emergence of international markets for research-based, differentiated, novel canola products raises the potential and need for more intrasectoral trade in canola. All of these trends increase the importance the industry places on an effective international trading system.

Evolution of the Multilateral Trading Rules

Although the General Agreement on Tariffs and Trade (GATT) was negotiated in 1947, it had only a minor impact on the flow of rapeseed and canola products through its first 48 years. Agricultural products were largely exempted from the increasingly tight rules that had been applied to manufactured goods as a result of eight successive negotiating sessions. Article 11.2c of the agreement allowed domestic support programmes and quantitative import restrictions as long as domestic production was controlled and no products were exported to international markets, while no countries bound their tariffs on primary agricultural products. As a result of the lack of coverage, the GATT dispute settlement system was never effectively used to manage international disagreements in agricultural trade. Instead, 'might was right'. The large countries, especially the US and EU, constructed policies to suit themselves and the other traders accommodated themselves as best they could. As with many agri-food products, canola tariff levels escalated based on the amount of processing involved, with relatively low tariff rates on the raw seed, somewhat higher levels for meal and the highest levels for processed oil (Table 13.7).

As a result, production and trade in rapeseed and canola, in common with most agri-food products in the post-war period, was increasingly influenced by domestic support measures. By the late 1980s the level of support for producers was quite large, equal to more than a 30% producer subsidy equivalent (Table 13.8), composed of price supports, income subsidies, quantitative import restrictions and export subsidies. Despite this level of distortion in the produc-

Table 13.7. Applied tariffs on canola seeds, oil and meal, by country (1997). (*Source:* Meilke, 1998.)

	Oilseed	Meal	Oil
Argentina	8	6	10
Australia	0	0	5
Bangladesh	8	0	30
Brazil	8	6	10
Canada	0	0	8
China	7	5	19
EU (15)	0	0	8
Former Soviet Union	9	5	15
Hungary	8	6	10
India	15	0	20
Japan	0	0	23
Mexico	13	15	10
Norway	0	0	0
Pakistan	10	21	47
Poland	14	3	67
Switzerland	50	0	184
Turkey	25	2	32
USA	0	3	20
Rest of world	11	6	18

tion and trade system, trade accelerated strongly during the 1960–1990 period.

After eight rounds of negotiation at the WTO, where agriculture was either ignored or where negotiations were unsuccessful, the condition in the industry changed enough to put the issue on the agenda. The combination of greater reliance on trade in the agri-food sector increased industry pressures for more liberal market access, while the industrialization of farming and the general rise in the welfare of farmers throughout the developed world reduced antagonism to reform.

The Uruguay Round of negotiations began in Punta del Este in 1986 with a decision to include agriculture as a major element of the process. Throughout the next 7 years the farm trade talks often set the pace and defined the critical issues and decision points in the negotiations. Following the stalemate at the Montreal mid-term meetings in 1990, the EU and US shifted some of the discussions to the bilateral level, finally coming to a compromise in London (the so-called Blair House Accord). Apart from the importance this agreement had for the entire talks, it also had a direct impact on the rapeseed/canola industry, as the EU agreed to cap its subsidies and support for oilseeds (soybeans, rapeseed and canola) at 6 million acres annual production. This has worked to slow the growth of canola production in the EU since then.

Table 13.8. Oilseeds/rapeseed producer subsidy equivalent support.

	1989–1991	1992–1994	1995	1996	1997
Australia	10	10	5	5	5
Canada	31	19	13	13	10
China	6	–10 (92)	na	na	na
Czech	49	7	–8	–11	–2
EU	68	58	54	41	48
Hungary	27	–19	10	19	17
India	41 (89–90)	na	na	na	na
Japan	78	31	37	38	33
Mexico	23	22	13	12	1
Poland	–35	27	14	28	7
Switzerland	95	92	91	89	89
Turkey	42	33	34	50	51
USA	10	9	8	6	8
Total PSE	31	26	22	15	20

Sources: OECD and USDA Economics and Statistics System.
na, not available.

Then in 1993, the negotiators came to an agreement, which was implemented in 1995. The key elements that have affected agriculture are the creation of the World Trade Organization (the GATT only had a secretariat before) and the mandating of the agency to proactively monitor trade policies, development of a new and more powerful dispute settlement system and extension of the multilateral undertakings with agreements on agriculture, intellectual property and sanitary and phytosanitary measures. The rapeseed/canola sector is affected by all of these measures.

The most immediate and direct impact of the deal was reductions in agricultural subsidies and protection (Table 13.9). After difficult negotiations all member states agreed to convert variable levies and import quotas into *ad valorem* or fixed tariffs, which they then agreed to reduce over the implementation period. The developed countries agreed to reduce the average tariff level on agricultural products by 36% in six equal annual increments; tariff cuts for individual products can vary, subject to the agreement that every tariff will drop by at least 15%. Developing countries agreed to a schedule of tariff reductions that is both less (–24% with a minimum of –10%) and slower (over 10 years). A complementary commitment was that the developed countries make minimum access commitments. In short, each country agreed to set for products not currently imported special, lower tariff rates for an import quota equal to an initial 3% of domestic consumption (1986–1990 reference period), rising to 5% of consumption by 2000. These tariff rate quotas, set for each product, were designed to open every market to at least some trade. For the canola sector, the lower tariffs are the most important development, with the tariff rate quotas creating some value in a few selected countries.

Table 13.9. The WTO Agriculture Agreement (%).

	Developed countries	Developing countries
Implementation period	6 years: 1995–2000	10 years: 1995–2004
Tariffs: average cut for all products	−36	−24
Tariffs: minimum cut per product	−15	−10
Domestic support: total AMS cuts for sector (base period: 1986–1988)	−20	−13
Exports: average cut in value of subsidies (base period 1986–1990)	−36	−24
Exports: cuts in subsidized quantities (base period: 1986–1990)	−21	−14
Minimum access (base period: 1986–1990)	3–5	0

Source: WTO.
AMS, aggregate measure of support.

The parties to the WTO agreement went well beyond traditional trade measures, with agreement both to reduce the level of distorting subsidies in the sector and to wind down export subsidies. The rationale was that if subsidies that distort production could be reduced, there would be less need for export subsidies. The agreement on domestic support was quite complex. The first step was a definition of which subsidies would be involved. They used a traffic-light system, categorizing subsidies whether they were green (not distorting production), red (distorting production), amber (possibly distorting production) and blue (distorting but provisionally allowed). Green subsidies – crop insurance replacing less than 70% of income losses, environmental subsidies, disaster programmes, R&D – were allowed without restriction. Amber subsidies either had to be 'decoupled' from production or would be treated as red programmes. The agreement therefore initially focused on red, production-distorting subsidies. The second step was to estimate the aggregate level of support in each country (called the aggregate measure of support, AMS), bind that level as the maximum allowable and then seek commitments to reduce the support over time. Developed countries finally agreed to reduce their AMS by 20% in six equal instalments. The only twist was that they used the period of 1986–1988, when subsidies were very high, as the base level. For many countries that had reduced their programmes in the interim, this provided significant 'water' in the commitment. Even so, by 2000 much of the water had been wrung out of the commitments and the bindings were beginning to bite. Developed countries only committed to a 13% reduction over 10 years. Once the base agreement was reached, there was one add-on. Both the EU and the US wanted to be allowed to provide compensation to their producers for winding down various domestic support programmes. They agreed jointly, and got concurrence with the rest of the member states, that they would be allowed provisional 'blue box' subsidies that might be based on production; the plan was that these subsidies would last

Table 13.10. PSE support for canola producers in key producing and exporting countries (%).

	Period	Market price support	Direct payments	Reduction of input costs	General services	Subnational and tax support
Australia	79–81	0	0	15	26	63
	86–88	0	0	0	27	74
	95–97	0	0	2	24	74
Canada	79–81	44	12	5	7	23
	86–88	23	48	1	4	24
	95–97	0	53	2	12	26
China	86–88	92	1	3	3	0
EU	79–81	0	83	0	17	0
	86–88	0	89	0	11	0
	95–97	0	91	0	9	0
India	86–88	66	0	34	0	0

Sources: For Canada, EU and Australia (OECD Producer and consumer subsidy equivalents database, 1998); for China and India (USDA, Producer and Consumer Subsidy Equivalents, 1982–1992 (95001) http://usda.mannlib.cornell.edu/data-sets/international/95001/).

at most 7 years and then would be eliminated or made green. The USA has already made its programme green but the EU has indicated it would like an extension of the blue box to facilitate its programme (Table 13.10).

Finally, the agriculture agreement addressed export subsidies, the most visible and troubling problem for most agricultural exporting nations. The countries made two commitments. First, the developed countries agreed to cap and then reduce the value of export subsidies provided to each product by 36% over 6 years. Unlike in the other areas, this commitment was by product. Developing countries agreed to reduce the subsidies they provided by 24% over 10 years. In order to make the commitment tighter, the countries also agreed to reduce the volume of subsidized exports by 21% over the 6 years (–14% over 10 years for the developing countries), using the 1986–1990 period as the base. The logic was that if the agreement worked, agricultural prices should rise, naturally reducing the value of subsidies. So, if volumes were also bound, that would guarantee lower competitive subsidies.

Four other parts of the agreement have significant potential to influence the future evolution of the canola market. First, the new dispute settlement system has both a bark and a bite. Under the GATT, disputes could be taken to Geneva but any panel decisions required unanimous approval of the council, which involved the disputing countries. The US, for example, as a matter of principle always voted against any decisions that found against US policy, thereby stalling their implementation. The new system is more judicial. Any member state with a problem related to the implementation of the agreement can initiate a dispute case, which goes through a period of consultation and adjudication. Ultimately, a dispute settlement panel (comprised of experts in trade economics and law) presents its findings and it is implemented. There are no opportunities for disputants to block adoption of the decisions. Between the start of the Agreement on 1 January 1995 and May 2000, there were 193 dispute cases initiated, more than 30 of which related to agri-food products. To date, most decisions – none related directly to canola so far – have been respected and implemented (if only tentatively and, at times, with poor grace).

Second, the agreement implements new commitments by member states to limit the use of sanitary and phytosanitary (SPS) rules as disguised trade measures. There were already well-established mechanisms to address legitimate concerns about public health and safety – in Canada, for instance, through the various regulatory systems now managed by the Canadian Food Inspection Agency – but there was considerable concern that domestically managed systems could be used to distort trade. The Agreement on Sanitary and Phytosanitary Measures aims to reduce the trade distortions that SPS measures can cause, by encouraging countries to base their SPS measures on existing international standards and to recognize other countries' standards, as long as they achieve the same degree of protection. The SPS Agreement also imposes certain obligations in determining what measures to adopt to safeguard health. Under the Agreement, SPS measures must be applied only to the extent necessary to protect human, animal or plant life, and be based on scientific principles and on an objective assessment of the risks posed to health. They may not discriminate unjustifiably between countries where the same conditions prevail, or be applied in a way that makes them a disguised barrier to trade. Disputes between contracting parties regarding the requirements of the SPS Agreement are to be dealt with under the established WTO consultation and dispute settlement procedures. For disputes involving scientific or technical issues, the panel ruling on the dispute seeks advice from experts, and may establish an advisory group of technical experts (USDA, 1997). This effort is strengthened by a notification and review system in the SPS Committee. This is likely to be a major focus for knowledge-based traded products, such as canola, as many domestic regulatory systems around the world seek ways to incorporate subjective consumer, citizen and environmental concerns into their processes.

Third, the technical barriers to trade (TBT) agreement, analogously to the SPS agreement, attempts to manage labelling and other technical measures to avoid them becoming unnecessary obstacles to trade. The key principles driving consideration of technical measures are non-discrimination and national treatment, harmonization or equivalence of technical regulations, mutual recognition of conformity assessment procedures, and transparency. A number of issues related to canola might at some time be considered by this committee, including potentially discriminatory consumer labelling requirements for 'genetically modified' products, country of origin labels and various environmental requirements stemming from the BioSafety Protocol.

Finally, the agreement involves provisions that manage trade related to intellectual property rights, which are of vital importance to those in the industry. These will be discussed below in the context of the evolving international intellectual property system.

On the face of it, the extension of trade rules into the agricultural sector with the implementation of the WTO agreement in 1995 would appear to solve many of the problems facing the industry. In practice, however, the system still has some kinks, with many countries just, if at all, complying with their commitments. More importantly, however, is the non-membership of key countries in the WTO. China, the single largest producer of canola in the world, is not currently a member, while potential key countries such as Russia are waiting to join. Although negotiations for accession have begun, it will take years for them to conclude and an even longer time for adjustments to be completed. In the interim, the international system will evolve on multiple tracks: developed country members of the WTO will increasingly liberalize, developing member states will slowly open and non-members will continue on their own route.

The Evolving International IPR System

Access to markets for producers of knowledge-based products is only useful if the companies can also exploit the value of their product. Especially with reproducible plant and animal products, few companies are willing to export into countries that will not protect their rights to the embodied knowledge. Hence companies and producers are seeking a more effective regulatory and trade system for products involving intellectual property.

As discussed in Chapter 4, innovation in plants is not limited to the development of plant varieties. Innovation also takes the form of inventive processes and downstream products, which qualify for patent and other IP protection, such as trademarks, trade secrets and geographic designations. Meanwhile, plant innovators may seek protection by means of patents for invention (where their innovation fulfils the criteria of the patent law of the relevant member state) or by plant breeders' rights (where their innovation fulfils the criteria of the UPOV Convention) or conceivably, in some countries, by both forms of protection. As a result, the array of protective measures for intellectual property rights is quite wide.

There are four main groupings of protections for intellectual property rights, which influence trade in agri-food. Before 1961, the only protection was from national countries. In the USA, the 1935 Plant Variety Protection Act subsequently granted plant variety protection for 18 years to asexually produced plant varieties (e.g. hybrid maize), but it was not until 1970 that the USA renewed the PVPA and granted the same property rights to sexually reproduced varieties, such as open or self-pollinated varieties. In The Netherlands, the Breeders' Ordinance of 1941 granted a very limited exclusive right for breeders of agriculturally important species to market the first generation of certified seed, but an exclusive right to market propagating material of other species. In Germany, after many years of limited protection for breeders based upon seed certification, the Law on the Protection of Varieties and the Seeds of Cultivated Plants of 1953 gave breeders the exclusive right to produce seed of their varieties for the purposes of the seed trade and to offer for sale and market such seed.

Between 1957 and 1961, a number of European countries that were concerned with the inconsistency of definitions and rights between countries and the absence of any rights for foreign plant breeders in national laws, met and negotiated the International Convention for the Protection of New Varieties of Plants (UPOV). The Convention was subsequently revised in 1972, 1978 and 1991. Currently, all but two signatory members use the 1978 Act. The Convention requires member states to provide protection for new varieties of plants, but also contains explicit and detailed rules on the conditions and arrangements for granting protection. Furthermore, it contains rules on the scope, the possible restrictions and exceptions, and the forfeiture of protection. It establishes, subject to certain limitations, the principle of national treatment for plant breeders from other member states; in any member state nationals or residents of another member state enjoy the same treatment as nationals or residents of that state. The agreement and the corresponding national laws grant breeders exclusive rights to market varieties for a set period (often 20 years), often with farmers' exemptions (to allow producers to save and reuse their seed) and with breeders' exemptions, which permit the use of protected varieties as genetic resources in further breeding. Finally, it introduces a right of priority. The 1978 Act provides that only States can be parties; the 1991 Act provides for the possibility of accession by any intergovernmental organization which has competence in the field of plant breeders' rights, with its own legislation providing for the granting and protection of breeders' rights binding upon all its member states. Currently 48 countries have some form of plant breeders' rights, representing 25% of the world. Of those, 38 have adopted UPOV standards; the other ten have national policies. In 1998, more than 23% of Canada's canola exports went to countries without any form of PBR or other IPR for plants (Strategis), up from an average of about 10% in the previous 4 years.

Since the advent of biotechnology, patent rights have been extended in a number of ways. As discussed in Chapter 11, the US patent office has extended patents to whole plants, which does not provide for farmers' or breeders' exemptions. Although other countries have extended patent protections less liberally, US predominance in the world markets for most products makes it the effective mediator of many rights issues. Theoretically, a research effort in Canada, or any other country that does not grant whole plant patents, could use genetic material that is patented in the US. But if it then tried to export any of the resulting product to the US, it would be open to litigation and seizure of the product and any resulting profits. So, practically speaking, the US law has extraterritorial application for export-based industries.

As intellectual property rights were being developed both nationally and internationally, the agri-food sector increasingly came under the aegis of the World Intellectual Property Office (WIPO) and the relevant treaties. There are seven key international treaties. The Paris Convention for the Protection of Industrial Property (1883), ratified by 151 countries, establishes the basic principle of national treatment and rules for facilitating granting of rights in multiple countries. The more recent Patent Co-operation Treaty (PCT) (1970), ratified by 100 countries, implements the concept of a single international patent application that is valid in many countries. Once such an application is filed, an applicant has time to decide in which of the countries to continue with the application, thereby streamlining procedures and reducing costs. The PCT system is expanding rapidly: the number of member states has more than doubled in the past 7 years, to 96, and the number of international applications has grown from 2625 in 1979 to 54,422 in 1997. Because each application extends to more than one country, those 54,422 applications represent nearly 3.5 million national applications for inventions. The Trademark Law Treaty (1994), establishing national treatment and rules for trademarks, is ratified by only 22 states. Other agreements, including the Madrid Agreement for the Repression of False or Deceptive Indications of Source on Goods (1891) and the Protocol Relating to the Madrid Agreement Concerning the International Registration of Marks (1989), are ratified by about 60 countries and deal with the international registration of marks and industrial designs. In 1997, there were 19,070 registrations of marks under the Madrid system, representing some 220,000 national applications and 6223 deposits, renewals and prolongations of industrial designs. The Budapest Treaty on the International Recognition of the Deposit of Microorganisms for the Purposes of Patent Procedure (1977) is ratified by about 30 countries. The Lisbon Agreement for the Protection of Appellations of Origin and their International Registration (1958) is ratified by 18 countries.

As WIPO had 171 members as of 30 June 1998, the commitment to the fair and equal treatment of intellectual property has expanded. Even so, delivery was somewhat spotty, as only a few countries have acceded to all of the agreements. Furthermore, although WIPO offers conciliation services, there are no formal dispute settlement systems within the various treaties.

More recently, in the 1986–1993 round of international trade negotiations, the member countries negotiated the Trade Related to Intellectual Property (TRIPS) Agreement to address the deficiencies of national, UPOV and WIPO systems. The Agreement, which came into effect on 1 January 1995 and has been ratified by 139 nations, covers a number of areas relevant to agri-food development: trademarks, including service marks; geographical indications, including appellations of origin; patents, including the protection of new varieties of plants; and undisclosed information, including trade secrets and test data. At its base, the Agreement provides for certain basic principles, such as national and most-favoured-nation treatment, and some general rules to ensure that procedural difficulties in acquiring or maintaining IPRs do not nullify the substantive benefits that should flow from the Agreement. The obligations under the Agreement will apply equally to all member countries, but developing countries will have a longer period to phase them in. In essence, the agreement provides three specific features.

First, the Agreement sets out the minimum standards of protection to be provided by each member. Each of the main elements of protection is defined, namely the subject matter to be protected, the rights to be conferred and permissible exceptions to those rights, and the minimum duration of protection. The Agreement sets these standards by requiring, first, compliance in the substantive obligations of the main conventions of the WIPO, the Paris Convention for the Protection of Industrial Property (Paris Convention) and the Berne Convention for the Protection of Literary and Artistic Works (Berne Convention) in their most recent versions. The main substantive provisions of these conventions are incorporated by reference and thus become obligations under the TRIPS Agreement between member countries. The relevant provisions are to be found in Articles 2.1 and 9.1 of the TRIPS Agreement, which relate, respectively, to the Paris Convention and to the Berne Convention. Second, the TRIPS Agreement adds a substantial number of additional obligations on matters where the pre-existing conventions are silent or were seen as being inadequate. The main exemption from this provision is that members may exclude from coverage plants and animals other than microorganisms (e.g. biological processes for the production of plants or animals other than non-biological and microbiological processes). However, any country excluding plant varieties from patent protection must provide an effective *sui generis* system of protection, such as plant breeders' rights. Moreover, the whole provision is subject to review 4 years after entry into force of the Agreement (Article 27.3b). Compulsory licensing and government use without the authorization of the rights holder are allowed, but are made subject to conditions aimed at protecting the legitimate interests of the rights holder. The conditions are contained mainly in Article 31. These include the obligation, as a general rule, to grant such licences only if an unsuccessful attempt has been made to acquire a voluntary licence on reasonable terms and conditions within a reasonable period of time; the requirement to pay adequate remuneration in the circumstances of each case, taking into account the economic value of the licence; and a requirement that decisions be subject to judicial or other independent review by a distinct higher authority. Certain of these conditions are relaxed where compulsory licences are employed to remedy practices that have been established as anticompetitive by a legal process. These conditions should be read together with the related provisions of Article 27.1, which require that patent rights shall be enjoyed without discrimination as to the field of technology, and whether products are imported or locally produced. Article 40 of the TRIPS Agreement recognizes that some licensing practices or conditions pertaining to intellectual property rights that restrain competition may have adverse effects on trade and may impede the transfer and dissemination of technology. Member countries may adopt, consistently with the other provisions of the Agreement, appropriate measures to prevent or control practices in the licensing of intellectual property rights which are abusive and anticompetitive.

Second, the agreement deals with domestic procedures and remedies for the enforcement of intellectual property rights. The Agreement lays down certain general principles applicable to all IPR enforcement procedures. In addition, it contains provisions on civil and administrative procedures and remedies, provisional measures, special requirements related to border measures and criminal procedures, which specify, in a certain amount of detail, the procedures and remedies that must be available so that rights holders can effectively enforce their rights.

Third, the Agreement makes disputes between WTO members about TRIPS obligations subject to the WTO's dispute settlement procedures.

The TRIPS Agreement gives all WTO members transitional periods so that they can meet their obligations under it. Developed country members have had to comply with all of the provisions of the TRIPS Agreement since 1 January 1996. For developing countries and transition economies, the general transitional period was 5 years, i.e. until 1 January 2000. However, all members, even those availing themselves of the longer transitional periods, have had to comply with the national treatment and most-favoured national treatment obligation as of 1 January 1996.

Even with the apparent international expansion of property rights to intellectual property, the system is not as simple as it might appear. As with the WTO trade rules, the absence of Russia and China from the agreement leaves a major hole in its coverage. In addition, many of the countries have yet to address the commitment to extend patent rights to whole organisms, or to implement an effective *sui generis* system. Finally, the dominant role of the US in the trade system has often worked to discourage private companies hoping to trade their products from using their intellectual property, thereby effectively extending US-style intellectual property rights to non-conforming countries.

Despite the unevenness, it is fair to say that the trade and intellectual property systems are working to liberalize trade. In contrast, some consumers and other groups are working feverishly to use other international forums to re-regulate and constrain trade in these products. In particular, consumers, environmentalists and other public advocacy groups have targeted the *Codex Alimentarius* and BioSafety Protocol processes as ways to assert their interests.

Labelling and the *Codex Alimentarius*

Although agri-food trade was outside the discipline of the GATT until the Uruguay Round's Agreement on Agriculture in 1994, multilateral efforts to discipline agri-food trade have been at the heart of the *Codex Alimentarius* Commission (CAC) since its establishment in 1962. Administratively, the CAC is a joint agency of the UN's Food and Agricultural Organization (FAO) and the World Health Organization (WHO) and it was created under the UN's Food Standards Program. As a result, all member countries of the United Nations may be members of the CAC.

The creation of the CAC was driven largely by desires to ensure food safety and consumer protection in the context of rapidly growing international food trade. Producers, in pursuit of economies of scale, increasingly exported food products, while imported products invariably composed part of domestic consumption bundles. Therefore, the objective of the CAC was to develop the *Codex Alimentarius* (which is Latin for Food Code) to provide internationally agreed standard food safety guidelines which may be universally applicable in order to ensure international consumer protection, while at the same time facilitating international trade in food products.

The importance of the *Codex Alimentarius* to international trade has grown steadily. Although the initial intent was to focus on food safety and consumer protection, the recent inclusion of *Codex* principles into the WTO's SPS and TBT Agreements inextricably links the *Codex Alimentarius* to international trade concerns (Buckingham *et al.*, 1999). Supporters of this development are encouraged that international agricultural trade concerns must defer to an established international food safety-standard-setting body. This preserves the dominance of food safety and international consumer protection objectives over commercial trade interests. Critics, on the other hand, perceive that the *Codex Alimentarius* has been captured by commercial trade interests and now sacrifices food safety and international consumer protection to meet trade objectives.

There are three separate *Codex* agencies which together work to develop the *Codex Alimentarius*. The first, already mentioned, is the CAC which meets every 2 years and is composed of all member countries to the United Nations (as of 2000 the membership was 165 countries). The CAC is composed of committees organized around commodities, general subjects and around expert groups which provide supporting advice and guidance. Committees are generally chaired by a member country and they may be active or dormant. There are 14 'vertical' commodity committees, including one for fats and oils, currently chaired by the UK. In addition, there are eight 'horizontal' *Codex* General Subject Committees. The ones relevant to the canola industry are: import/export inspection and certification; food additives and contaminants; general principles; pesticide residues; food labelling; and analysis, sampling and food hygiene. Two other joint FAO/WHO agencies provide expert advice and consultation to the CAC and the various committees – the Joint Expert Committee on Food Additives (JECFA) and the Joint Meeting on Pesticides (JMP).

The second agency is the permanent *Codex* Secretariat that is located in Rome and administered by the FAO. The purpose of the Secretariat is to provide day-to-day support for member countries as they attempt to interpret, develop and implement national food regulation congruent with the *Codex Alimentarius*.

The third *Codex* agency is the *Codex* Executive Committee. The Executive Committee meets yearly and, unlike the CAC, is organized according to principal regions: Europe, Africa, Asia, the South Pacific, Latin America and North America. Hence, the *Codex* Executive Committee is the regional coordinating committee, which provides the regional perspective on food safety, consumer protection, and, increasingly, agri-food trade.

As the *Codex Alimentarius* attempts to develop universal food safety and consumer protection principles, it should come as no surprise that the administrative process is lengthy and subject to many iterative review processes. A *Codex* food standard is adopted only after eight stages or steps of consultation have been completed. First, a food safety issue is identified by the CAC, the *Codex* Secretariat of the *Codex* Executive Committee and presented at a CAC plenary session (every 2 years), where, if it is determined that a *Codex* food standard ought to be elaborated, the CAC assigns the issue to either a commodity or a general subject committee. Second, the committee presents its elaboration, based on *Codex* food standard elements, to the *Codex* Secretariat who produces a Proposed Draft Standard. Third, the Proposed Draft Standard is sent to all member governments and identified non-governmental international organizations for review and comments. Fourth, comments from step 3 are returned to the Committee who initially elaborated the food standard. Fifth, the committee amends the Proposed Draft Standard subject to the review and comments and the amended Proposed Draft Standard is presented to the CAC by the Secretariat at a plenary session where it may be adopted as a Draft Standard. Sixth, the adopted Draft Standard is sent to all member governments and identified non-governmental institutional organizations for further comment. Seventh, comments are returned to the Committee through the Secretariat for amendments to the Draft Standard. Eighth, the amended Draft Standard is presented to the CAC for adoption as a *Codex* Standard to be sent to member governments for acceptance.

A final *Codex* food standard includes several elements. First, it includes a description of the product and the essential composition and quality factors, which identify the product from close substitutes. Second, the standard includes both identification and analysis of any additives and potential contaminants in the food product. Third, the food standard incorporates established *Codex* requirements such as the *Codex* product hygiene requirements and the *Codex* labelling requirements. Fourth, the standard includes a complete description of the scientific procedures used to sample and analyse the product during review. Determination of the safety of the food product is based on scientific risk analysis and toxicological studies of pesticide residues, microbial contaminants, chemical additives and veterinary biologicals.

Generally it takes about 7 years to develop a *Codex* food standard (i.e. one-half of a year for each of steps one to six, while steps seven and eight take 2 years each). There is a fast-track procedure that can be employed, however, if the proposed standard is relatively uncontroversial. Under the fast-track approach, it is possible for the Proposed Draft Standard to be adopted at step 6 as a *Codex* Food standard, instead of being sent for further review, if consensus has been achieved. The *Codex* decision-making process is preferably done by consensus, although it can be by vote in the CAC if consensus cannot be achieved on very controversial issues, such as the use of hormones in beef production.

Essentially, *Codex* is a top-down approach to developing universally acceptable food standards through its elaboration and consultation procedures at the multilateral level. Once a *Codex* food standard is adopted, member countries are encouraged to incorporate the standard into any relevant domestic standards and legislation. However, under the principles of *Codex*, member states retain the right to unilaterally impose more stringent food safety regulations that may be deemed necessary to ensure domestic consumer protection. It is anticipated that when countries deviate from the *Codex* food standard, they do so in a scientifically justifiable manner.

The *Codex Alimentarius* plays an important role in agri-food trade because its principles are enshrined in the SPS and TBT Agreements of the WTO. The SPS Agreement requests that all parties harmonize their domestic standards with *Codex* standards, guidelines and other *Codex* recommendations, where such exist. In the case of trade disputes, *Codex* standards, guidelines and recommendations are to be employed under the SPS for WTO dispute resolution procedures (Buckingham *et al.*, 1999).

There are some important challenges on the horizon for the *Codex Alimentarius*. One crucial challenge is the debate over scientific risk assessment. The 22nd Session of the *Codex Alimentarius* Commission (23–28 June 1997, Geneva, Switzerland) considered amendments to the *Codex* Procedural Manual including four statements of principles concerning the role of science in the *Codex* decision-making process. It has been the traditional stance of the *Codex Alimentarius* that risk should be based on scientific evidence of risk to human health, based on the reasonably certain interpretation of the precautionary principle. However, recent efforts by the EU to include 'risk to other legitimate factors' such as 'social dimensions' have created controversy. The pressure has been to create an opt-out situation so that a country may abstain from the *Codex* food standard decision, and the standard may still be adopted. This is directly against the original intent of adopting standards by consensus and promoting the international harmonization of food standards.

With respect to biotechnology-based canola varieties, the *Codex Alimentarius*, like other international institutions such as the OECD, deals with biotechnology on a product basis, not on a technology basis. In this sense, there is no horizontal biotechnology committee. However, through the General Subject Committee on Labelling, *Codex* may have an impact upon the marketing of GM canola-based products. The precedence for labelling based on process or production method has been set by the *Codex* guidelines on labelling irradiated meat products in order to increase consumer information. By extension, labelling guidelines may be established to identify whether the product contains any genetically modified components such as inputs, processing aids, re-works, etc. From an industry perspective, developing labelling guidelines at the *Codex* may be commercially advantageous. This is because many countries have expressed an intention of imposing unilateral labelling policies for genetically modified products (Phillips and Foster, 2000). In order to prevent the possible divergence in labelling policies and the subsequent market fragmentation, *Codex* label standards would provide a truly international approach to labelling GM products.

The Convention on Biological Diversity and the BioSafety Protocol

In 1992, the UN convened a conference on global biological diversity in Rio de Janeiro, known as the Earth Summit. The result was a Convention on Biological Diversity which has since been signed by 171 countries (about 88% of all the countries in the world) and ratified by 134 countries. All of the countries that are markets for Canadian canola products have signed the protocol, although not all have ratified it. The most notable exception is the US, which although not formally bound by the agreement has remained a key actor in the negotiations related to greenhouse gas emissions and the BioSafety Protocol. Nicaragua is the only other market of any size that has not ratified the convention.

Internationally, the effort to manage genetic resources began in the immediate post-war period with the development of international research centres. Over the following years 12 international commodity research centres were established to undertake crop, livestock and fish development and to conserve genetic resources. These centres – which along with the International Plant Genetics Research Institute (IPGRI), the International Food Policy Research Institute (IFPRI) and the International Service for National Agricultural Research (ISNAR) were informally united in the Consultative Group for International Agricultural Research (CGIAR) in 1971 – gathered samples from various countries around the world, which they stored and used selectively to improve the commercial breeding stock. The collections were viewed to be held 'in trust' and resulting improvements were provided to national and private breeding programmes without any charge. Since 1983 the FAO has worked with the CGIAR to develop a global system for managing their resources and to ensure that the returns to their research flow as much as possible to poorer people and farmers in developing countries (FAO, 1998). That effort got a boost from the Convention on Biological Diversity (1992), which provided the base for an agreement between FAO and the CGIAR centres to place most of their accessions into an 'in-trust' international network of collections managed by FAO. Approximately 443,000 accessions in the centres, equal to 75% of the total germplasm collections in the centres, have been designated as part of the network (material may not have been designated because it is duplicate or because the rights to the material are in doubt). More than 30 other countries with more than 775,000 accessions have expressed willingness to join the network. One condition of using the materials henceforward will be explicit agreements that prohibit national or private breeders from seeking property rights to the germplasm. Combined, these genebanks are the single largest repositories of genetic material in the world, which will work to impede the efforts of national and private research programmes to capture returns for research involving that material.

The big push in the late 1990s is to use the BioSafety Protocol process to compensate for the limited domestic regulation in many countries. In addition, some developed countries with extensive domestic regulations are seeking to insert in the Protocol measures that would re-regulate trade by requiring monitoring, notification and labelling for non-environmental effects, such as food safety. Currently 75% of the countries in the world, and almost all of those that are likely targets for deliberate commercial release of foreign or genetically modified plants or animals, have national quarantine rules. Almost three-quarters of those countries (103) are members of the International Plant Protection Convention and their quarantine rules conform to the convention. Furthermore, more than 68% of the countries in the world have some form of controls on the seeds industry; most countries regulate the seed quality while the rest certify new seeds before they are allowed to enter the market (Table 13.11). Finally, approximately three-quarters of the countries in the world have some type of domestic crop production programme. Most countries have a genebank while most developing countries have basic or developing research programmes while most developed countries have advanced breeding programmes. Furthermore, about 75% of those countries with national research programmes were also active members of a subregional research network, with the opportunity to access a wider level of technological support (FAO, 1998). The one group at the greatest risk of indiscriminate genetic introductions is the poorest and least developed nations, especially those in central Asia and Africa. Nevertheless, the vast majority of countries and regions that are likely to be deliberate targets for new crops have the basic mechanisms to protect their ecosystems.

The BioSafety Protocol (BSP) is an effort, under the auspices of the Convention on Biological Diversity, to create a legally binding international agreement that governs the transboundary movement of living products of modern biotechnology in order to protect and conserve biological diversity. Originally, the Protocol was scheduled to be signed at a Conference of the Parties in February 1999. The talks were officially suspended in March 1999 when the 'Miami group' of countries (Canada, the USA, Australia, Argentina, Chile and Uruguay) rejected European efforts that were supported by the other 140 negotiating countries, to extend the coverage of the Protocol to include risks to human health and to agri-food shipments intended for processing. Agreement was finally reached in January 2000 and the BSP will come into force after 50 signatory countries ratify the Protocol domestically. Regardless of the specific issues in the Protocol, the proposed compliance system has significant potential to seriously limit canola trade because the bulk of the production in Canada and the US is now genetically modified, while research efforts in the EU, Australia, China and India are all targeted to develop transgenic varieties.

The agreement requires exporters to acquire advanced informed agreement (AIA) for all first-time transboundary movements of living modified organisms destined for commercial release. Exporting countries will be required to inform the importing country, which will then accept or reject a shipment based on results of a scientific risk assessment (SRA). Movements for breeding programmes are exempt, while commodity shipments must be labelled but are not subject to SRA.

The potential economic and trade impact is associated with the presence or

Table 13.11. National and international undertakings on genetic resources.

		National legislation used					Status of national crop programmes			
	Number of states	CBD	Quarantine rules	PBR	Seed certification	Seed quality	Basic	Developing	Advanced	With gene-bank
Africa	48	46	33	3	17	16	8	23	2	37
Americas	36	34	35	13	22	5	10	13	9	27
Asia/Pacific	37	31	25	4	12	7	5	9	7	29
E. Europe	21	19	13	12	11	8	0	13	4	19
W. Europe	22	20	18	15	17	0	0	0	17	16
Near East	30	21	22	1	14	2	3	16	2	20
Totals	194	171	146	48	93	38	26	74	41	148
% total		88%	75%	25%	48%	20%	13%	38%	21%	76%

Source: FAO, 'State of the world's plant genetic resources for food and agriculture', Appendix 1 (Rome 1998).
CBD, Convention on Biological Diversity; PBR, plant breeders' rights.

absence of an identity-preservation production (IPP) system. A recent study of the potential impact of the BSP on the canola industry (Isaac and Phillips, 1999) concluded that the range of trade impacts for canola could be as high as 100% or Can$1.2 billion worth of exports (without an IPP system) to as low as 0.5% or Can$6 million (with an IPP system if only plants with novel traits intended for deliberate release are regulated). The annual private sector incremental cost of complying with the new rules for canola would correspondingly range from Can$3 million to as low as Can$400,000 (based on an estimated 59 genetically modified (GM) varieties and eight plants with novel traits released per year, and assuming that the cost of generating a regulatory portfolio for each variety is Can$50,000). This assumes that regulators in different countries would accept the data and trials undertaken in Canada. Without an IPP system the 45% of Canadian canola production that moves across international boundaries (to between 13 and 22 countries annually, involving 107 and 385 transboundary movements) would all be affected by the new rules. In practice, however, little of the canola exported is intended for deliberate environmental release. In 1996, according to CSTA Export Statistics, canola seed for deliberate environmental release accounted for only 0.5% of total exports, to five destinations (the USA, Australia, Belgium, Finland and South Africa). In addition to the compliance costs, either the government or the industry would be required to pay the costs of monitoring the estimated 42 to 1770 separate notifications of transboundary movements that would be required annually (estimated to cost between Can$400K and Can$7 million annually, depending on the definition of LMO).

If genetic transformations were one-time events, the new rules would be less onerous. The trend, however, is for exponential growth in the use of biotechnology. Up to and including 1997, 138 varieties of canola have been registered in Canada (approximately 100 more varieties were registered in 1998–2000), including 31 varieties which draw on four novel, single traits. Field-trial data for 1997 indicates that double- and triple-trait stacking categories are a focus of significant research effort. There are 25 categories of double-trait stacking while triple-trait stacking is being employed in over nine categories. The past biotechnology effort indicated that, on average, four transgenic modifications were possible for each trait-stacking category and that subsequently three new GM varieties were required to 'cover the market' for each new transgenic modification. Therefore, based on the research in the field as of 1998, there is the potential for 408 new GM varieties over the next 7 years, involving 54 novel traits.

The key issue to producers and exporters is ensuring market access. If IPP systems are not possible, then all the production must be considered as GM if approval for the unconfined production of even one GMO variety has been granted in Canada. In the short term, participants in the Canadian grains and oilseeds industry insist that the present Canadian distribution system makes it 100% logistically impossible to segregate GM product from non-GM product (Hart *et al.*, 1997), a view shared by both US and European industry participants (Agrevo, Nov. 1997; GAFTA, May 1997; Central Soya, Dec. 1996; NOPA,

Dec. 1996; ASA, Dec. 1996 and Sparks Companies, Sept. 1996). Therefore, in the short term, any commodities that have GM varieties that fall under the BSP's Notification and AIA (advanced informed agreement) procedures face the problem that without the ability to segregate, all exports would have to be monitored.

A potential solution to the segregation issue is to develop and implement strategies to create an identity-preserved production system for GM grains and oilseeds. However, the short-term economic feasibility of such a system is seriously questioned by industry. An experimental IPP system for GM canola varieties was implemented in Canada in 1995 and 1996. From this experiment, it was identified that an IPP system created incremental costs of between Can$34 and Can$37 per Mt for grains and oilseeds (Manitoba Pool Elevators, 1997). That cost was assessed against only about 60,000 tonnes of produce. If all the 2.8 million tonnes of export product required such an expensive IPP system, the industry would need to absorb more than Can$95 million of additional costs, which is probably more than the current profit margin of all parts of the supply chain. Other estimates suggest that developing and implementing an international IPP system would require a commodity price rise of between 140 and 180% (EuropaBio, June 1997). Further, even if segregation could be ensured on the production or supply side and the price rise of commodities could be absorbed, industry participants argue that many importing countries simply lack the capacity to ensure no co-mingling when the commodity is imported.

If the estimated costs of notification, regulatory oversight and segregation are accurate, the competitiveness of Canadian exports could be drastically affected by the Protocol. In the smaller markets, the increased burden may exceed the margins being earned in those markets, essentially ending mutually beneficial trade. This would be especially true in those markets where domestic regulatory capacity was weak. Canada currently ships 20% of its canola to developing countries (*Strategis Trade Data*, 1997), many of which lack the capacity necessary to conduct a scientific risk assessment (SRA) to determine advanced informed consent. Using Kenya as a proxy for other less-developed countries (LDCs), a recent study found that it 'is still in the first stages of development of biotechnology research'. The available resources seem poorly targeted because Kenya has 'a low technical support staff to researcher ratio' while it has a 'high manager to researcher ratio' (ISNAR, 1998). The capacity of like countries to undertake assessments is limited. This implies that the burden of assessment will fall upon Canada, either the exporters or the government.

Combining exporter notification with the lack of capacity in many of the potential export countries creates an incentive to not notify a shipment which may have varieties considered to be LMOs under the BSP. Since visual segregation is often not possible, and the capacity to conduct genetic tests on the shipment in the importing country may be limited, it may be possible for exporters to claim LMO-free shipments in order to avoid the costs of notification. In the end, given that the objective of the BSP is to protect and conserve biodiversity, creating such an incentive would be counterproductive.

Conclusions

In the past, there was little or no effective regulation of international trade in knowledge-based agri-food products. With increasing concentration of production and the rise in international and intrasectoral trade, the need for new rules became obvious. The array of relevant rules has mushroomed, at first working to liberalize markets and more recently to limit trade. When consumer and citizen concerns could no longer be satisfied within domestic regulatory systems, many of those groups sought more effective international regulation. As the trade system had moved significantly towards an objective, science-based system of regulation, consumers sought new venues to influence, finding opportunities through the SPS talks at *Codex Alimentarius* and the environmental discussions through the BioSafety Protocol. Continued market access now is more uncertain than in previous years, with weaknesses in the domestic regulatory systems causing significant friction in various multilateral forums. The root of the current challenge lies in the fact that rapid technological and economic change has surpassed the absorptive capacity of most consumers and their regulatory agents.

Winners and Losers

V

The Theory of the Gains from Research   14

Peter W.B. Phillips

Introduction

Two key issues will determine the sustainability of a knowledge-based industry. First, there must be adequate returns to justify the investments and outlays in the research. Second, the returns must be distributed in such a way as to sustain the participation of all the necessary actors in the system: private companies, public research agencies, farmers and consumers. This section provides an examination of the methodologies for evaluating returns, summarizes the literature and offers new estimates of the impacts on the canola industry.

Methodologies

Economists estimate the returns to research in a somewhat different way than many others. The basic approach is to evaluate the costs and benefits of innovation in separate product markets over the effective life of a project, calculating an internal rate of return on the investment by discounting the stream of costs and benefits.

In the simplest case where there are competitive supply and demand markets, the gains to any innovation are the resulting increased consumer or producer surplus. Consumer surplus is the area on a graph bounded by the demand curve, the y-axis and the horizontal line through the market-clearing price. Producer surplus is the area on a graph bounded by the supply curve, the y-axis and the horizontal line through the market-clearing prices. In short, consumer surplus is the area below the demand curve, which represents the amount consumers would have been willing to pay to consume the good; producer surplus is the area above the supply curve, which represents the rent earned through lower cost production as prices rise.

Innovations to production technologies within existing product markets are assumed to increase productivity and therefore to shift the supply curve. Some innovations can reduce costs regardless of the volume produced, which would shift the supply curve down but parallel to the original supply curve (Fig. 14.1a). In the canola story, the introduction of higher-yielding public varieties represents this case. Other innovations reduce costs in proportion to the volume produced, which tends to rotate the supply curve down and to the right by a proportionate amount, with the old and new supply curves intersecting at the y-axis (Fig. 14.1b). The introduction of herbicide-tolerant canolas represents this case, as not all farmers will use the new technology; the more who do, the greater the pivot in the curve. Without any shift in the demand conditions, the shifts in the supply curve will both increase the equilibrium supply and lower the equilibrium price. The gains to the innovation in any period are the sum total of the change in the area bounded by the y-axis, and the two curves (area $b+c$ in both Fig. 14.1a and b). Both consumers and producers have the potential to gain some. Consumers generally gain because they consume more at a lower average price: their 'consumer surplus' rises by the area bounded by the y-axis between the original and new equilibrium prices and the demand curve (area $a+b$ in Fig. 14.1a and b). Producers can either gain or lose from an innovation. Producers gain surplus between the two supply curves and the new equilibrium price (area c on each figure) but lose the area bounded by the demand curve and the old and new market clearing prices (i.e. area a in each figure). Parallel shifts in the supply curve tend to yield higher returns to producers than pivotal shifts in the curve, because pivotal shifts lower the equilibrium price faster than parallel shifts (i.e. area $c-a$ in Fig. 14.1a is definitely positive, while area $c-a$ in Fig. 14.1b is negative).

Ultimately the share of the returns producers and consumers receive depends on the relative elasticities of demand and supply. Three discrete outcomes are possible. First, if the supply curve is flat, with constant returns to scale in the production of the product, all of the benefits of innovation will go to consumers (Fig. 14.2a). Second, if the demand were perfectly elastic (e.g. producers are price takers as in commodity markets) then all of the returns to innovation would go to producers (Fig. 14.2b). If, as is more normal, there are decreasing returns to scale and a negatively sloped demand curve (as in Fig. 14.1), then the benefits will be shared between producers and consumers.

For each period in time there is a supply–demand situation yielding returns. Against this, one must account for the investments and other outlays. Over the life of a project all of the investments and all of the returns can be tabulated and then discounted to a set point in time to calculate the internal rate of return for the project. If the discounted stream of benefits (the sum of the producer and consumer surpluses) exceeds the discounted cost of investments, the project is deemed to yield a positive internal rate of return.

Of course, the simple model seldom reflects reality. For a starter, the con-

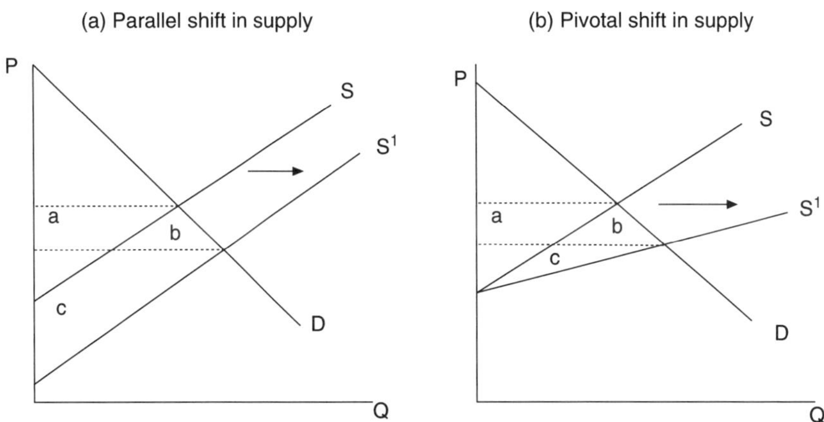

Fig. 14.1. Gains from research. (a) Parallel shift in supply; (b) pivotal shift in supply.

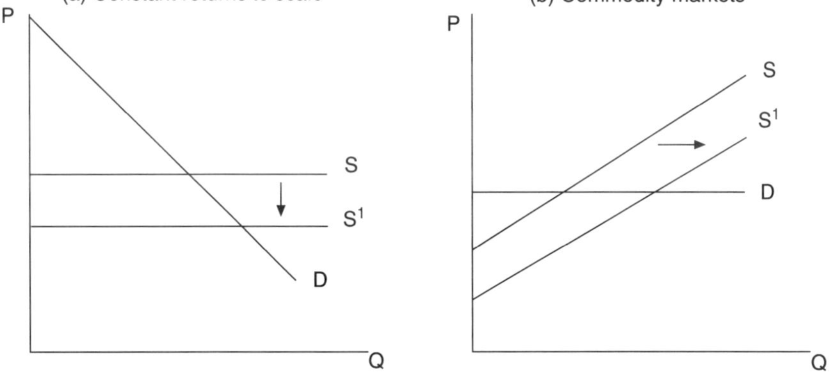

Fig. 14.2. Impact on distribution of gains to research of different supply and demand conditions. (a) Constant returns to scale; (b) commodity markets.

sumers often are scattered about the world, so if one were interested in the national returns of a project, it would be necessary to separate consumer benefits that accrue within the home country from those gained offshore. In the canola case, about 80% of Canada's production is exported, so much of the benefit flows abroad. A second twist on the consumer side is that many of the innovations change the nature of the good or service and do not simply produce the same good at a lower price. This will tend to shift or rotate the demand curve, with the potential for consumers to gain or lose from that change. The introduction of novel oil canolas represents this case. On the supply side, there are many potential differences from the simple model. First, the model assumes that production comes from a single production process with competitive market conditions. In practice, goods and services are produced within a supply chain, often with some perfectly competitive links and some with significant market power. Thus, not all producers within the chain will capture part of the producer surplus. High industrial concentration in the chemicals and processing sector relative to the farm level suggest that whatever gains accrue to the production system may not accrue to farmers. Second, imperfect competition within the supply market creates the potential for producers to act oligopolistically or monopolistically, extracting a portion of the consumer surplus by withholding production and enforcing a monopoly price, which both reduces the overall returns and shifts the returns from consumers to producers. Although the canola oil and meal markets do not exhibit many imperfections, the agrochemical, novel oil, nutraceutical and pharmaceutical markets certainly do.

Returns to Innovation: the Literature

A succession of studies since the late 1950s has estimated the costs and benefits of innovation in the agri-food industry. Alston *et al.* (1998) surveyed more than 294 studies in the agri-food sector and found that the 1821 estimates of returns ranged from −100% (e.g. the project yielded no benefits for the investment) to +724,000% (a suspect return). Ignoring the extreme results at both the top and bottom, they estimated that the mean internal rate of return was 72%. One way to examine the studies is to look at them over time.

One set of studies between 1958 and 1986 (Table 14.1) examined returns to research in a variety of developed and developing countries for a mix of products under conditions of perfect competition and a closed economy. They estimated that the internal rate of return (IRR) ranged from 20 to 95%. Ulrich *et*

Table 14.1. Recent estimates of the returns to research in the agri-food sector.

Authors	Sector	IRR
Griliches (1958)	Hybrid maize in USA	35–40%
Peterson (1967)	Poultry in USA	20–30%
Schmitz and Seckler (1970)	Mechanical harvesting of tomatoes	30%[a]
Ayer and Schuh (1972)	Cotton in Brazil	89%
Hertford *et al.* (1977)	Rice in Colombia	60–82%
Ulrich *et al.* (1986)	Barley research in Canada	16–75%
Unnevehr (1986)	Improved rice quality in Indonesia and the Philippines	29–61%

[a]Schmitz and Seckler estimated the gross and net social rate of return was in the range of 1000% but did not calculate the IRR. The estimate of 30% was supplied based on comparable studies which showed external rates of return of approximately 1000% yield IRR of about 30%.
IRR, internal rate of return.

al. (1986) further estimated, at least for barley in Canada, that public returns were generally higher than private returns due to the externalities of much of the research. Schmitz and Seckler (1970) added that welfare economics suggests compensation should have been paid to those affected negatively by innovations (in their case farm workers in the tomato sector). They concluded that high rates of return to research (in their case about 30%) should be more than adequate to compensate the losers and still yield an above-average return to farmers.

Beginning in 1975, many economists extended their analyses to examine the distribution of the benefits and costs among farmers and consumers, and in some cases between domestic and foreign actors (Table 14.2). Early studies of rice in Asia (Akino and Hayami, 1975; Hayami and Herdt, 1977; Scobie and Posada, 1978) estimated that yield-enhancing innovations do not benefit producers – instead consumers gain all of the benefits. In contrast, Nagy and Furtan (1978), Mullen *et al.* (1988) and Lemieux and Wohlgenant (1989) show that producers could gain from innovations that improve yield (canola, beef and pork, respectively) but that consumers still gain more than producers; Mullen *et al.* (1988) show that in the long run, consumers gain almost 90% of the return to the innovation. A succession of more recent studies of the US and Australia by Voon and Edwards show that innovations that improve the quality aspects of specific products (in their studies pigs and wheat) generally yield high returns, a greater share of which is captured by farmers than for yield-enhancing innovations. Alston and Mullen (1992) go further to investigate the international distribution of returns to yield-enhancing innovations in Australia, concluding that the Australian share of benefits was 58% from farm-level technology improvements and 27% from processing technology innovations.

Finally, a number of recent studies have looked at the role of market distortions and government interventions. Alston *et al.* (1988) concluded that net world benefits would be reduced if there were output quotas, increased if there were equalization schemes and unchanged with import taxes or export subsidies. Murphy *et al.* (1993) further estimated that if a country has domestic price supports and export subsidies, the returns to innovation could be negative if the effective export subsidy is increased. Freebairn *et al.* (1982) examined the gains from research in a multistage competitive market system, concluding that research benefited all participants in the system. In contrast, Alston and Scobie (1983) show that producers could lose from innovations that effectively increase the role of non-farm inputs. Most recently, Alston *et al.* (1997) estimated that the greater the degree of imperfect competition in an industry, the more producers would lose. Their analysis of the Taiwanese tomato sector showed that as the degree of imperfect competition in the processing industry rose, farmers' share of the benefits of innovations dropped from almost 95% to only 32%, with the processors capturing almost all the farmers' losses. Their analysis of the US beef industry showed that at the extreme, as meat processing moved to a monopoly, farmers could become net losers.

Table 14.2. Distribution of the benefits of research.

Authors	Sector	IRR	Distribution of costs and benefits
Akino and Hayami (1975)	Rice in Japan	75%	Producers = net losers; consumers = 100%
Hayami and Herdt (1977)	Rice in Philippines	na	Producers = net loss to 33% of gain; consumers = 67–100%
Scobie and Posada (1978)	Irrigated rice in Colombia	79–96%	Producers = net losers; consumers = 100%; poorest 50% of consumers captured 70% of benefits
Nagy and Furtan (1978)	Rapeseed in Canada	101%	Producers = 47%; consumers = 53%
Mullen *et al.* (1988)	Substitution of inputs in US beef production	na	Short run: producer = 57–72%; consumer = 28–43%; long run: producer = 9%; consumer = 88%
Lemieux and Wohlgenant (1989)	PST hormone in US pork	na	Producers = 22–27%; consumers = 73–78%
Voon and Edwards (1992)	10% reduction in backfat in Australian pigs	na	Farmers: 67–80% (down to 58% with perfect elasticity of substitution between pigs and marketers); marketers: 0–13%; consumers: 16–20% (up to 42% with perfect sub)
Voon (1991)	1% reduction in PSE syndrome in pork in Australia	na	Producers: gain 85%; consumers: gain 15%
Voon and Edwards (1992)	Protein in Australian wheat		Producers: 90%; consumers: 10%
Alston and Mullen (1992)	1% cost saving for wool in Australia	na	Topmaking: Australia = 24%; ROW = 86% Textile processing: Australia = 27%; ROW = 73% Farm-level savings: Australia = 40–58%; ROW = 42–60%

IRR, internal rate of return; na, not applicable; PST, porcine somatotropin; PSE, pale, soft and exudative pork; ROW, rest of world.

In summary, the studies show that innovations in agriculture have high private returns and even higher public returns, but that farmers get a smaller share of the returns on innovations that improve yield rather than quality and when the related processing sector is imperfectly competitive. Surprisingly, given the variety of products, mix of countries, and differing assumptions about market conditions, the results have been fairly consistent.

Returns to Canola Research

When the economic theory and literature is juxtaposed with the rise of private investment (often for biotechnology-based effort) in the canola sector, a number of tentative hypotheses can be drawn and tested.

First, gains for research, estimated to yield an internal rate of return between 20 and 95%, may actually be larger for specific biotechnology-based developments because of the reduced cost of the research and the increased array of attributes that can be bred into the seed, which add new value to consumers.

Second, the increase in private funding has the potential to lower the overall internal rate of return, first, because there are likely localized decreasing returns to scale within the industry and, second, because much of the investment will be competitive and, with intellectual property rights, only one of the efforts will yield a commercially exploitable innovation.

Third, given that past studies have shown that the gains from yield-enhancing innovations tend to be captured by consumers, one can tentatively conclude that the development of HT varieties and other agronomic-related research will probably not benefit producers over the long term. The presence of intellectual property rights, however, may allow the breeding institutions to capture a portion of the surplus that historically has been captured by consumers.

Fourth, the conclusion of some studies that quality-enhancing innovations benefit farmers somewhat more than yield improvements suggests that the novel attributes being bred into canola could yield a higher return to farmers, partly due to the new value being created through market segmentation and partly because high non-separability will make farmers more valued partners in the production system than previously.

Finally, many recent studies suggest that producers and consumers should be concerned about the imperfectly competitive state of the agri-food chain. For canola, the four-company concentration ratio is 57% in the canola seed industry (Table 8.1), 65% in the agrochemicals industry (Just and Hueth, 1993) and 100% for the western Canadian processing sector (Wensley, 1996). Furthermore, the end-users of many of the novel oil products are monopsonists (i.e. monopoly buyers). Taken together, there would appear to be significant potential for actors in the supply chain to reduce both farmers' and consumers' share of the returns from innovation. Nevertheless, Green (1997) suggests that the innovative nature of the industry will limit the market power of many of the actors. The short breeding cycle and the short life span of each new variety (approximately 3 years) may require the holders of intellectual property rights to share a larger percentage than otherwise.

Chapter 15 examines the global industry and calculates the total internal rate of return to research and development, estimating that the IRR has dropped from more than 40% in the 1970s, when the industry was dominated by public breeding programmes, to under 10% in the 1990s, when the industry was dominated by private firms. Chapter 16 looks at the distribution of the returns within the supply chain and between producers and consumers.

The Aggregate Gains from Research

15

Stavroula T. Malla, Richard S. Gray and Peter W.B. Phillips

Introduction

Past assessments have shown a high rate of return for canola research. Government, and particularly private industry, have continued to increase research expenditure with the hope of capturing future dividends from oilseed research. Nevertheless, as discussed in Chapter 6, the funding of canola research in Canada has undergone many changes since its inception in the mid 1950s, when Agriculture Canada began a programme to improve rapeseed-processing methods. Over time, research has shifted from a modest public research programme to a large research industry dominated by private-sector participation. In 1970, 83% of research spending was public investment. By 1997 the private sector's share had grown to 80% of the total (Canola Research Survey, 1997). This funding shift is evident in the registration of new varieties. Prior to 1973 all varieties were public, while in the 1990–1998 period 86% of the varieties were private (Chapter 6). This large shift in emphasis from public to private research is due to the large increase in private-sector investment rather than a reduction in public research.

The change in the private funding of research has coincided with a change in the ownership of the property rights for the research and, implicitly, who benefits from the resulting returns to the investment. In 1987, virtually all of the canola varieties were open-pollinated and non-transgenic, and there were no Plant Breeder's Rights until 1990. This meant that virtually all of the acreage was grown without a production agreement, giving producers the right to retain seeds for future use and to sell non-registered seed to their neighbours. In contrast, by 1999, almost 70% of the acreage was planted to herbicide-tolerant (HT) varieties, with producers required to sign a technology agreement or to purchase a specific herbicide–seed package. An estimated 30% of the acreage was seeded to hybrid varieties and much of the remaining acreage was seeded to varieties with Plant Breeder's Rights. These changes have put plant breeders in a far better position to capture value from genetic innovation.

Previous research has shown very high rates of return for canola research. An evaluation of public investment in canola research and development (R&D) was first published in 1978 by Nagy and Furtan. For the period 1960–1974 they calculated the internal rate of return (IRR) from improved yield research to be 101%. Ulrich *et al.* (1984) updated the estimates of IRR in canola research for period 1951 to 1982 and calculated the IRR from research into improved yields to be 51%. Ulrich and Furtan (1985) incorporated trade effects and found the estimated Canadian IRR from higher-yielding varieties to be 50%. Despite the dramatic changes in the industry since 1982, there has not been a more recent comprehensive analysis.

There have also been some recent advances in the estimation of the returns to research that have not previously been applied to canola. While many studies used econometrics to examine the effect of R&D investment on agricultural productivity (e.g. Thirtle and Bottomley, 1988; Pardey and Craig, 1989; Huffman and Evenson, 1989, 1992, 1993; Leiby and Adams, 1991; Chavas and Cox, 1992; Alston and Carter, 1994; Evenson, 1996), those studies imposed an assumed shape and length of adoption lag to calculate the returns. Some of the more recent econometric studies have instead estimated statistically the shape and length of adoption lag and generally have found lower rates of return (e.g. Akgüngör *et al.*, 1996; Makki *et al.*, 1996). Alston *et al.* (1998) also dealt explicitly with the concept of knowledge depreciation, which is not common in the agricultural R&D literature. These new approaches have relevance for estimating the IRR for canola research.

Given the dramatic changes that have recently occurred in the canola industry, there is a need to re-examine the returns to research in the sector. In particular, the entrance of private industry, the change in property rights, the introduction of biotechnologies, and the changed role of the public institutions have influenced the benefits created. This chapter provides new estimates of the returns to yield-increasing canola research over time. In addition, recent information suggests there is significant potential for both positive and negative externalities from the adoption of new technologies. The chapter concludes with a short discussion of the rigour of the estimates, both in the context of new theoretical approaches being examined, and in terms of the potential for misestimation due to spillovers or unmeasured benefits and costs.

An Economic Model to Examine the Effect to Property Rights and the Return to Research

The establishment of enforceable property rights for products of genetic crop research has significant implications for the amount of research that the private sector will provide. In the absence of enforceable property rights, many of the products of research can be copied or reproduced. While all firms that use the research output may benefit, without property rights there is no way for the market to fully remunerate any firm for doing research. This creates a 'public-good' market failure, resulting in underinvestment in research activities. As shown in Fig. 15.1, in the absence of complete property rights the private marginal benefit (MBp) that can be captured from the marketplace is less than the public or social benefits (MBs) of the research. A private research firm will equate the marginal cost (MC) of doing research with the private demand (or private marginal benefit) for the research and produce a quantity of research Qp. At this amount of research the social marginal benefit of research is far greater than the marginal cost of doing research. In this case, the marketplace fails to produce the socially optimal amount of research, Qs, where the marginal cost of research is equal marginal social benefit. If the government provides a quantity of research $Qg-Qp$, this research creates a social benefit equal to the additional area under the social benefit curve while incurring costs equal to the much smaller area under the marginal cost curve. In this instance there is a high rate of return to public research, which has been found in many empirical studies. This illustration may characterize the situation in canola research until the mid 1980s before the private sector played a major role.

Government has also addressed the incomplete set of property rights for research goods by providing assistance to private firms doing research. As discussed in Chapter 7, this assistance has come in many forms. Research tax credits have been used in many countries and in many sectors to stimulate research. Recently, grants have been provided to match private expenditures on research. Infrastructure has also been provided at a reduced cost in many jurisdictions.

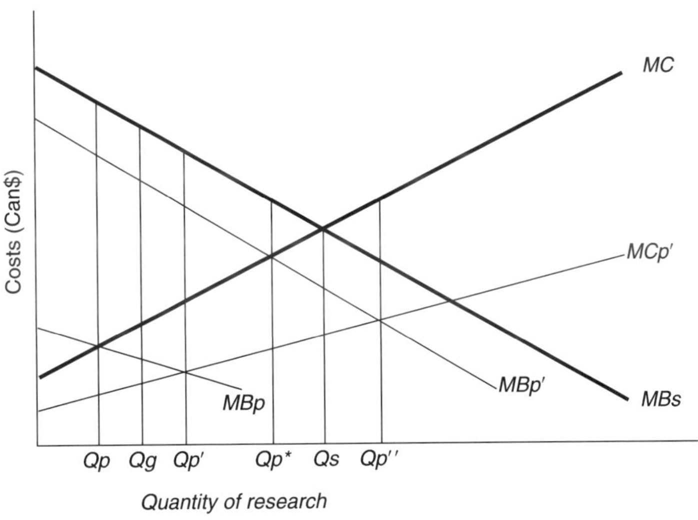

Fig. 15.1. The marginal private and social costs and benefits of research.

Indirect support from the public sector has also come in the form of the public education of research scientists and the free provision of the output of public research. This public assistance to private research has lowered the private cost of doing research, allowing the private sector to do more research. In Canada the public sector has always provided some direct support for private-sector research on canola. Investment tax credits and infrastructure grants have existed for some time. Since 1995 the government has also offered matching investment initiatives (MII), which match private research expenditure in approved projects. This has allowed the private sector to play a greater role in research provision. In Fig. 15.1 this is equivalent to lowering the marginal cost of private research to MCp', which, in the absence of complete property rights, moves the private investment toward the socially optimal level Qs to a level of Qp'.

Recently, governments, and to some extent the private sector, have addressed the 'public-good' market failure in research by establishing effective property rights over the products of research. As outlined by Malla *et al.* (1998), the assignment of intellectual property rights provided some of the added ability to capture value from research. Although the Seeds Act has always protected the name of registered varieties, it wasn't until the adoption of Plant Breeder's Rights in Canada that breeders could forbid the sale of registered varieties without royalty payments. This assignment followed a number of milestones, including the US Patent Office decision of 1985 to grant patents for whole plants. As discussed, many of the seeds produced during the 1990s had very specific attributes. Herbicide-tolerant canola requires the use of a specific herbicide in order to be useful. Similarly, canola with particular oil characteristics needed specialized processing and marketing in order to be viable. The development of hybrid varieties has given private firms a greater ability to capture value from their genetic material. The first hybrid variety was introduced in 1989. Although often protected with Plant Breeder's Rights and production contracts, hybrids do not require the enforcement of contracts to maintain control over the use of the genetics.

The establishment of enforceable property rights has the effect of moving the private marginal benefit (market demand) curve toward the social demand curve. As was discussed earlier, this has had the effect of increasing the demand for private research and the amount of private research provided by the private sector, thus partially addressing the market failure. In the absence of government support for research this moves the private investment from Qp to Qp^* in Fig. 15.1. The establishment of property rights changes the optimal role for government. If government provides support for private research, once property rights are established this further increases the private incentive to do research. If the property rights are nearly complete and public-research support is significant, then the private sector can provide more research than is socially desirable, as represented by point Qp'' in Fig. 15.1. Thus, both correcting the public-good failure and subsidizing research can result in excessive research, as noted in Chapter 8.

A few other points apparent in Fig. 15.1 are worth noting. The highest benefit-to-cost ratio, which will generate the highest IRR, will be at some level of research less than the social optimum. The socially optimal quantity of research occurs where the marginal social benefit is equal to the marginal social cost. At this point total net benefits of research arc maximized. Additional research beyond this point is socially wasteful, costing more on the margin than what is produced. At these excessive levels of research (anything greater than Q_s in Fig. 15.1) the total benefits can still be greater than total costs, and the IRR can still be above market rates. Importantly, a positive overall return to research, or an IRR greater than market rates, does not imply that more research is socially desirable – rather, it suggests that the research programme taken as whole has produced net benefits.

This simple economic model presented in Fig. 15.1 illustrates several important concepts for research policy. The first is that in the absence of enforceable property rights, the private sector will under-invest in research, creating a role for government to address the research shortage. Second, the assignment of property rights to research products can increase the amount of private investment toward the socially optimal amount. Third, if enforceable property rights have been established, the subsidization of private research could lead to socially wasteful overinvestment in research. Finally, neither an assessment of total research benefits nor of the rate of return on total investment are good indicators that on the margin more research is socially desirable.

Estimating the Returns to Agricultural Research

By its very nature, investment requires a commitment of resources at a point in time, with the benefits flowing in future time periods. This dynamic nature of investment precludes the simple comparison of benefits and costs. Specifically the time value of money must be considered in the appraisal of an investment. Fortunately methodologies for valuing investments are well developed and can be used to appraise the investment in canola research. This section contains a brief description of the conceptual framework used to estimate the returns to canola research.

The process of creating new crop varieties can be described in four phases, as shown in Fig. 15.2. During the first phase, or the research phase, research resources are spent to develop a crop variety that has commercially desirable characteristics. This production process depends very much on the stocks of human capital, knowledge and germplasm as inputs into the creation of a new variety. The attribution of the cost of creating these important stocks is difficult. As a result, the creation of these stocks are often considered to be sunk costs, independent of the particular research programme. The whole study of research spillovers would be important if these costs were to be attributed. At the end of the research phase a new variety is created.

There are many years between the research expenditure to develop a new

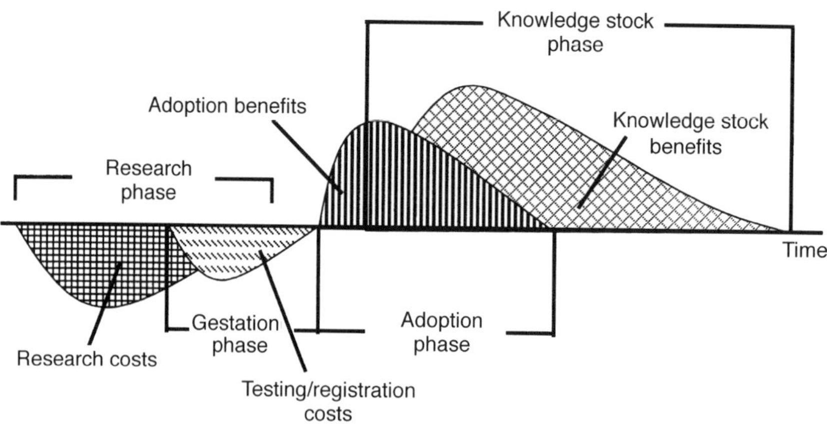

Fig. 15.2. Four phases of crop development and the path for R&D cost and benefit (adapted from Alston *et al.*, 1997).

variety and the variety reaching any end user. Research itself takes a number of years to produce any tangible product. Even after a variety with potential has been created it must be tested both internally and by external regulatory agencies before it can be licensed for sale. This period of waiting is referred to as the 'gestation lag of research', and is defined as the number of years between making the investment and generating new technology or useful knowledge. In practice, the gestation lag is difficult to estimate, because the expenditure to create a new variety is often spread over many years. For instance, a new variety released in year T may have involved research and development expenditure in years $T–2$ to $T–8$. To get around this problem of multiple gestation lags, most studies have estimated a single gestation lag, which represents the lag between the weighted mean time of expenditure and the commercialization of a licensed variety.

The third relevant period for estimating the returns to research is the adoption phase. During this phase the new variety is adopted and then replaced by other varieties. The typical pattern is low adoption in the first year of introduction, growing to peak adoption in 2 or more years, then slowly being replaced by other, newer varieties. In terms of economic impact the variety has it largest annual impact in the year when the adoption rate reaches it peak.

The final research stage is the depreciation phase. Research often creates a new process or new germplasm. These innovations provide a very important base on to which subsequent research is built. Thus, innovations in the form of new varieties contribute to the stock of knowledge or germplasm, which continue to play a role long after the particular innovation has been supplanted by newer innovations. For instance, the first semi-dwarf wheat varieties are no longer used but some of the germplasm from these varieties continues to be in many of the varieties grown today. Although durable, the contribution to the stock of germplasm is not permanent and depreciates over time. One of the common reasons cited for depreciation is that pests in the environment eventually adapt themselves to attack a particular germplasm and new germplasm is required.

The expenditure and benefits for a typical variety through the four phases of research impacts are depicted in Fig. 15.2. The period of research is when the investment is made and typically only involves costs. The benefits reflect the adoption curve after a period of gestation and represent the current benefits from growing the new variety. Finally, the cross-hatched area to the right of the adoption benefits represents the benefits from the contribution to knowledge stock which depreciates over time. These four phases of crop variety development, and the long lags between investment and output, have made the estimation of the returns to research difficult and a subject of considerable discussion.

As discussed in Chapter 14, the estimation of net economic benefits has a relatively long history. The main outcome of almost all of the studies was that the net social rate of returns was high and hence there was evidence of under-investment. In Canada, Nagy and Furtan (1977) first performed the evaluation of public investment on R&D in canola research for the period 1960–1974. They calculated the IRR from improved yield research to be 101%. In their analysis they used a 4- to 5-year lag between the initial investment and the first commercial effect of this expenditure. Furtan and Nagy assumed a pivotal shift in the supply curve (as in Fig. 14.1b). For the calculation of future benefits the annual future returns were set at the 1972 level until the year 1995, with no depreciation but with an annual maintenance cost of 35% of the 1974 expenditure charged until the year 1995. Furthermore, their estimates of cost excluded all research not directly and immediately related to the development of the specific new varieties. As a result, all of the academic and much of the competitive but unsuccessful applied effort was not included.

Ulrich *et al.* (1984) updated the estimates of IRR in canola research for the period 1951–1982. Specifically, they calculated the IRR from improved yield research to be 51%. They assumed a 10-year lag between research expenditure and research benefits. They also revised the cost calculations, using Zentner's (1982) estimates of the annual cost of a professional person-year of research. On the benefit side, they assumed a proportional divergent shift of the supply curve due to technical change, used the same supply and demand elasticities as Nagy and Furtan (1977) and employed a modified formula to estimate the annual producer and consumer surplus. They further assumed that the annual future returns to be the average of 1980, 1981 and 1982 level until the year 2002. Meanwhile, the costs after 1971 were assumed to be 35% of the average cost of 1969, 1970 and 1971 until the year 2002.

Finally, Ulrich and Furtan (1985) estimated the Canadian IRR from higher yield varieties equal to 50%, and the total IRR from higher yielding varieties equal to 51% for a period 1951 to 1983. They used the same research cost estimates for professional person-years as in the 1984 study and the same lag (10 years) between research investment and benefits. On the benefit side, they assumed a parallel shift of supply curve due to technical change and they modified Rose's (1980) formula for estimating the annual producer and consumer surplus. They also assumed that the supply and demand elasticities of canola are similar to those of wheat and barley, and used cited elasticities of wheat and barley. Finally, the annual future returns from 1984 to 2003 were set equal to the average of 1981, 1982 and 1983 levels. This was the most recent estimate of the IRR of canola research in Canada.

Estimating the Relationship between Expenditure and Yield Increase

The rest of the chapter examines the gains to research in the canola sector since the beginning. The empirical procedure begins estimating the relationship between research expenditures and yield increases. So far almost all of the research produced since the 1980s has, in one way or another, improved the yield. This and the next chapter will also discuss the implications of quality-trait improvements.

The relationship between research expenditures and yield increases is estimated through regression. The first step was to construct a yield index of different canola varieties to the same base variety (Torch = 100). The annual yield index was created from an average of the yield index for varieties grown each year, weighted by the seeded acreage. The data, on the relative yield of different varieties, were obtained from various issues of Saskatchewan Agriculture and Food, *Varieties of Grain Crops in Saskatchewan*. These data are based on the research station yield trials at a number of locations across Saskatchewan, which were designed to measure varietal performance due to genetic causes. In short, the data hold constant other agronomic factors, such as quality of land, agronomic practices, variable use of herbicides and fertilizers and weather. The data on the percentage acreage of each canola variety were obtained from three sources: the 1978 study by Nagy and Furtan, which covered the period 1960 –1976; the Prairie Pools Inc., *Prairie Grain Variety Survey* surveys for 1977 to 1992; and Manitoba Crop Insurance Corporation's varietal survey data.

In addition to the genetic stock there are two other factors that influence the yield index of the varieties grown. First, there are two types of rapeseed/canola grown in Canada: Argentine species (*Brassica napus* L.) and Polish species (*Brassica rapa* L.). Argentine varieties are higher yielding than Polish varieties (15–20%) while Polish varieties mature faster and have a lower erucic acid content. The area grown to each variety varies from year to year, with more Polish varieties seeded when spring is late. In order to capture the effect of planting Argentine versus Polish varieties, a variable indicating the area seeded to Argentine varieties is included in the regression. The other yield factor that needs to be accounted for is the switch from rapeseed to canola varieties. The selection for low erucic acid and glucosinolates in canola quality was attained at the expense of seed yield. From 1978 to 1984, while the changeover was under way, the annual weighed average yield fell about 9%. This result is similar to other findings, which have shown that the combined yield of Argentine- and Polish-type canola varieties decreased from the middle 1970s to the beginning of the 1980s (e.g. Forhan, 1993; Malla, 1997). To account for the yield effect of the conversion of rapeseed to canola, a variable was created that represents the percentage of total rapeseed/canola varieties seeded that were canola varieties. This variable is also included in the regression.

The total research expenditure per year was calculated by multiplying the total person-years invested per year in the research by the total variable and fixed research cost (Canola Research Survey, 1997–1998), as reported in Chapter 6. A person-year is used to define both a professional person-year and a technical person-year, where a professional person-year corresponds to full-time annual work dedicated to professional research, and a technical person-year corresponds to full-time annual work on technical research (as reported in the Inventory of Canadian Agri-Food Research). The data on canola research professional and technical person-years were obtained from five sources: Canola Research Survey 1997–1998; Nagy and Furtan (1977); ISI (Institute for Scientific Investigation); ICAR (Inventory of Canadian Agri-Food Research); and Phillips (1997). Where there were discrepancies in the overlapping periods from the data source, the earlier estimates were indexed and adjusted (usually upward) to reflect later estimates.

An average adoption curve for canola varieties was estimated rather than assuming a specific adoption lag structure. The individual adoption rate of each rapeseed/canola variety was calculated by dividing the acreage sown of each variety by the maximum acreage sown of that variety for each year after the year of introduction. These adoption rates were then averaged for all varieties and weighted to sum to one. The weighted average adoption rate of rape-seed/canola varieties was applied to the cost data to create a variable of the weighted lag research expenditure. The adoption curve means that on average the acreage planted today to specific varieties is a function of when in the past the varieties were introduced. The annual weighted yield index is therefore an average of the yield of varieties previously introduced weighted by the respec-tive coefficients on the adoption curve. Given this relationship it follows that the annual change in the weighted yield index is a weighted average of the changes in the yield of the varieties introduced. If the change in new variety yield is pro-portional to the lagged expenditures on research, then the change in the annual weighted yield index is proportional to the adoption curve weighted yield expen-ditures (see Gray *et al.*, 1999, for mathematical presentation and regression results).

A regression was specified and estimated to calculate the effect of a research expenditure on the average yield index. The change in the annual weighted average yield index was fitted with the annual adoption-lag-weighted research expenditure (lagged by the number of years between making an investment and its effect on yield), the percentage of the total canola/rapeseed varieties that are canola grade (a value between 0 and 1) and the percentage of the total canola/rapeseed varieties that are Argentine (*B. napus*) varieties (which takes a value between 0 and 1). The regression showed that a complete switch from Polish to Argentine varieties would increase yield by 17.86 index points, while the complete switch in the early 1980s from rapeseed to canola varieties reduced the average yield by 17.09 index points. These large effects have implications for the value of non-yield traits in canola. Furthermore, holding all other variables constant, the regression suggests that a 1% increase in the annual lag-weighted research expenditure in year $T-4$ increases the yield index level by 0.00425 index points. Given that the yield index was 127 in 1997, a 1% increase in the annual lag-weighted research expenditure in year $(T-4)$ increases, on average, the yield index by approximately 0.0033% at 1997 yields.

Net Returns to Canola Research

To calculate the social return from the yield-increasing research, the econometric estimates of the yield increase due to research expenditure were applied to the historical production of canola. The regression was used to predict the amount of yield increase due to research expenditure in each year. As an approximation, it is assumed that the additional yield due to genetic improvement came at no resource cost and thus benefits are in direct proportion to revenue each year. Benefits in 1997 dollars are estimated by multiplying the quantity of canola seed (production) by the price of canola seed, and deflating by the consumer price index.

The present value of research benefits are estimated by first calculating all future yield increases due to the yield increases in a particular year. This uses the notion that there is a stock of knowledge that is subsequently built upon. This calculation was made using various rates of depreciation. These future yield increases are then applied to the revenue in each future year to calculate the future benefits. For 1997 and beyond it was assumed that 1997 revenue would continue indefinitely. Once the future stream of benefits is calculated, the figures are then brought back to the present value in the year of introduction, using a discount rate. The present value of costs for varieties grown in year T is calculated from the present value in year T, of the weighted expenditures lagged by the gestation period and the adoption curve. The net present value of research is calculated from the difference between the present value of the benefits and the cost of the research using a number of depreciation and real discount rates. The results over time are shown in Fig. 15.3.

What is most striking from the analysis is that, regardless of the discount rates and depreciation rates used, the net present value (NPV) peaks in the 1980s and declines thereafter. For instance, with 5% depreciation and a 6% real discount rate, the increase in the research expenditures resulted in an increase

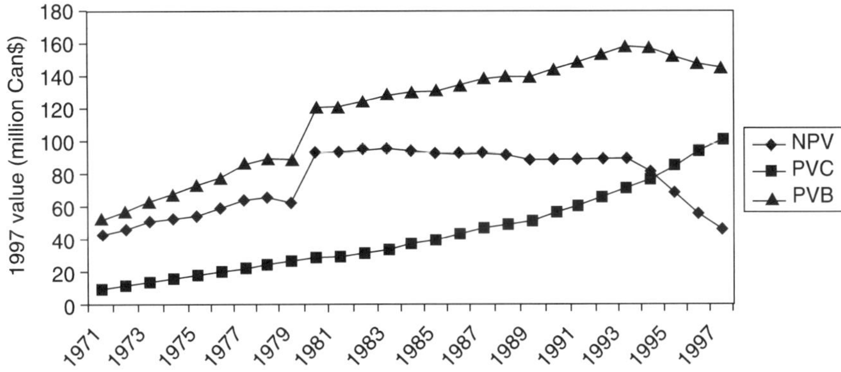

Fig. 15.3. Present value of benefits and costs of yield-increasing canola research 1970–1997. NPV, net present value; PVC, present value of costs; PVB, present value of benefits.

in net present value from Can\$41 million in 1971 to a peak of Can\$88 million in 1983, which then began to decline as the increase in expenditures exceeded the growth in benefits. By 1997 the net present value of yield increases had declined to Can\$25 million (all real 1997 dollars).

The model was also estimated under different assumptions about depreciation rates and lag structures. The internal rate of return was calculated using a 4-year gestation lag and depreciation rates of 0%, 1%, 5% and 10%, respectively. Not surprisingly, the IRR declined as the depreciation rate increased because the investment is not durable. The rate was also calculated with a 1-year gestation period. As expected, this increases the internal rate of return, confirming that the rate of return is sensitive to assumptions about the gestation lag. What is most striking about the results is that the return declines in each case as the level of investment has increased. While the internal rate of return was clearly excessive in the early 1970s, it declined to or below market levels by the mid 1990s.

Given that many biotechnologies became predominant during the 1980s, the declining NPV and IRR provide little support to the notion that biotechnology has led to significant increases in the returns to research. However, in 1997 about 35% of area was sown to herbicide-tolerant varieties and thus the research may have produced other benefits not measured as a yield increase. Yet, given the very recent introduction of herbicide-tolerant varieties, this phenomenon does not explain the decline in the NPV prior to 1996.

The increase in the level of expenditures and the declining internal rate of return approaching market rates suggest that the assignment of property rights and matching grants has largely corrected the public-good market failure. Further extrapolation of these rates of return would suggest that overinvestment in the sector may be causing low private and social internal rates of return for investment. The decline in the net present value of research as expenditure

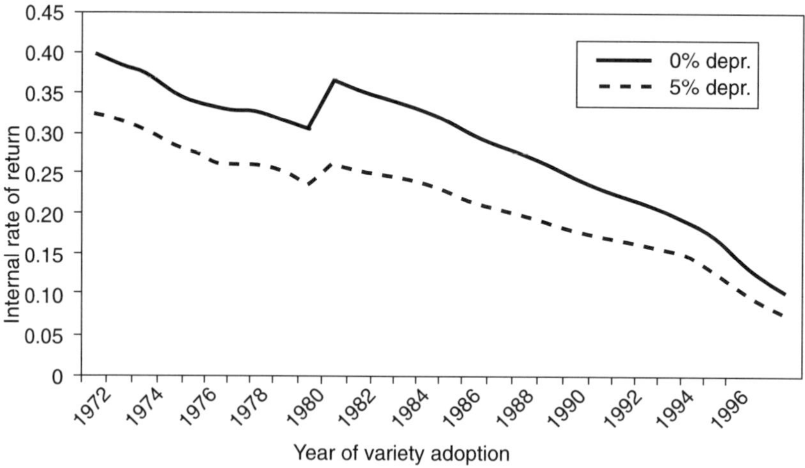

Fig. 15.4. Estimated IRR for canola yield-increasing research (1967–1993).

increases is consistent with moving beyond the optimal amount of research (Q_S in Fig. 15.1).

Figure 15.4 shows the internal rate of return with 1% and 5% depreciation rates. The rate drops by 1997 to 10% or below in each scenario examined. One interesting feature of these series is the increase in the rate of return in the late 1970s when canola acreage surged in response to a producer shift into canola precipitated by poor wheat markets. Note however, both acreage and revenue during the mid 1990s was near record levels, and despite this, the rate of return is low.

Extending the Analysis

While the above analysis is state-of-the-art, there are a number of extensions that should be possible with better data and methods.

First, with the built-in lags between research and production, it is possible that part of the tail-down in returns is the result of a build-up of research capacity that has yet to yield new products. Nevertheless, this build-up began in the 1980s and the first wave of products entered the market in 1995, yet there was a continued slide in the rate of return. As competitive private programmes proliferate, there is significant potential for wasted effort and correspondingly lower returns. Furthermore, this analysis has not included the discovery and commercialization costs of isolating the patented gene constructs which are being inserted into canola to make herbicide-tolerant plants. This was a major cost to all of the companies (estimated to be in the range of US$10–15 million per gene) and could legitimately be partially booked against canola. Currently there are four herbicide-tolerant canola varieties. The companies that isolated the active genes report that in many cases they developed and patented two or more genes that would do the same function. So there is somewhere between US$60 million and US$100 million invested in genes of direct impact on canola production today. As of 1999, 72% of the global transgenic acreage was for herbicide-tolerant plants (e.g. maize, cotton, soybeans and canola) and canola accounted for approximately 1 in 7 of those acres (James, 1999). So one could reasonably assign approximately 15% of the cost of developing the genes to the canola research effort, which would further depress the rate of return. By the same token, the costs of developing and improving the *Agrobacterium* transformation system could be partially assigned to canola (as it is the preferred technique for genetic transformation in this species). As with any study, it is hard to determine where to stop.

Second, the analysis does not capture all the potential spillovers. Although the above analysis uses constructed cost estimates based on global research effort, the estimated returns are based on production in Canada, which represents only approximately 20% of the global production of rapeseed or canola. This is justified on a number of grounds. First, approximately half of the global research is happening in Canada (Table 9.1). More importantly, all of the multinational companies surveyed in 1998, regardless of whether they did their research in Canada or elsewhere, reported that their primary market for new canola innovations was Canada. Furthermore, yield gains in the rest of the world are to a great extent based on government policies encouraging intensive practices (e.g. in the EU, China and India) rather than transfer of research-based germplasm. In 1998 seed for deliberate release is estimated to have been exported from Canada to only five regions (the US, Chile, Australia, the EU and South Africa), and most of that seed was used for contraseason multiplication for Canada rather than for local production. Upon further examination of the seed variety registration records for Canada, the US, Australia, India, China, Germany, the UK and Sweden, there was little evidence of rapid cross-border transfer of new varieties. India and China, which together represent approximately 47% of global canola production, have not imported any new varieties since the 1960s and 1970s, when some germplasm was transferred. Australia and Germany, representing 10% of global production, had only a few Canadian varieties recorded as parents of registered varieties or as direct imports. The UK, Sweden and the US each reported significantly more import of Canadian germplasm, but they represent only about 6% of global production (by acreage). Nevertheless, there is potential that a better estimate of the spillovers to the EU, the US and Australia could raise the rate of return estimates.

Third, the modified traditional approach to estimating the gains to research tends to ignore unquantifiable externalities. Those will be discussed in more detail in Chapter 16.

Finally, Moschini and Lapan (1997) and Moschini *et al.* (1999) argue that given the existence of IPRs, it is important to estimate the monopoly rents to the innovators. In their 1999 study, they calculate that the adoption of Round-Up

Ready™ soybeans as of 1998 yielded approximately US$480 million of incremental social welfare to the US, of which US$355 million, or 75%, is captured by monopolists. They further estimate that the global return was approximately US$790 million, of which 45% was monopolists profit. This might hold for the soybean sector, but is unlikely to hold for canola. In 1991, before the introduction of HT canola, more than 90% of the acreage seeded to canola in western Canada used herbicides (Statistics Canada). Just and Hueth (1993) estimate that in 1994 the four-company concentration ratio for the agrochemical industry was 65%, implying significant market power in the hands of the main companies. For the most part, the introduction of new herbicide-tolerant canola crops has simply replaced one set of chemicals with another set, both of which were priced oligopolistically. Hence, there is unlikely to be any net monopoly rent created through these innovations that is not already counted in the above analysis. Chapter 16 will discuss the implications of this approach further on the distribution of the returns.

Conclusions

Many changes took place in the industry during 1960–1997. A small but very successful public research programme eventually became dominated by a large influx of private research investment, encouraged by property rights and technologies that provided a greater opportunity to capture the benefits from research. During the private growth period, the technologies used for genetic improvement shifted from traditional breeding to the use of many biotechnologies.

The rate of return from canola research has been on the decline throughout the study period. Specifically, the internal rate of return declined from a high rate in the 1960s and 1970s to a much lower rate in the 1990s. Moreover, the total net present value of yield-increasing research peaked some time during the early 1980s and subsequently declined – suggesting a possble overinvestment in research. This result indicates that the increase in private research and development efforts did not actually yield as much net benefit as one would expect when witnessing a large amount of private investment flowing into an otherwise public-funded research area. Therefore, further investment in canola research and development may not be as profitable a venture as the investment stampede would lead us to believe.

This study challenges the current government policy in canola research. The canola research industry is heavily subsidized and property rights for canola seed are well established. Given that property rights allow private firms to capture most of the social benefit of investment, capital is attracted, driving down the rate of return toward normal levels. If government also subsidizes the costs of private research, then it is certainly possible to create overinvestment in an industry. The present analysis shows declining net present value of investment. The industry might actually already be operating beyond the point where the marginal social benefit is equal to the marginal social cost. Hence, this study indicates a need for a much closer examination of policy in this industry.

The general result that the new biotechnologies have yet to produce measurable high social returns in the canola sector raises some very important questions. Clearly, genetic traits other than yield have economic value and, if incorporated into the analysis, could change the general conclusion. If the net present value has fallen, does the age of a crop, to a large extent, dictate the rate of return to research? Is there a natural cycle to crop development, which has an increasing and then a decreasing return to research investment? If this is true, then should public investment be targeted to crops on the basis of age rather than historic rates of return? Answering these questions could provide important insight into the best policies to govern the rapidly expanding biotechnology industry.

This research also begs the question of what has been happening to the rate of the return in other crops and in other sectors. Of particular importance is whether the falling rate of return in canola is the result of the assignment of IPRs, which has become general to all crops, or do market failures continue to exist in other crops where hybrids or other physical reproductive barriers do not exist?

Distributing the Gains: Producers, Consumers and Others

16

Peter W.B. Phillips, Murray E. Fulton,
Lynette Keyowski, Stavroula T. Malla and
Richard S. Gray

Introduction

The ultimate challenge of any economic and policy analysis is to determine *qui bono* (who benefits). One can gain significant insight into an industry and its evolution by looking not only at the changes and the aggregate economic impact, but also by looking at the winners and losers. The canola industry is no different. Although significant public and private funds have been attracted into both research and corporate development, Chapter 15 demonstrated that the aggregate returns in the late 1990s may not be adequate to justify the accelerated level of activity. Perhaps more crucial from the perspective of medium- to long-run sustainability, it is important that each of the key actors in the system – research companies, farmers, processors, traders and, last but not least, consumers – be no worse off than without the activity. If one or more of them are adversely affected, they may opt out of the supply chain, bringing it to a standstill and causing the economy to lose the benefits that could otherwise have been obtained.

As noted in Chapter 14, both the theory and the past studies suggest that past innovation benefits have been shared between producers and consumers, with consumers gaining a larger share of innovations that increase yield, while producers gain a larger share when quality traits are involved. The distribution of the returns ultimately depends on market conditions.

As mentioned in Chapter 15, there have only been a few studies that undertook evaluations of the returns to research in the canola sector. The last comprehensive one was completed in 1985. Since then a number of developments have occurred, with significant potential to change the distribution of returns. First, there has been a revaluation of the relative value of canola. Health studies in the mid 1980s were just beginning to hint at the substantial health benefits of consuming the mono-unsaturated fats in canola rather than polyunsaturated fats found in coconut and animal-based oils. As the evidence became clearer, consumer attention to canola rose. A significant change came in 1985 when the USA granted GRAS status to the oil. As adoption rose, the price of canola oil rose from a perpetual discount to soybean oil during the 1980s to approximate parity in the 1990s.

Second, during the 1980s there was a large, rapid infusion of private capital into canola breeding and development, which both depressed marginal rates of return (see Chapter 15) and shifted the focus of much of the research away from broad yield-enhancing research into narrower niche areas, such as herbicide tolerance and novel oil attributes. At the same time, the results of this private research effort were protected through strengthened intellectual property rights. As a result, it is no longer clear that the traditional distribution of benefits continues today. It now is important to disaggregate the returns to the production chain into shares held by farmers vs. shares captured by either the input suppliers (e.g. research companies) or the processors.

This chapter examines the historical and contemporary evidence on the distribution of benefits between input suppliers, producers, processors, consumers and taxpayers. The assessment demonstrates the difficulty of a priori identifying winners or losers, and the importance of considering the type of innovation and the market circumstances for the innovation. In brief, it is no longer possible to use the old rule of thumb that benefits are large and that they generally flow to consumers and producers.

Producer Returns

Given the relative importance of canola to Canada, it is somewhat surprising that more work has not been done to determine the benefits from the crop. As discussed in Chapter 15, there were a number of studies carried out in earlier years that showed large gross returns. This section looks at those studies and how they estimated the distribution of those returns. More recently, these analyses have been extended through a number of working papers and research projects, using both more advanced theory and more current data. The result has been a refinement of the estimates and evidence that the distribution of returns is shifting.

Nagy and Furtan (1978) were the first to estimate the returns to canola research. They estimated that producers gained about 47% of the estimated 101% return resulting from research undertaken between 1960 and 1975. Consumers gained the rest. Low erucic acid varieties had been developed by 1978 but little adoption had occurred, so the authors were forced to make an educated guess about the extent of the adoption of the research. This assessment is consistent with the situation portrayed in Fig. 14.1a. Ulrich *et al.* (1984) updated that study to include the research effort between 1951 and 1982, which incorporated both a longer period and the almost complete conversion between rapeseed and canola-quality seed. They estimated that the internal rate of return for that period was 51% and calculated that approximately 68% of the benefit went to producers. Although they did not formally calculate the distribution of the returns in that study, Ulrich and Furtan returned to the subject the next year and produced a more extensive analysis for the period 1951–1983. In that study they estimated the Canadian internal rate of return was approximately 50%, and that Canadian producers gained about 68% of the benefit (Table 16.1). This higher share of the return, relative to other studies (Table 14.1), can be explained by the fact that canola had become a premium oil, which tended to raise its price relative to rapeseed, at least partially offsetting the price-depressing effect of higher yields. As rapeseed/canola is a traded commodity, Ulrich and Furtan (1985) extended the analysis to include welfare effects in the rest of the world. Given that little or none of the new technologies embedded in canola had been transferred to producers abroad at that time, the main impact of including the rest of the world was on consumers. They estimated Canadian producers gained a slightly smaller share of the benefits (63% versus 68% in the domestic case), and that none of the benefits went to foreign producers.

Table 16.1. Studies of the distribution of gains to research into rapeseed and canola, 1979–1999.

Study	Period	Focus	Social IRR	Distribution of returns
Nagy and Furtan (1978)	1960–1975	Canada	101%	Producers = 47%; consumers = 53%
Ulrich *et al.* (1984)	1951–1982	Canada	51%	Private IRR was 35%
Ulrich and Furtan (1985)	1951–1983	Canada	50%	Producers = 68%; consumers = 32%
		World	51%	Canadian producers = 65%; Canadian consumers = 31%; foreign consumers = 4%
Gray *et al.* (1999)	1951–1997	Canada	32% in 1971 to 7.5% in 1997	NA
Gray and Malla (2000)	1960–1992	World	NA	Canadian producers = 8%; Canadian consumers = 2%; ROW producers = 48%; ROW consumers = 42%

NA, not applicable; IRR, internal rate of return; ROW, rest of world.

That was the end of the analysis for more than 10 years. In the past 2 years, however, interest has rekindled in the subject because of the sharp rise in the canola research effort and the commercialization of the first transgenic canola in 1995. Three studies have addressed different aspects of the issue.

Gray and Malla (2000) undertook an analysis of the decision made in the late 1970s to improve the quality of canola. The removal of erucic acid and glucosinolates from rapeseed came at both some resource cost and loss in yield. In genetic selection, higher quality usually comes at the expense of yield, which is no longer the sole consideration of research. A variety with both higher quality and higher yield would clearly dominate, but is seldom found. If yield is reduced (or the cost of production rises), the supply curve shifts up (not down as shown in Figs 14.1 and 14.2). This means less of the higher-quality product will be produced at any given price. The quality improvement will also increase the willingness to pay for a product, which is reflected in an upward shift in the demand curve. That is, consumers are willing to pay more for any given quantity of product. The upward shift in both the supply and demand curves yields a higher market price for the higher-quality product.

The actual distribution of benefits, and ultimately the decision on whether to pursue quality attributes, depends on the relative shifts in the supply and demand curves. If consumers are willing to pay more for the improved quality than it cost to produce, then the quantity demanded would rise and both producers and consumers would benefit. The investments necessary to improve quality would be made. In contrast, if the costs of producing the better-quality product rise faster than the willingness of consumers to pay, the volume of product being produced and consumed would decline, leaving both consumers and producers worse off. This net loss would preclude the investment.

Gray and Malla (2000) applied this approach to determine the impact of the earlier decision to breed canola-quality rapeseed. On the supply side, they calculated that the yield for canola-quality seed in 1990 was approximately 9% below the recent trend for rapeseed yield. In other words, the changeover from rapeseed to canola raised the supply curve and raised costs. On the demand side, the new canola quality product was demonstrated to have high levels of mono-unsaturated fats, which were linked to lower LDL (low-density lipoprotein) cholesterol levels; palm and coconut oils were found to be high in polyunsaturated fats, which contributed to higher cholesterol and a greater incidence of coronary heart disease. To gain some measure of the impact on demand, they compared and estimated the change in the price premium between rapeseed/canola oil and soybean oil. They concluded that the changeover to canola from rapeseed and the resulting increased interest in it as an edible oil shifted the demand curve up by approximately Can$32 per tonne. Putting these calculations into a three-region global model for rapeseed/canola (Canada, Japan and the rest of the world), they calculated that although the cost of production went up, the rise in consumer demand translated into a net positive price effect of Can$26 per tonne for canola oil and Can$8 per tonne for canola seed. As a result, the quality improvement resulted in an overall gain in economic surplus. Producer surplus increased in all markets while consumer surplus rose in both Canada and the rest of the world, but not Japan. In Japan higher production costs raised prices and reduced quantity demanded because there was no offsetting demand shift – the new quality attributes in canola were judged to be of little consumer interest, given the low rate of coronary heart disease in Japan.

Gray and Malla's analysis has particular relevance for the industry in the late 1990s as many of the private companies are investing heavily in the search for new novel oil attributes to breed into canola. Calgene, for example, bred and commercialized Laurical™ canola, which has transgenes for expression of laurate oil. It first commercialized this product in the US but brought it to Canada for production in 1997. After 2 crop years of production, Calgene had stockpiled enough seed to supply the market for an extended period. Laurical™ canola has both a lower yield than traditional canola and the added cost of a contract-registration-imposed identity-preserved production and marketing system. Both as a result of oversupply and competition from the natural laurate market (the second crush of palm and coconut), the price premiums were seen to be too low to offset the higher costs of producing Laurical™ canola. As a result, Calgene, and its Canadian partner Saskatchewan Wheat Pool, did not offer any production contracts to Canadian producers in 1999. In short, the technology worked but the economics did not. The higher costs of producing Laurical™ canola were not offset by a large enough rise in demand to compensate. There are mixed views in the industry about whether this was an anomaly due to competition from Third World producers of laurate oil, or whether this is likely to be the case for most modified oil canola varieties. The theory and evidence suggests that the only novel oils in canola that will be commercially successful will be those that have significantly higher value to consumers than laurate.

Gray *et al.* (1999) provide some further evidence on the benefits of research. Although they did not examine the distribution of the returns from recent investments in canola, Gray *et al.* (1999) did find relatively low internal rates of return in recent years. One implication of this result is that it is no longer possible for all participants in the supply chain to gain significant returns. In earlier periods, when there were large estimated returns to innovation, even small shares of the benefits would translate into noticeable improvements in welfare. Now, with low rates of return, moderate shares of the net returns will translate into minor benefits to many of the participants in the supply chain.

Notable features of the seed industry in recent years are its increasing integration with the chemical industry, the growing corporate concentration in the sector, and the increased importance of private R&D funding. These features raise important issues when determining the distribution of research benefits.

The growing concentration in the seed and chemical industry and the rise in privately funded R&D are closely linked. The key aspect of this link is intellectual property rights. As Fulton (1997) and Moschini and Lapan (1997) point out, intellectual property rights were introduced to encourage private funding of R&D. However, one consequence of introducing intellectual property rights is to convey some degree of market power on the firms operating in the industry. Moschini and Lapan (1997) explore some of the economic issues associated with estimating the distribution of the benefits of the increasingly private R&D activity when the innovating firms have market power. A key concept in their analysis is the notion of a drastic innovation. An innovation is drastic if it is priced lower than the existing technology, thus completely taking over the market, while an innovation is non-drastic if it is priced competitively with the existing technology.

The notions of drastic and non-drastic innovation provide a relatively easy method of determining the distribution of benefits. Drastic innovations provide benefits to the agricultural production sector and the rest of the supply chain, since they fundamentally shift the supply curve. Non-drastic innovations, however, are different. If the existing technology was being provided competitively and farmers are homogeneous, then the introduction of a new proprietary technology that is non-drastic will, at the limits, provide virtually no net benefit to the agricultural production sector, nor to other downstream sectors or consumers (Moschini and Lapan, 1997). In the case of herbicide-tolerant canola, where the selected genes and the varieties are fully protected by intellectual property rights, the theory suggests the owner of the IPRs will choose to price their seed (which embodies the technology) at some profit-maximizing price. If they price the technology to fully capture the return, farmers are indifferent to the innovation and would not have any economic incentive to adopt the technology (i.e. the demand curve for HT seed would be horizontal at that price). At any price below 100% value capture, farmers would have an economic incentive to adopt the technology and the demand curve would begin to slope downward.

Moschini *et al.* (1999) used the theoretical observations from Moschini and Lapan (1997) to evaluate the creation and distribution of economic welfare due to the introduction of Round-Up Ready™ (RR) soybeans by Monsanto. Rather than assuming the innovation to be drastic or non-drastic, they relied on observed pricing to carry out the welfare calculations. They developed a stylized three-region world model for the soybean complex (involving the USA, South America and rest of the world and the seed, oil and meal sectors) which involved a monopolist supplier of the new technology in an otherwise competitive input market. Their baseline analysis, using observed elasticities and prices, showed that if 55% of US acreage and 32% of South American acreage is planted with RR soybeans in 1999–2000 (current expectations), the global efficiency gain of RR soybeans would be US$789 million, 45% of which is captured by Monsanto, the innovator-monopolist. As output is expected to rise more than 2% and prices decline by more than 6%, producers in the USA, South America and the rest of the world would lose; collectively their losses would be about the same magnitude as the monopoly profits. Because of the price decline, consumers around the world gain; their collective gains equal more than 100% of the net welfare gain. They looked at a variety of adoption scenarios, including with technology being adopted only in the US, only in the US and South America and being adopted universally. US farmers gained only when they were the only producers able to adopt the technology, suggesting that there may be real first-adopter benefits of many of these technologies. As the technology is dispersed globally, however, the first-adopter benefits are eroded. Moschini *et al.* (1999) calculated that as the technology spreads, world soybean prices fall and US producer losses occur; worldwide consumer benefits correspondingly rise. The innovator-monopolist profits rise as adoption increases, reaching a maximum with total global adoption.

Given the similarities between canola and soybeans, many of the results of Moschini *et al.* (1999) have particular relevance. First, their model shows that the US consumers do not gain as much as US industry; the US economy only gains because it is the home for Monsanto, which is presumed to repatriate its monopoly profits to its headquarters in St Louis. In the case of canola, even though the majority of the canola-directed research is located in Canada, the companies developing the technology are, for the most part, multinational firms, so that profits likely would be repatriated to the home countries. If their analysis holds, new yield-enhancing technologies, such as herbicide-tolerant canola, might not benefit Canada as a whole.

Secondly, they modelled the input sector (e.g. chemicals and seeds) as perfectly competitive before the presence of Monsanto's product. Hence, any monopoly profits generated are net additions to social welfare. While that may or may not hold in the soybean industry, it is not true for canola. Before the introduction of herbicide-tolerant canola varieties, between 85% and 95% of all western Canadian farmers used herbicides on their crops, so the introduction of herbicide-tolerant varieties did not generate new demand. Rather, they simply reallocated market shares. Just and Hueth (1993) characterize the herbicide market as oligopolistic, with a global four-company concentration ratio of 65%. Thus, oligopolistic rents were already being extracted from that market. With the introduction of three competing HT varieties of canola in 1995, new oligopolistic rents were likely created, but for the most part they simply replaced the rents forgone, often by the same companies, as older chemicals were replaced by this new technology. In short, the agrochemical industry cannibalized their own products. Thus, it is not clear that herbicide-tolerant canola varieties would generate as large a return as Moschini *et al.* (1999) would suggest.

Thirdly, the adoption rates and agronomic benefits for canola do not support the view that HT canola is a drastic innovation. Although HT canola was rapidly adopted – acreage of herbicide-tolerant canola increased an average of 22% per year between 1995 and 1999 – the adoption rate of HT has not approached 100%. For instance, in 1999, the adoption rate was approximately 70% in Saskatchewan (Fulton and Keyowski, 1999). This levelling out of the adoption rate is consistent with the agronomic benefits of HT canola. As the

Saskatchewan Canola Development Commission notes in an article, the relative yields of HT and conventional varieties depended critically on the weed infestations present and on the geographical location. Where weed infestations were large, the HT varieties performed better. However, where weed infestations were less of a problem, the conventional canola produced similar yields to the HT varieties. A comparison of the yields of conventional canola varieties (Argentine) and HT varieties by Alberta Agriculture, Food and Rural Development shows considerable variation in the yields of both types of canola. Overall, the comparison data suggest that in the summer of 2000, neither HT nor conventional canola can be said to have an absolute yield advantage. Given the higher cost of the chemical for some of the HT varieties or the presence of Technology Use Agreement fees, HT varieties are thus not always economically superior to conventional varieties (Table 16.2).

Fulton and Keyowski (1999) further observe that the concept of a drastic innovation is only relevant when all producers of the product face the same costs and agronomic factors. However, if the products produced are not the same and/or if the production factors differ, then generally both technologies will coexist. The argument that producers benefit if the relative price of growing new varieties falls, depends critically on the belief that all farmers are identical in the agronomic factors they face, the management skills they possess and the technology they have adopted. If farmers are different in these characteristics, no such easy test of producer benefit is available. Instead, the determination of farmer benefits requires a detailed examination of the agronomic, management and technology factors facing different farmers.

Table 16.2. 1999 Canola product line, system comparison. (*Source:* Pioneer Grain Company Limited, reported by Fulton and Keyowski, 1999.)

	Round-Up Ready[TM]	Smart[TM] Open Polish	Liberty[TM] Hybrid	Conventional Open Polish
Seed cost per acre[a]	$18.70	18.70	24.75[c]	$13.47
Herbicide cost per acre	$5.00	$26.20	$22.75	$30.00
TUA	$15.00	None	None	None
Total cost per acre	$38.70	$44.90	$47.50	$43.47
Bundle price per acre	$38.70	$38.25	$47.50	$43.47
Yield potential/check	110%	105%	119%	119%
Yield – bushels per acre[b]	33.0	31.5	35.7	35.7
Commodity price per bushel	$8.00	$8.00	$8.00	$8.00
Expected gross	$264.00	$252.00	$285.60	$285.60
Less system costs per acre	$(38.70)	$(38.25)	$(47.50)	$(43.47)
Gross per acre	$225.30	$213.75	$238.10	$242.13

[a]Seed cost was calculated assuming a seeding rate of 5.5 lb per acre.
[b]Yield was calculated assuming the average yield of the check is 30 bushels per acre.
[c]Recommended seeding rate is 5 lb per acre for Liberty[TM] Hybrids.

Fulton and Keyowski (1999) argue that herbicide-tolerant canola is most appealing for farmers who have adopted conservation systems of land management. In conservation systems, farmers make fewer passes over the land with tillage equipment. In the zero tillage system, for instance, farmers place the seed and fertilizer directly into the undisturbed soil using an air seeder. No other tillage takes place. Conservation systems are being used to maintain higher levels of soil organic matter and to minimize soil erosion. Because tillage is minimized in these systems, weed control must be carried out solely with chemicals rather than with a combination of chemicals and cultivation, the usual method under traditional land-management systems. When producing conventional crops, weed control is done prior to seeding using cultivation or chemical spraying, or after seeding using spraying. In most cases only a limited spectrum of weeds can be controlled using either pre- or post-emergent chemicals. The availability of a canola resistant to a chemical that can control the entire spectrum of weeds gives farmers much more flexibility in terms of the timing and type of weed control. The one-pass chemical operation characteristic of herbicide-resistant canola systems not only improves the yield potential of the crop by removing competition for moisture and nutrients, but also eliminates the cost of additional machine operations over the field. Farmers that benefit most from the technology are those that have fully adopted reduced tillage.

A number of observations can be made from the approach of Fulton and Keyowski (1999). First, their model shows that some producers benefit even if only a portion of the market switches to the new technology. Second, the price of traditional seed is a key factor in determining the benefits of the traditional seed, and the benefits of the new technology. Specifically, all else being equal, decreases in the price of traditional seed result in an increase in the benefits of traditional canola, a fall in the share of farmers that use HT canola and a fall in the benefits of HT canola. Third, the price of HT seed is an important factor determining the benefits of the new technology. An increase in the price of the HT seed results in a smaller portion of farmers adopting the new technology and a smaller producer benefit.

The data tend to conform to their theoretical results. Indeed, farmer diversity is a hallmark of modern agriculture. As noted in Chapter 9, farmers differ substantially in terms of age, education, farm size, product specialization, farm management skills and the geographical location of their farming operation. Canola yields and the percentages of acreage devoted to canola differ substantially across crop districts. Generally speaking, canola production occurs in the dark brown and black soil zones, areas where the average yield is greater. As noted previously (Table 9.10), farmers also differ in the degree to which they have adopted conservation tillage practices. In Manitoba in 1996, 37% of canola farmers reported using conservation tillage and 17% reported using no-till practices, in Saskatchewan, the numbers were 48% and 25%, respectively, while in Alberta 41% reported using conservation tillage and 16% reported using no-till practices. Conservation tillage requires a different set of equipment than does conventional tillage practices, and the cost of this equipment is substantial. The cost of adopting this new technology, along with factors such as degree of risk aversion, farm size, age and management skills, all contribute to this technology not being adopted by all farmers. Generally speaking, farmers who have not adopted conservation practices do not have the same agronomic and economic benefits of herbicide-resistant canola as do those farmers who have adopted conservation practices.

Perillat and Phillips (1999) undertook to survey a number of farmers in Saskatchewan about their experiences with various herbicide-tolerant canola systems. Although the sample was not statistically valid, the results supported Fulton and Keyowski's conclusion that the benefits vary depending on the farmer. Each farmer surveyed had different returns from using the HT canola varieties. The skills of the farmer, the quality of the land and the agronomic practices largely determined the returns to producers. Some farmers gained from using different HT canola varieties while others lost. Gianessi and Carpenter (1999) produced similar results, looking at the use of the *Bt* gene in maize, cotton and potatoes – uneven and unpredictable levels of insect infestation yielded wide variations in actual net returns to the technology, with both winners and losers in the late 1990s.

Zatylny (1998) noted a related benefit of the new technology – the new herbicide-tolerant canola can be planted earlier. The standard recommendation since the early 1970s is that canola should be seeded into a warm, firm, moist seed bed that is weed free and only after all danger of frost has passed. This recommendation had little to do with the requirements of the plant itself – the plant can withstand temperatures of $-9°C$ in the early stages. The main reason was for weed control – winter annual weeds, such as stinkweed and shepherd's purse, traditionally could only be controlled with tillage and pre-seeding herbicide applications. To do this, the soil needed to be warm to fully activate the soil-incorporated herbicides. This pushed the recommended seeding date into the second half of May. Zatylny reported recent research that showed that earlier seeding (i.e. in late April and early May) with HT canola was profitable. Earlier-seeded canola produced a clear yield advantage as the plant is able to take advantage of a longer cool period to set the seed. Zatylny estimated that if earlier seeding were widely adopted, total average yields for Canada could rise by 1.5 bushels per acre, which would translate into more than Can$150 million increased revenue to producers.

Falck-Zepeda *et al.* (2000) introduce many of the factors outlined above into their analysis of the impact of the introduction of Bt cotton in the US in 1996. In their analysis they focus solely on the distribution of the benefits of Bt cotton among four groups – consumers, producers, the gene developer (Monsanto) and the germplasm supplier (Delta and Pine Land Company). The analysis does not examine the rate of return to the R&D activity that gave rise to the innovation. Using a simulation model, they examine the distribution of the benefits of introducing Bt cotton into a large open economy (the US) with no technology spillovers (i.e. other countries were not able to adopt the technology that was introduced in the US). The model assumes linear supply and demand curves.

The introduction of Bt cotton is assumed to result in a parallel downward shift of the supply curve for cotton. A downward-sloping demand curve for cotton seed reflects producer heterogeneity in such things as lepidopteran pest pressure, yields and seeding rates. Their analysis indicates that the biotechnology innovation generated a worldwide increase in economic surplus of US$240.3 million. Of this total, 59% went to US farmers, with the gene developer obtaining 21% and the germplasm developer receiving 5%. US consumers received 9%, while the rest of the world received 6%. A key assumption in this model is that of no technology spillovers. Since 1996, the *Bt* technology has spread to other countries. As Moschini *et al.* (1999) show (see the discussion above), allowing other countries to adopt the technology changes the distribution of the analysis considerably.

It is important to note that all of these studies have modelled the processing sector as competitive, so that none of the returns to innovation accrue to processors. Given that the four company concentration ratio is 100% for canola crushing in Canada (Wensley, 1997), there is a possibility that producers would need to share with these processors a portion of any returns that reside in the supply chain.

One issue that none of the studies have directly examined is the cross-product impacts of technological change. The results of an international conference in 1999 on biotechnology and the poor, presented in Persley and Lantin (1999), offer some insights into how this technology could influence many of the LDCs. Development groups have for quite a while pointed out that technological change tends to be concentrated and commercialized first in more developed nations, leading to lower per capita incomes for many developing countries. The Rural Advancement Foundation International (RAFI) has used Calgene's Laurical™ canola as an example of the adverse effects of technological change on developing countries. Laurate is an oil that occurs naturally in a wide number of oil-producing crops, including palm, coconut and rapeseed. The oil is used as a sudsing agent in industrial detergents as well as a stiffener and whitener in confectionery and baked goods. Most of the laurate used until recently was produced in the palm or coconut sectors as a by-product of the primary oil processing procedure; it was the result of the second crush of those products. Naturally occurring rapeseed has only trace amounts of laurate oil. As discussed in Chapter 8, in the mid-1980s Procter & Gamble went in search of a new source of laurate and helped to finance Calgene's research into the product. To make a long story short, Calgene bred a gene that expressed for laurate in canola and commercialized the product beginning in 1994. This new, higher quality supply of laurate has both depressed prices and displaced exports from a number of developing countries. Although many developing countries produced coconut oil, the Philippines is perhaps the most dependent on the product. RAFI reports that the coconut industry accounts for 44% of the country's total agricultural export earnings, 7% of total export earnings and 30% of total employment. Globally, annual lauric oil consumption is about 4.6 million metric tonnes; US imports of tropical lauric oil were valued at

US$350 million in 1992. Although laurical canola production in 1994–1999 only amounted to a small share of this total, it has the potential to have a significant effect if it grows in size. A full costing of the returns to innovation should ideally include all of Schumpeter's creative destruction to get a net welfare gain and to determine the distribution of the returns. Only then will we know whether an innovation is a potential Pareto improvement and thereby worthy of either direct or indirect public support.

In conclusion, there would appear to be a number of developments that influence the share of benefits that producers can expect. Perhaps most fundamental of all, the sharp increase in competitive investment appears to have depressed the net returns on new innovations, which simply means there is less to go around. At the same time, the proprietary nature of almost all the commercially valuable innovations creates a new countervailing factor in the system, which makes it less likely that producers in aggregate will gain a significant share of the returns. The innovations freely delivered by the public sector previously are now gone. Furthermore, the increasingly heterogeneous nature of farming has the potential to enable a proliferation of new technologies that in aggregate may not benefit all producers but will benefit at least some, while some technologies have the potential to destroy significant value for other producers.

Consumer and Citizen Benefits

Agricultural and industrial analyses tend to focus predominantly on the returns accruing to farmers. As noted already, the impact on consumers, and by implications on taxpayers, is often many times larger and more important than the producer returns.

Consumers get whatever returns are not captured in the production chain. As shown in Table 16.1, that means that consumers, over the past 30 years, have gained somewhere between 0% and 53% of the total returns from research into rapeseed and canola. Ulrich and Furtan (1985) and Gray and Malla (1998) make an important distinction when dealing with the distribution of returns. Given that Canada produces approximately 20% of world's annual output of rapeseed/canola but accounts for only 4% of its consumption, the distribution of benefits and costs between those two groups has significant implications for Canada. If an innovation is not transferable to other competing jurisdictions, then, as Fig. 16.1 shows, the Canadian supply curve would shift outwards (from S to S*) and Canada's exportable surplus would shift accordingly (from ES to ES*). World prices would drop, trade would expand and foreign producers would lose (due to lower prices) while foreign consumers would gain. In the canola case, given that more than 80% of Canadian production is exported annually, the majority of consumer benefits should flow abroad. Gray and Malla (1998) estimated that about 95% of the net consumer benefits flowed offshore.

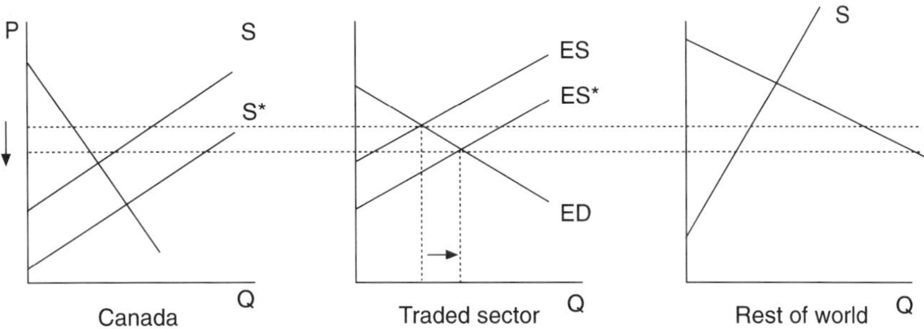

Fig. 16.1. International impact of biotechnology innovation.

Traditional studies of the gains to research have identified the potential for externalities of new technologies but seldom have quantified them. The history of canola presents two distinct examples of externalities that result from the new technology. In the first instance, the discovery in the 1980s that the oil profile in canola helps to reduce coronary heart disease created an unanticipated benefit for both consumers and taxpayers – research and development of the crop increased consumption, which led to less heart disease and reduced healthcare costs. Medical and nutritional studies suggested that replacing 10 g per day (3.65 kg per year) of saturated fats with canola oil would result on average in a 2.3% reduction in total blood cholesterol and on average a 4.6% reduction in the incidence of coronary heart disease. Assuming a one-for-one relationship between coronary heart disease and healthcare costs (which cost an average Can$265 per capita per year in 1993), the shift from saturated fats to canola was estimated to yield a health expenditure saving of Can$12 per capita. Thus, every kilogram of canola oil consumed would reduce health costs by Can$3.34. Although 15 million kilograms of canola oil were sold in Canada in 1993, in practice only about half the oil purchased was actually consumed. Thus, Malla (1995) estimated that the shift to canola oil yielded a Can$25 million per year saving in Canadian healthcare outlays annually. Even without putting any positive value on the reduced suffering and loss of human life, the present value of this healthcare saving was more than triple their highest estimate of welfare gain to overall consumption.

Meanwhile, the taxpayer faces mixed results. Apart from any externalities due to health benefits, the public treasury, at least in the first instance, is funding a large share of the research with only limited direct returns through royalties or other rents on innovations. In the longer term, however, taxpayers have the potential to offset some or all of those up-front costs as new technologies are commercialized, creating new incremental value for both consumers and producers. This ultimately leads to a larger tax base and greater fiscal capacity. It is unclear when the break-even point between outlays and returns will be reached, but it probably has not occurred yet.

Finally, there is significant potential for environmental impacts. On the plus side, the adoption of herbicide-tolerant varieties has changed agronomic practices, with the result that newer, less toxic herbicides are being used, which has some environmental benefits. Monsanto commissioned Sparks Companies Inc. (1997) to assess the impact of Round-Up Ready™ canola. They concluded that producers using Round-Up Ready™ canola used 54% less active herbicide ingredient per acre. Furthermore, glyphosate, the active ingredient, has a half-life of 60 days or less in the soil, and after 6 months approximately 90% of the compound will have been degraded by microorganisms – leaving carbon dioxide, nitrogen, water and phosphate – while the surfactants added to aid the absorption of Round-Up™ into plants has a half-life of less than 1 week. Tests showed that Round-Up™, which has been used for more than 20 years, had no weed-killing activity once in contact with the soil, as both the active ingredient and the surfactants bind tightly to most types of soil particles. As a result, water passed continually through columns of soil treated with Round-Up™ contained no chemical residue. Test results showed that glyphosate does not accumulate in animals, birds or aquatic species such as fish, clams and shrimp, which indicates that it does not accumulate in the food chain. In short, the company-sponsored study shows that the adoption of herbicide-tolerant canola varieties has generally reduced the number of herbicide applications and the volume of active ingredient. Other herbicides linked to new varieties have similar traits. As *Bt* genes are expressed in canola (field trials were under way in 1999 on *Bt* varieties), pesticide use could also fall; Gianessi and Carpenter (1999) conclude that the introduction of *Bt* maize varieties in the USA led to 2 million fewer acres being sprayed against European corn borer in 1998 and reduced the number of applications on cotton from a previous average of approximately five applications per season to an average of only about one application per season. Ultimately, these benefits will flow to the larger society in terms of less environmental degradation and lower risks to plant, animal and human health.

On the negative side, the wider application of selected herbicides, not only for canola but for other field crops, has the potential to increase the risk of outcrossing with weedy relatives, creating 'super weeds'. Volunteer canola, resistant to Round-Up™, Liberty™ and Smart™ chemicals, has already been found in Alberta (*Western Producer*, 2000c). Further study will be needed to estimate the economic costs of that event. There is also significant debate in western Canada about the potential for genetic material to 'drift' between crops of the same species or between species or products, thereby 'contaminating' non-genetically modified products and creating a negative externality. There are already a few cases of this happening. Scientists have shown that canola pollen can blow more than 200 m (recent tests have found that some pollen travelled more than 30 km), fertilizing other non-transgenic fields. This is a special concern for producers of organic crops, but could also act as a barrier to containing and developing GM-free products. There is one case where the EU rejected Canadian honey when it found in the honey traces of canola pollen that contained transgenes not approved for general release in the EU. Furthermore, there is some limited information that GM traits could cross the species barrier. While preliminary tests show that GM traits in animal feeds do not affect milk (*Western Producer*, 2000a) or meat (*Western Producer*, 2000b), there is some evidence that bacterial DNA in herbicide-tolerant canola has transferred to a honey bee's digestive tract. Although the researchers concluded that the DNA does not affect the bees, its presence raises a host of new questions about possible externalities. All of the externalities – both costs and benefits – should be incorporated into the analysis of the gains for research.

On the whole, consumers and citizens may have little need to worry about most aspects of technological change in agriculture. The real challenge for consumers and governments, as discussed in Chapters 12 and 13, is managing the trade-offs between measurable consumer benefits of change and the uncertainties of the unknown future. Setting aside any credence factors involved in biotechnology-based products, all of the measurable risks, and a significant share of the direct and indirect returns to research, flow to consumers. Furthermore, although there definitely are uncertainties about the adoption of these new technologies, many of the potential impacts are more appropriately judged as risks, with at least some subjective probability of outcome. For many of those possibilities, the impacts will be largely internalized within the farm sector. On the positive side, less herbicide residue in the soil, for instance, should help to support the returns to producers as they have greater choice of rotational plantings. On the negative side, greater herbicide tolerance in crops increases the potential of crops becoming weeds in the rotation system. Furthermore, much of the cost of increased herbicide tolerance in weeds will be passed back to producers, as they are forced to find new, often more expensive weed-control systems. Over the longer term, however, many of the environmental effects of intensive agriculture are not now properly priced in the production system; evidence suggests that the new systems may reduce those impacts, which will benefit the broader society.

The recent consumer backlash against GM foods creates some doubts about the above analysis. Economists have begun to investigate the role of trust and confidence in the creation and operation of markets (Fukuyama, 1995; Stiglitz, 1999). Markets for many products are often not able to create, by themselves, the conditions of trust that generate the socially optimal quantities of goods and services produced and consumed. Hence, there is a much more explicit role for public and private regulation in markets than neoclassical theory generally suggests. This is especially true for GM agri-food products, where perceived risks and public uncertainties abound. Tirole (1988) identifies three types of goods: search goods, where consumers can visually identify attributes before consumption; experience goods, which require consumption to determine the attributes; and credence goods, where the unaided consumer cannot know the full attributes of consuming a good, at least for some period after consumption. Markets for search goods are for the most part able to function efficiently based on simple transactions – primitive barter economies and street markets all thrive with little or no government intervention. Experience and credence goods, such as GM foods, require a greater element of trust, which must involve active communication about the product's attributes. Those product factors that involve probabilistic or hypothetical public health and safety risks are usually regulated by the state. Formal communication of this regulation is sometimes signalled through labels (e.g. Canada Choice meats) but more often is simply implied by the presence of the product in the food chain. GM foods, however, also involve a wide array of speculative risks that the state does not handle (Phillips and Isaac, 1998). Following on Ackerlof's work (1970) on the market for lemons, Bureau *et al.* (1997), Giannakas and Fulton (2000) and Plunkett and Gaisford (2000) suggest that in some instances where consumer fears are high enough, the presence of unlabelled GM foods could result in global welfare losses.

Conclusions

Winners and losers are the real drivers of any change. Winners push for more change while losers either opt out or stall adoption of changes. The canola story runs true to form.

The rise in privately funded and directed research in the canola sector has probably fundamentally altered the distribution of returns. Perhaps most importantly, the large increase in private and public outlays has dramatically lowered the returns to research, to the point where it is no longer clear that there are adequate returns to create a Pareto improvement. Meanwhile, the combination of private research and proprietary strategies for innovations has shifted some of the producer returns away from farmers in aggregate and toward the input industry. But the heterogeneity of Canadian farmers virtually assures that the technologies commercialized so far have created some winners among farmers. Offshore producers, on the other hand, are certain to be net losers from the new technology. Looked at another way, if the same level of canola research was being undertaken globally, but without any activity in Canada, Canadian producers and society would be net losers. This is perhaps the strongest argument for searching for new and more effective ways to support and promote local research and development.

On the consumer side, we see the basis for the current policy conundrum facing agricultural biotechnology. While, by all conventional measures, consumers are likely the winners from the application of biotechnologies in the agri-food sector, doubts remain. While a comparative statics analysis would probably show that biotechnology raises productivity, increases supply and lowers market-clearing price, consumers might not notice. Given that commodities such as canola often represent a small percentage of any processed product and an even smaller percentage of the consumer's food basket, the per capita changes are often extremely small and often ignored by consumers. Meanwhile, consumer and producer perceptions of risk and uncertainties are asymmetrical. Whereas producers see most of the risks as small and manageable, consumers focus on the uncertainties of the new products, assigning high probabilities and large personal costs to many of their fears. As a result, consumers in many markets appear to have looked the gift horse in the mouth and rejected it.

Policy Implications 

Lessons for the Future

Peter W.B. Phillips and George G. Khachatourians

Introduction

On one level, this book is about the past and future for canola. As described in the previous chapters, oilseed brassicas (and canola in particular) are now a major source of both vegetable oils for human food and meal for animal feed. After soybean and palm, canola is the major source of oil for human consumption. Canola has achieved this position in a relatively short time because of the availability of planned and fortuitous technological, instrumental and capital investment and effective institutions. This story is far from done. From the beginning, canola has been relatively widely adaptable. Nowadays, because of our ability to genetically manipulate this plant into more adaptable varieties that can be grown from subtropical to cool temperate regions, in the Americas, Australia, Europe and Asia, the potential for further adoption is significant. Some changes in *Brassica* traits will need only conventional breeding, while others will require sophisticated genetic engineering. As a result there will be many varieties of these oilseeds that involve higher yields, new tolerances to abiotic stresses and new oil characteristics, as well as novel medicinal, nutraceutical and other functional properties. Meanwhile, on the industrial-use side, canola plants with oil profiles matching industrial needs for lubrication and diesel oils will continue.

Furthermore, the canola case provides a comprehensive overview of the economic, commercial and public policy implications of the use of biotechnology processes and commercialization of genetically modified products on the global agri-food system. The canola industry experience sheds some light on the potential trends and implications of biotechnology in the broader agri-food industry. Although the commercial canola seed business represents less than

1% of the US$23 billion global commercial seed industry, it is one of the earliest and most extensively transformed of all the products so far targeted for genetic transformation, with adoption of new biotechnology-based varieties in Canada estimated at above 70% of total acreage in 1999. As such, it provides some early evidence of the impact of biotechnology on the global agri-food system.

Economic Trends

The introduction and use of biotechnology in the agri-food industry globally has caused five identifiable economic impacts, which together will influence the volume, pace and location of research-based agriculture.

First, the new technologies have led to a change in the innovation process. Historically in the agri-food sector, the innovation process has been quite linear, often starting with curiosity-based research and leading through development, production and, ultimately, marketing the new product (Fig. 2.1). As a result, the research process has been relatively narrowly defined and self-contained. The advent of biotechnology and its close connection with public and private funding of research, with obvious expectations of a return on investment, did two things. First, it created the potential to more finely target the research to specific market needs, providing a very real incentive for processors and end users to engage in the research effort and direct some of the research. Secondly, it made the research process far more integrated, with no individual or small group of individuals able to undertake the entire research process. As well, individual researchers no longer have effective monopolies on research areas. Rapidly changing knowledge and technologies have changed the innovation process itself, making it operate more like a 'chain-link model' (Fig. 2.2). The basically linear process spanning design, adaptation and adoption, is now quicker start-to-finish, with feedback loops from each stage to previous stages and with the potential for the innovator to seek out existing knowledge or to undertake or commission research to solve problems in the innovation process.

This has altered fundamentally the research effort related to canola and has the potential to alter the rest of the agri-food research system. The codification of know-why and know-what technologies appears to have made them accessible and mobile (subject to constraints imposed by intellectual property rights), opening up the potential for greater competition in the research area. Meanwhile, the increasing complexity of the science has increased the importance of know-how and know-who knowledge, which, as discussed in Chapters 3, 6 and 8, tend to be located in limited sites. In the canola sector, this has allowed the rise of Canada as an 'entrepôt' of research, development and commercialization. For the most part, the basic science and the patentable technologies have been developed at isolated locations around the world. Canada, and particularly Saskatchewan, has found a niche in the research and production chain, assembling the basic scientific knowledge and proprietary technologies and using them to breed in novel traits, which are then planted, grown and crushed to a first stage of processing, before being exported to global markets. Both the upstream and downstream capacity is largely elsewhere. Canada and Saskatchewan have found a competitive role as the providers of know-how and know-who, the vital connective elements that make knowledge flow. Both types of knowledge are often learned by doing, which makes them more difficult to transfer to others and, hence, more difficult to codify and transfer. The growing public and private research and development capacity in Saskatchewan and Canada created the base for the increasingly important networks and alliances that make this new model of innovation work.

The changing nature of the innovation system is also changing the targets for research. In the early years of rapeseed research, when a linear innovation system dominated, the focus tended to be on yield and improving the agronomic performance of the plant. There was little or no focus on output traits. The rapeseed research effort went through a partial change, as consumers and sophisticated intermediate processors entered the research effort in the 1970s and began to direct effort towards output traits – particularly the reduction of erucic acid and glucosinolates. This was one of the first examples of a chain-link innovation system in the agri-food sector. As technologies were improved, first with the half-seed process and the use of the GLC and, more recently, with biotechnologies, the outcomes of research became more predictable and the incentive for end users to engage in the research system increased accordingly. Now much of the research is directed or financed by one or more sophisticated input suppliers or buyers seeking innovations to market to their key clients. Table 17.1 illustrates the array of efforts under way in the canola sector, both on the input and output side. The input trait research has small to medium economic value and relatively early commercialization dates, while the output trait

Table 17.1. Potential economic opportunities involving canola (1997–2007). (*Source*, KPMG, 1997.)

Opportunity	Expected date of commercialization	Relative economic value of opportunity	Number of participants
Yield increase	1997–2000	Medium–large	>5
Modified oils	1998–2001	Medium	5
Disease resistance	1999–2002	Small	3
Insect resistance	1999–2002	Small	2
Adaptation to US climate	2000–2002	Small	3
Increased oil content	2002–2005	Large	3
Meal enhancements	2002–2005	Medium	2
Nutraceuticals	2002–2005	Medium	2
New oil profiles (e.g. low-fat oil)	2003–2007	Very large	4
New meals	2004–2007	Large	2

research, with a larger expected economic return, is probably a few years from full-scale commercialization.

Advances in genomics increase the likelihood of more customer-driven research. A number of efforts are under way around the world. In Canada, the federal government, in partnership with the Saskatoon AAFC Research Station and NRC's Plant Biotechnology Institute in Saskatoon, announced in 1999 a Can$20 million initiative to complete the genetic coding of *Arabidopsis thaliana*, a close relative of canola. They will then develop a concordance of that genome to canola to enable them to target breeding programmes with increased precision (Parkin and Lydiate, 2001). In theory, once the DNA genome is mapped (e.g. allele, phenotype and stock data) and all the functional attributes of the genes are known (gene, protein and transcript data), it should be possible to manipulate very precisely the genetic code of canola to produce almost any attribute that anyone could want. At that point, canola should attract significantly increased attention from those interested in its lubricant and energy uses, its edible-oil properties, its animal nutritional value, its functional food properties, or its potential as a host plant for pharmaceutical proteins or enzymes or for use in environmental remediation (Palmer and Keller, 2001). One area of specific interest would be its use as a fat-soluble-vitamin delivery system, which could have significant health, nutrition and welfare benefits for consumers around the world.

The biology and genetics of canola make it an ideal species for modification (Palmer and Keller, 2001). Having developed a multiplicity of genetic and metabolic engineering capabilities, we should be able to apply many of these technologies to other oilseed crops. Already many of the concepts that have been applied to the biotechnology of canola have been applied to several other oilseed plants (Khachatourians *et al.*, 2001). Although a number of other species may have similar traits, many of the large acreage crops are different and may not generate as large an interest relative to their acreage base. Three features in particular are worth noting. First, breeding of canola is relatively new, which leads some to believe that there may be a large untapped genetic potential. In contrast, some argue that crops such as wheat, which have gone through extensive genetic modification over the past 6000 years through both farmer selection and intensive public breeding programmes, may be nearing their genetic potential. If true, that could limit commercial interest in those crops. Second, the *Brassica* species has a relatively simple genome (Parkin and Lydiate, 2001), which makes it very attractive to work on. Compared to complex genomes such as wheat, canola is both relatively easy to modify and yields significantly more predictable outcomes. The greater the predictability, the more attractive the species. Third, being an edible oilseed that is intensively cropped, canola is both a relatively high-value plant, which has historically commanded a premium in food markets, and an attractive market for input suppliers. Other plants such as wheat and barley, which have a lower value per acre and are much less intensively cropped, are correspondingly of lower interest. The one mitigating factor for crops such as wheat is that large acreages present larger markets than smaller acreage crops, such as barley. Furthermore, canola is a good plant in which to express other proteins and enzymes because of its oil properties and the relatively simple processing systems that can extract novel attributes from the oil. Although many other commercial crops exhibit some of these characteristics, few of them conform to all of them, which will tend to limit research on them.

Accepting that caveat, the trends in the canola sector suggest that the research related to other crops may undergo significant specialization based on core capacities in coming years. Know-why and know-what research is likely to continue to be produced in the public sector and commercialized or disseminated through private action, while the assembly of the technologies to create new agri-food products will tend to be privately directed at those locations which have achieved a critical mass of know-how and know-who capacity. This will likely vary, based on the product and the technologies involved, and will not always be consistent with current production or trade patterns. It may be possible through government action to invest in facilities that provide the critical know-how and know-who to attract the attention of private companies to areas that have not traditionally been centres of agri-food research.

Biotechnology is also likely to have a somewhat different impact on the many other crops that cannot economically or effectively be modified to express new output traits. In these instances, the focus will be on input-traits, which tend to be specific to specific regions. This may work to offset some of the tendencies to clustering, as the research programmes will need to focus on narrower traits that have only local value, and likely will be forced to locate at least part of the industrial activity in those markets. Output traits tend to be more transferable, which enables firms and the industry to cluster their activity for crops like canola in a few locations.

Perhaps equally important, the effective hybrid technologies for canola have enhanced its attractiveness as an investment target. Over and above the yield bounce that hybrids tend to provide, such systems also help to protect and facilitate exploitation of new plant-based technologies, because farmers are forced to purchase new seed each year. The economics of hybrids for canola make it especially attractive. Given average seeding rates of 5.6 kg per ha, 1 ha of F1 hybrid seed plants approximately 88 ha. In contrast, unless an effective hybrid system for wheat is developed that yields a substantial yield bounce, the economics are not favourable – 1 ha of F1 hybrid seed would only plant about 10 ha. In short, the hybrid approach to value capture is not universally available. Other crops therefore may have somewhat limited attractiveness as investment targets.

Secondly, the fragmentation of the innovation system into different types of knowledge development and use has fundamentally affected the economies of scale and scope in the industry. As discussed in Chapter 4, basic, know-why research does not appear to exhibit any significant economies of scale in the short run. Given that, except in a few locations, researchers at universities and public laboratories collaborate most often with those outside their home base rather than with other local scientists, the economies of scope appear to be limited. Similarly, there are no obvious cases where biotechnology research programmes exhibit significant economies of scale in the development of know-what knowledge. Most of the fundamental patentable technologies that have been developed have come from disparate research programmes around the world and, once 80–95% of the research is done, are assembled, 'developed' and commercialized by private companies. The economics of lotteries may be a more appropriate way of looking at this research. A region cannot win without playing, but even for the players the odds of winning are slim and cannot be easily or significantly altered by the degree of effort. In contrast, there do appear to be significant economies of scale and scope in the know-how and know-who stages of development, if only because imperfect markets make smaller ventures invest relatively more of their resources on transactions, which tends to favour larger, more integrated firms.

There is some evidence of increasing returns to scale in the industry, at least partly due to the high fixed costs to enter the business. With high entry costs, firms face declining average costs over the feasible operating range. Each unit of input yields successively higher volume of output. This is more likely in a knowledge-based world where the high cost to enter is not the purchase of lumpy capital equipment but rather is the result of an investment in 'learning by doing.' This investment is a real barrier to entry as there is significant uncertainty that the resulting output from the research effort will have any commercial value. It is not like making and selling a standardized product. Nevertheless, those entering the research business report increasing returns to scale as they find that they have acquired a capacity to learn – each new research effort becomes easier as it builds upon the experience of the researchers and developers. In crops-based agriculture, for example, regulatory compliance adds a significant barrier to new entrants. New varieties are first reviewed by regulators to ensure that the new product is both safe and conforms to industry standards. The first developers of herbicide-tolerant and Bt-resistant crops learned that convincing the regulators of the safety of their product imposes significant costs. Then, the developer must convince producers that the new variety is worth introducing to their enterprise. Finally, increasingly the developer must convince consumers to buy the product. All of these impose significant costs on developing new products. Companies that do this more than once learn from their efforts and do it more efficiently and effectively.

The canola evidence presented in Chapter 9 showed that while the total research budget for canola varieties rose more than tenfold in real terms over the past 30 years, the real average cost per variety has dropped sharply since the early 1980s. Throughout the period, however, the marginal cost of new varieties has remained relatively small. Beyond the basic economies of scale in the canola industry, there is also evidence that externalities or economies of scope exist and have influenced the industry. On the positive side, the sharp drop in the average cost per variety is at least partly due to 'mysteries being in the air' (Marshall, 1980), thereby disseminating advancements and improvements throughout the research and development community. A survey in early 1998 of firms undertaking research into canola suggests that the flow of knowledge has a real impact on their operations. Half of all the respondents, representing the majority of larger private companies responding, acknowledged the importance of proximity to either collaborators or competitors as important factors in locating their research efforts. About 40% recognized the importance of being close to their collaborators, particularly the National Research Council and Agriculture and Agri-food Canada research centres in Saskatoon. In a knowledge cluster, the gradients of diffusion of knowledge from the originator or source to the early users and adopters are crucial. If the capacity is not linked or in close proximity, knowledge will not flow effectively. Beyond the core, normal diffusion takes care of the rest.

There has been extensive research in other countries and in other product areas to determine the elements that influence the creation and location of these economies of scope. Chapter 9 examined the Saskatoon research community to determine whether there is any evidence of economies of scope developing. There was relatively strong evidence of a thickening of the labour market in Saskatoon and Canada and of a tightly integrated research and production system extending from the applied research stage through to marketing the intermediate market. As an entrepôt, the community imports the vast majority of its inputs (know-why published knowledge and patented know-how technologies) and assembles them with local germplasm, know-how and know-who. The resulting innovations are commercialized in western Canada first, in partnership with a large number of sophisticated farmers. The produce is then largely sold elsewhere. In short, Saskatoon offers a very efficient and economical import–export business for the biotechnology industry. As with any entrepôt, this effort is supported and nurtured by both industry and government. The Canola Council of Canada and Ag-West Biotech Inc., in particular, encourage and support collaboration and cluster building at the local level, while the federal and provincial governments have acted as collaborators, promoters and financiers, providing public research infrastructure, responsive regulatory systems and attractive fiscal regimes. Thus, Saskatoon has become a global competitor for canola-related R&D.

This model poses a number of possible scenarios for other agri-food research efforts. While canola is somewhat unique, in that its development effort has always been led from Canada, there nevertheless remain a number of features of the canola experience that may match other products. At least for commercial crops in developed regions, there likely would be significant potential for economies of scale and scope if research and development clusters develop in one or a few sites in the world for each key product. Although the immediate benefits of agglomerations would flow to the sites themselves, the broader agri-food sector and global consumers would ultimately gain from such a development, as greater innovation leads to more returns. In a perfect world, the knowledge that the benefits will flow widely should reduce the tendency for regions to compete to attract these research clusters. But given past practice of governments intervening in agri-food policy, it is very likely that competition for investment attraction will accelerate. The result could be higher investment in research, excessive 'creative destruction', lower returns to the public treasury (due to payments of subsidies) and mixed returns to research companies, producers and consumers.

Thirdly, the introduction of new technologies has precipitated a major industrial restructuring in the agri-food sector. The economies of scale in the agricultural biotechnology industry, due to the high cost of entry and extensive regulatory hurdles during product commercialization, have led to an industrial structure somewhat different from the conventional view of knowledge-based industry. Each year the agricultural biotechnology sector has a smaller number of increasingly large multinational enterprises, which have integrated vertically and absorbed many of the smaller research and service companies created over recent years. At the other end of the scale, there are very few entrepreneurial start-ups annually and few independent on-going small or medium-sized ventures. This is almost diametrically opposed to the experience in the information technology and computer software industries, where the barriers to entry are relatively small and there are few regulatory hurdles. That industry is, for the most part, driven by entrepreneurial start-ups, with a large number of small and medium-sized companies and a few large enterprises.

The opportunity presented by biotechnology to manage an integrated research process to deliver custom products has presented attractive investment opportunities for some private companies while creating a threat or risk to many others. The first-generation biotechnology innovations lowered the cost of production (e.g. herbicide tolerance and Bt resistance) and created significant overall returns to the industry and society, while second- and third-generation products (which change the value to the end consumer) have even greater potential to generate profit. Just and Hueth (1993) were among the first to point out that the chemical companies, in particular, have significant potential to gain or lose. The companies themselves were almost as quick to realize that the new ability to manipulate the genetic coding of crops changed both their markets and their business itself. The new technologies created the opportunity for them to develop new varieties tolerant to their patented herbicides, thereby opening new markets for their product. At the same time, some companies recognized that biotechnology-based products unrelated to chemicals had the potential to produce significantly higher profits than many of their traditional lines of business. Many large multinational companies that were upstream or downstream of the farmer in the food system have restructured to capture this business, often divesting themselves of lower-return activities and replacing those with direct investments or acquisitions of knowledge-based parts of the production system. More recently, the presence of significant economies of scale and scope in the product-development end of the business has tended to cause firms and research facilities to merge or ally themselves into larger groups. The difficulties in protecting and transferring technologies among partners, combined with the imperative to quality-assure the product through the production and marketing chain, have also driven the firms to integrate vertically, so that the production system from basic genetics right through to the consumer is increasingly tightly managed and controlled within a small group of industrial networks (some call them the 'life-science companies'). The role of intermediate processors in directing and funding research in search of novel output traits is further encouraging consolidation. In many cases these companies are the sole buyers of the resulting product and have the potential to exert significant influence over market conditions. The larger life-science companies have more capacity both to manage and weather risk than smaller enterprises.

This trend has been clear in canola. As discussed in Chapter 8, these large life-science companies have acquired most of the independent seed and breeding programmes. Meanwhile, mergers and acquisitions in the global agri-business market have caused significant consolidations in the research, breeding and seed industry. The canola industry provides a unique and revealing perspective on how the companies have been positioning themselves. These companies have integrated through mergers, acquisitions or expansion to manage a larger portion of the vertical supply chain (by buying genetics, seed and technology companies) and to control a larger share of horizontal markets at key stages in the production chain (through such mergers as AgrEvo/Rhône Poulenc). The desire to control greater parts of the canola research and development business clearly was a significant part of the rationale for Monsanto purchasing Calgene, for BASF undertaking a joint venture with Svalof, for the AgrEvo/Rhône Poulenc merger, for Cargill acquiring InterMountain Canola and for Dow purchasing Mycogen. The same logic has been and will continue to drive mergers, acquisitions and strategic investments in other parts of the agro-biotechnology industry.

Nevertheless, the economies of scale and scope observed and sought in the biotechnology industry are not likely great enough to drive the industry to the levels of concentration seen in the traditional agrochemical business. Many of the steps in the research and breeding process do not exhibit economies of scale (e.g. seed multiplication and farming). While there may be some natural monopolies (e.g. transformation systems), it is not yet clear how governments will respond. In other sectors, governments have either expropriated and offered the technology as a public service or have regulated the approved economic returns and competitive access to the monopolistic technology. A well-regulated natural monopoly could have significant potential to offset the market power of the life-science companies.

Fourthly, biotechnology is both changing national and regional comparative advantages and raising the trade dependence of the new technologies. Agri-food production, which has been relatively capital intensive for almost a generation, is now also relatively research intensive, due to the sharp rise in private investment in the sector since 1980. As discussed in Chapter 1, economic trade theory suggests that production will tend to locate in regions of the world where the local factor endowments favourably match with the factor requirements of a good. Although theory suggests technology should flow as easily as products, in practice technologies do not disperse rapidly or completely – there is an increasing technology or productivity gap developing between low-technology countries and research-intensive countries (Romer, 1990). Coincident with the rising research intensity of agriculture has been a significant push from the WTO agreement in 1995 to reduce barriers to agri-food trade. Now, agricultural production is being driven by factor endowments, the key one of which is research. As a result, production of research-intensive products is tending to concentrate in and around research centres, mostly in developed countries. As Chapter 13 notes, canola production, in particular, now is significantly more concentrated in research-endowed countries than it was in the 1960s. Two other related trends are increasing the dependence on trade. Many of the new types of agriculture produce being developed, such as designer canola oils, are superior goods, with rising per capita consumption strongly positively correlated with per capita incomes. Yet the factor intensities of producing these goods do not tend to correlate well with the market areas. Hence, trade is being encouraged. At the same time, many second- and third-generation biotechnology innovations exhibit strong 'product' attributes. In many cases these products, which may have relatively short product lives, often will be produced only in the country near the appropriate research centre and then traded globally; other producing countries will concentrate on other goods, which also may be niche products. As a result, traditional net exporters now both export and import. All of these trends – in economies of scale, consumer demands and product attributes – increase the dependence of the canola industry on international trade and markets, and are likely to be matched in other agri-food product lines.

The fifth, and perhaps most important economic impact of the changes under way is that there are both winners and losers. Economic studies in the past 30 years show that research in agriculture has traditionally yielded relatively high total private returns and even higher public returns. Furthermore, farmers have been estimated to get a smaller share of the returns on innovations that improve yield rather than quality and their share is depressed further whenever the related input or processing sectors are imperfectly competitive. When juxtaposed with biotechnology-based production, four general conclusions can be drawn. First, gains to research, which have been estimated to yield an internal rate of return between 20 and 95% for agri-food research (Alston and Pardey, 1998), are likely lower for biotechnology-based developments. While some selected projects to breed high-value attributes into plants and animals may still have high yields, in aggregate the action of competitive private research efforts, subsidized by public programmes, is likely to lead to excessive creative destruction and relatively low social returns. The overall rate of return for canola research is estimated already to have dropped into the single-digit range. If the developments seen in the canola sector are repeated in other agri-food areas, similar results can be expected. Second, past studies have shown that the gains from yield-enhancing innovations are often bid away by competitive farmers and translate into lower retail prices and higher consumer welfare;

hence first-generation biotechnology products (e.g. HT and Bt-resistant crops) should not benefit farmers over the long-term. Nevertheless, as discussed in Chapter 16, some individual farmers may gain as their agronomic and management circumstances allow them to profit from the new technologies. This will become even more relevant as new input and output traits are inserted or stacked in individual crops, making seeds more tailored to specific agronomic circumstances. As noted in Table 17.1, there is a wide variety of different input traits being worked on for canola alone, each of which will have a different application for different farmers. Table 17.2 lists an array of output traits currently being researched. Third, even though the research and supply side of the agri-food industry is not perfectly competitive, it is not at all certain that it will gain. Many studies suggest that high concentration ratios in the input and output sectors allow firms to capture a higher share of the return. Given that many of the first-generation innovations link chemical products to plants, some individual chemical companies will certainly gain market share and profits. Although early research by Moschini and Lapan (1997) suggests that intellectual property rights should enable research firms to price their innovations to extract a portion of the surplus that historically has been captured by consumers, there are convincing arguments that they may not be able to do so. As discussed, incomplete property rights ensure that at least part of the benefit is not captured. Monsanto has estimated that by using open-pollinated platforms, approximately 25% of their benefit would be lost to the bin-run or brown-bagged seed markets; even with their technology-use agreements, they estimate they still lose 10% of the benefit (Roth, 1999). Furthermore, as noted, Green (1997) argues that the relatively short life cycles of the new varieties and products force the innovators to share some of the returns to ensure rapid market adoption. In aggregate, it is also not clear that industry is gaining. Given the high usage of chemicals in the canola sector even before the HT crops, gains by one company were offset by lost market share by others, so that the chemical industry may have gained little from these innovations when research expenses were deducted. Fourth, a variety of studies, including two for canola, conclude that quality-enhancing innovations benefit the production system relatively more than yield improvements because they enable producers to segment the market and increase demand for their product, thereby offsetting any price-dampening effects. Second- and third-generation biotechnology products could therefore yield a higher return to the production system, some of which could accrue to farmers because they will need to be paid to produce and market in a way that protects the quality of the product.

Finally, it is important to remember that the winners and losers are distributed widely about the world. With much of the canola produced in Canada exported to other markets, any consumer benefits would flow accordingly. Similarly, in many cases the novel traits being bred into canola replace or compete with other sources of supply, with the result that producers in those sectors would likely lose due to any innovations in this sector. Table 17.2 presents some of the current sources of supply for specific novel oils currently being

Table 17.2. Potential oil properties to exploit using canola.

Class	Example	Uses	Current source
Saturated medium- and long-chain oils	Lauric and palmitic acid	Detergents, soaps, margarine	Coconut and palm oil
Mono-unsaturated oils	Oleic and erucic acid	Cooking and salad oils; lubricants, nylon, plasticizers	Olive, rapeseed, crambe, *Brassica* spp.
Polyunsaturated oils	Alpha-linolenic and gamma-linolenic acids	Paints, varnishes, cosmetics, health foods, pharmaceuticals	Flax, borage, evening primrose
Epoxy	Vernolic acid	Plasticizers	Epoxidized soybean oil, vernonia
Hydroxy	Ricinoleic acid	Polyurethane coatings, lubricants, plasticizers	Castor
Low-melting solids		Confections	Cocoa butter

Source: MacKenzie and Taylor,
http://www.pbi.nrc.ca/bulletin/may96/may96.html#seedoils

researched. Many of the producers for such products as coconuts, palm oil and cocoa are in developing countries. The distribution of winners and losers from the development of knowledge-based canola products is likely to be replicated to a greater or lesser degree in most other agri-food markets, with the impacts varying based on the industrial structure, the nature of the innovations, the location of competing suppliers and the whereabouts of consumers.

In conclusion, the agri-food industry has been fundamentally changed by the introduction of biotechnology. The ability to selectively breed and grow crops with targeted traits has opened the door for new innovation structures, a rapidly industrializing agriculture, and shifting production within and between countries. As a result, there are major winners and losers.

Political Challenges

As with any economic event where there is potential for either significant gains or losses, there is pressure for the state to respond and intervene in the market. This is especially true for the agri-food sector, which has traditionally been viewed as a strategic sector. Many assert that agricultural policy has been always at least 50% politics. The agri-food sector has been and remains a politically important economic, social, regional and electoral constituency, which produces a strategic necessity that at times has tremendous value as a geo-political tool. That historical relationship continues in the biotechnology-based agri-food industry.

There have so far been four key political responses to the advent of agricultural biotechnology which, for the most part, relate to the fundamental economic changes brought by the new technology. Governments have two conflicting goals when addressing biotechnology. On the one hand, they would like to gain for their citizens the economic benefits flowing from the new technologies. To that end, they have introduced new intellectual property rights regimes, designed supportive domestic and international trade rules and engaged as a partner and promoter of research activity. On the other hand, they worry about managing the costs of technological change (e.g. environmental, health and safety) and ensuring an equitable distribution of the benefits. This at times has led to regulations, processes and policies that limit unrestricted development. Governments have been challenged to balance the two sides – often within the same policy envelope – and have had only mixed success.

First, the most significant government response to the introduction of biotechnology was the extension of intellectual property rights to products of biotechnology. In the first instance, governments throughout the developed world extended rights to new technologies, genes and germplasm in an effort to encourage greater private investment. This started in the United States, where at least three types of intellectual property protection relating to plants are in place, but extended rapidly to new domestic rules in Canada, the EU, Australia and many other developed countries. That effort was globalized in the 1970s, with efforts to reinvigorate and extend private rights to new plant varieties through the UPOV system. That effort was continued in the 1990s, with the adoption of the Trade Related to Intellectual Property Agreement at the WTO in 1995. Now, private rights to parts or all of the biotechnology processes and products are ubiquitous. Although most economists, industry and policy advisors accept that intellectual property rights are necessary to correct market failure, there are some concerns that the specific forms of protection being extended are inappropriate for the agri-food sector. Patents are, by definition, a second-best solution. They grant a monopoly right to use an innovation for a set period of time as an incentive to encourage both research and disclosure of results. In essence, the state has judged that the long-term benefits of innovation and knowledge more than compensates for the monopoly profits innovators are allowed to extract during the life of their patents. There is a rising debate about whether the nature and structure of the protections offered are appropriate.

From the industry side, there are some concerns that the level of protection still may not be high enough. Even with the apparent international expansion of property rights to intellectual property, the system is not as simple or complete as might appear. As with the WTO trade rules, the absence of Russia and China from the agreement leaves a major hole in its coverage. In addition, many countries have yet to address the commitment to extend patent rights to whole organisms, or to implement an effective *sui generis* system for plant varieties. Canada and many other countries have so far refused to grant patents for living organisms. In the absence of a comprehensive international system, many innovators pursue US patents and rely upon the dominant role of the US in the trade system to protect their rights. The extraterritorial nature of US patent law – prohibiting imports of products produced using unlicensed technologies – has for the most part effectively extended US-style intellectual property rights to non-conforming countries.

Opinion is mixed in the industry, however. Many smaller research firms, commodity groups with research programmes and public researchers, in particular, are concerned that the concentration of intellectual property in a few firms jeopardizes research on smaller crops and agronomic traits and threatens competition. They all raise, in one fashion or other, the fear that their freedom to operate is being limited or removed. At one level the cost of searching, negotiating contracts and enforcing arrangements imposes significant costs to accessing protected property, which limit its use for marginal products or markets. In addition, there is some evidence that firms are using their property strategically, either providing access on profit-maximizing terms or limiting access to competing projects. All of these concerns suggest that more work is needed to determine whether there are real impediments to accessing protected technology.

Many worry that the unique nature of genetic material and whole organisms make them inappropriate matter for protection. With traditional utility patents on processes, disclosure through the patent increases the likelihood that other innovators could either improve or invent around the patent. Genes and whole organisms, in contrast, are unique and it not easy to see how they can be invented around. As a result, patents on them appear to be valued much more than traditional utility patents. Many point to the large premiums the life-science companies have paid for companies with germplasm collections as evidence of the undesirable influence of patents on this matter.

Some have suggested that governments may need to address these concerns, either by imposing anti-trust measures in national competition acts or by using compulsory licensing provisions in the IPR acts to force use of the technology. Some suggest that one alternative is for the state to purchase at fair market value from private companies those key enabling technologies and genes that have greater public than private value. Either way, more can and should be done to understand the impact and implications of the current system and its alternatives.

Secondly, governments have been challenged to refashion their role in agri-food research and development. Agriculture has been one area where the public sector has historically contributed a significant share of research resources and undertaken a large share of the research effort. Except for those agri-food products with effective hybrids (e.g. maize), most of the effort has been undertaken by governments, publicly funded universities or by private companies funded by public grants. Until the 1980s, that relationship held true in most product markets. Since then, new, proprietary technologies have been developed and most of the resulting crop innovations have been commercialized by private companies. As the germplasm, technologies, genes and seeds industries have been privatized, the public sector's historical role as proprietor or lead innovator has been challenged. Now, the state acts more often as a partner and promoter, creating the basic economic structure for public and private investment in R&D through direct investment and through a selection of fiscal measures targeted on the industry. This has involved a shift away from doing all of the varietal development in public or university laboratories to doing more custom work and collaborations, often on pre-commercial or non-competitive projects. Although it is next to impossible to determine explicitly the exact impact of these policies on private research, industry competitiveness and industrial location, there are a number of examples that demonstrate that, at least at the margin, these policies can and do influence private decisions about location of research effort. Governments believe the benefits of action are significant. If knowledge spillovers (e.g. know-how related to genetic transformations) are limited to a specific location (perhaps because the diffusion of the knowledge requires face-to-face interactions), then any scale economies that result will be captured by the region that undertakes that activity. Grossman and Helpman (1991) argue that 'comparative advantage evolves over time', so that if the final product of biotechnology is tradable but the innovation-based knowledge is a non-transferable intermediate factor of production, then the fact that innovation begins or is supported in one jurisdiction could indefinitely put that site on a higher trajectory of R&D and new product development. As a result, the high-technology share of GDP and of exports will be greater than otherwise, and society will be better off.

Whereas economic assessments suggest that the public sector should re-examine its role in agri-food research, historical practices suggest that governments are likely to compete even more to attract research in order to gain promising benefits. Economists would argue that this would simply lead to excessive creative destruction and dilution of the social benefits of this research. Clearly, there needs to be more investigation of the gains to research and the related incentive structures – both through intellectual property regimes and fiscal programmes – to determine the appropriate role for the state in the agri-food research sector.

Thirdly, governments are being forced to examine the international trade regime and its impact on biotechnology-based agri-food development (see Buckingham *et al.*, 1999; Buckingham and Phillips, 2001). After almost 50 years of negotiation at the GATT, where agriculture was either ignored or where negotiations were unsuccessful, the last round finally yielded an agricultural agreement. Greater reliance on trade in the agri-food sector increased industry pressures for more liberal market access, while the industrialization of farming and the general rise in the welfare of farmers throughout the developed world reduced antagonism to reform. The World Trade Organization agreement, implemented in 1995, brought agriculture under the aegis of the trade rules, including the most favoured nation and national treatment provisions and the dispute settlement system. The agreement involved commitments to liberalize market access, reduce domestic and export subsidies, extend intellectual property protection to agri-food innovations, adopt science-based sanitary and phytosanitary measures and establish international rules for labelling foods and food products. Given that 139 countries are currently members of the WTO, the extension of trade rules into the agricultural sector with the implementation of the WTO agreement in 1995 would appear to solve many of the problems facing the industry. In practice, however, the system relating to biotechnology still has some kinks, with many countries barely, if at all, complying with their commitments. In addition, China, one of the largest users of biotechnology, is not currently a member, while potential key markets, such as Russia, are waiting to join. Although negotiations for accession have begun, it will take years for them to conclude and an even longer time for adjustments to be completed. In the interim, the international system will evolve on multiple tracks: developed-country members of the WTO will increasingly liberalize, developing member states will slowly open and non-members will continue on their own route. Perhaps most importantly, the international agreements have not resolved conclusively a number of key issues around biotechnology. Two industry concerns remain on the supply side. First, although NAFTA includes rules protecting investors' rights, these rights are not currently covered by any multilateral agreement; and, second, the rapid restructuring of the industry is raising concerns about the lack of international competition rules. On the demand side, both environmental and consumer concerns, while technically addressed in the agreement, have evolved, with the potential that they may impede trade in biotechnology products.

Although there are no current efforts to resolve the concerns of producers, there is a significant effort to address environmental and consumer concerns. The BioSafety Protocol, negotiated between 1996 and 2000 by 138 countries under the auspices of the 1992 Convention on BioDiversity, provides rules for transboundary movement of GM organisms intended for environmental release and for those destined for the food chain. For living GM organisms (e.g. seeds for propagation, seedlings, fish for release), exporters will be required to obtain approval from importing countries. Within 15 days of approving a new GM variety, a country would notify a BioSafety Clearing House with information about the traits and evaluations. The first time that new GM variety is to be exported as seed, the exporting country would notify the importing country. The importing country would then decide whether to approve the shipment or decline the shipment because of risks identified through a science-based risk assessment. This process is called 'advanced informed agreement' (AIA). Transboundary movements of genetically modified organisms intended for food, feed and processing (e.g. commodities) will be exempt from the advanced informed agreement provisions. Nevertheless, exporters must label shipments with GM varieties as 'may contain' GMOs, and countries can then decide whether to import those commodities based on a scientific risk assessment. Furthermore, GMOs intended for 'contained use' (e.g. national breeding programmes and research) and GMOs in transit through other countries will not require AIAs. Although this seems straightforward, the protocol includes two features that may raise conflict in coming years. First, the text indicates that countries may, in their reviews of GMOs, consider 'socio-economic factors' (e.g. the impact on local farmers), provided they respect their other international obligations. Second, the protocol includes a so-called 'precautionary principle', whereby countries do not have to have complete scientific certainty to block imports of a GMO that they fear could be harmful to biological diversity. It is likely, given the reference in the preamble to other international obligations, that any import bans that are not based on scientific risk assessments will be constrained. As with the SPS Agreement under the WTO, temporary bans may be permitted, but it is likely that countries will need to make real efforts to undertake the scientific research to validate (or refute) the concern. Meanwhile, a similar debate is being held at the FAO-based *Codex Alimentarius*, where the EU, in cooperation with developing countries, is suggesting that any country be allowed to require labelling for a wide variety of subjective risk factors – such as the use of biotechnology – that could impede or stop international trade in biotechnology-based agri-food products. So far, these issues have not been addressed at the international level and there is no sign that they will be resolved soon.

Fourthly, in the absence of an effective set of international rules, governments around the world have been pressed to develop and adapt their domestic regulatory systems to handle concerns related to biotechnology-based research and production processes. As discussed in Chapters 10–13, there is a wide array of competing public objectives for the regulatory powers of the state. In domestic research, production and marketing systems, states attempt to mediate between public and private goals by providing a set of rules and norms that ultimately determines the extent and scope of private initiative, including research and development policy and tax incentives, intellectual property rights, competition policy, regulation of the seeds industry, regulation of the agricultural and food markets (e.g. laws governing market structure and contracts), and laws relating to the environment and public safety in the production and marketing systems. In each case, the state attempts to balance the public interests of the general citizenry with the private interests of both domestic and foreign firms. Although two alternative approaches to the problem of regulating biotechnology have evolved, neither of them adequately addresses the concerns of industry and the public. The USA and Canada, for instance, have focused on the potential risks of biotechnology in the resulting products and have adopted the 'substantial equivalency' standard, such that if the genetically modified product that is consumed has the same molecular structure as the non-GMO, no incremental labelling or regulatory restrictions are applied. As a result, those countries have, for the most part, used existing legislation, regulation and agencies to review the risks of biotechnology, which has yielded a relatively efficient system for both governments and the companies. That system – which has relatively high public confidence – assumes that consumers will accept the outcome of the process. It is not clear, however, that the North American system would be capable of handling the level of consumer antagonism to biotechnology seen in the EU. The EU, at least partly because of consumer unease, has developed a regulatory system that focuses on the technology and not on the end product. As a result, they have developed entirely new legislation and regulations, which have effectively stalled regulatory approvals. While this system appears to satisfy many consumers, the environmental movement and many advocacy groups, it has not been popular with either EU-based or foreign industry. Clearly, neither system has really addressed the gap in confidence between consumers and producers; each has simply sided with one of the perspectives. Meanwhile, most of the rest of the countries in the world have not adopted any specific measures to regulate biotechnology. This incomplete, and at times conflicting, regulatory system has created pressures for change.

Governments have to some extent failed by not finding a process to resolve the wide differences of opinion over biotechnology. Both the weak regulatory system in Europe and the 'efficient' regulatory process in North America have failed to handle consumer or citizen concerns effectively. If consumer concerns about the health and safety, environmental, economic and ethical implications of biotechnology in the agri-food system are not managed, both systems could fail, jeopardizing the potential economic benefits of biotechnology-based agriculture. The large information gap that exists between producers and consumers, especially for biotechnology-based products, is a major market failure that necessitates government action. For any resolution, domestic regulatory systems must become more credible. But more are likely to be needed. Industry will also need to accept more responsibility to manage consumer fears. Already there are signs that some efforts will be taken (Phillips and Foster, 2000). In the EU, a number of grocers have taken the lead from the Commission labelling policy and adopted corporate strategies to develop GM-free products and to label the presence of genetically modified elements in all products, in order to let the consumer decide. In the US and Canada, there are a number of industry-led efforts to develop voluntary labelling programmes to provide more transparency and thereby more choice to domestic consumers. These corporate measures have the potential to both improve the situation in one way and to make matters worse in others. While labelling will address some consumer concerns, industry-developed standards have the potential to create non-competitive conditions in many markets, with a corresponding damage to social welfare. Governments everywhere are pondering how to manage this challenge to the system.

Although public policy and the domestic and international regulatory systems have been designed largely to nurture and support development of biotechnology-based commerce, they have been challenged to respond to citizen and consumer agendas. It is too early to say how governments and international institutions will address those concerns.

Risk of Outside/Exogenous Shocks

The economic and political analysis above tends to suggest that the future of the biotechnology industry can be determined by looking at the trends. If there is one thing we know, it is that exogenous shocks are likely. Three unpredictable but conceivable developments could shock the sector away from its current development path.

First, science itself could change course. There remain many imponderables because many of the details of the metabolic engineering through genetic engineering remain to be refined. There is still much to learn to improve our understanding of how cells, plants and their organs (seeds or flowers) function or produce new polymeric compounds (Poirier, 1999). To date, for example, no transgenic oilseed brassicas have been constructed to express polygenic traits (e.g. flowering time and plant architecture). Our understanding of transgene stability, transgenic metabolism and epigenetics and the relationship between the environment and genes are still outstanding. Although the stability of gene expression in transgenics and occurrence of gene silencing and inactivation exists in transgenic oilseed rape, new research suggests that such gene silencing can be controlled (Kasschau and Carrington, 1998; Vain *et al.*, 1999). Perhaps with the emphasis on genomics and proteomics research we will have quicker access to the application of knowledge to additional innovations to canola. Genomics research from other plants, especially model plants such as *Arabidopsis thaliana*, should make available the DNA sequences for these traits. But again, as such genomic exercises are already commercially spoken for, once genes are located and their sequences are identified, their commercial exploitation may be licensed to new players, new innovators and maybe even for species other than canola. On the one hand, new breakthroughs could make genetic manipulation more predictable and hence less risky, both economically and socially. It is impossible to say what these developments could be, but the potential is always there. Perhaps the major event of the earlier history of canola, the splitting of a single newly bred seed in two halves for oil analysis and growth, again may have a context. The availability of nano-technological tools and detection tools for scaling down the volumes of a sample for measurement may put us at the dawn of a new era. Concurrent and high-thoroughput robotic analysis could allow for the analysis of nanolitres of liquids and under a dozen molecules from a cell. Scientists and marketers in the agricultural biotechnology industry wax eloquent about a utopian future where essential amino acids will be produced in rice, where antioxidants will be in wheat used for bread and where plants will be used for environmental remediation of the filthiest industrial cesspools. As Shumpeter (1954) noted, breakthrough innovations destroy much of the value of existing systems and create new winners. If new, more valuable genes are identified, new transformation processes developed or new output attributes expressed, different products might be the target of research, new actors would emerge and some existing actors could disappear. Alternatively, science could make new links between the genetic base of various food products and disease or environmental damage, causing a re-evaluation of our mix of products and processes. Already we have seen that finer testing systems lead to new correlations between our environment and diet, and our health and changing patterns of consumption.

Secondly, the human side of the industry could fail. The research, production, marketing and regulatory systems that manage the development and commercialization of new science-based products are, in the end, only a human construct. As history tells us, humans are fallible and mistakes happen. Even if the science is above reproach, it is entirely possible that poor management systems could jeopardize the industry. Already we have an example in the canola area where Limagrain inadvertently bred an unapproved Round-Up Ready™ gene construct into a variety. That material got into the marketing channel and, in a few instances, was sold to and planted by farmers before the mistake was caught. In this case there was no lasting damage, except to a few companies. Similarly, there have been a number of instances where producers have not followed the agronomic advice, thereby increasing the risk of gene drift into other fields or to weedy relatives. We also have seen farmers and others act opportunistically in the marketing channel, substituting one product for another. In the canola case, so far, no lasting damage has been done. When novel oil canola varieties with potential allergens or non-edible oils and proteins reach the market, the risks of co-mingling, either deliberately or inadvertently, will rise. Finally, as discussed above, regulatory systems in most countries may not be adequate to the needs of this developing industry. As we have seen in Europe, failures in regulatory processes can impose punitive costs on industry and consumers. In the UK, the mad-cow beef problem in the mid-1990s almost wiped out the local beef industry, as a large proportion of cattle herds had to be slaughtered. Beef consumption plummeted on the news of the risk factors and was slow to recover. Consumers remain skittish about trusting regulators, especially when they approve genetically modified foods as safe. Another exogenous risk comes from the restricted availability of genetic resources and access. Denial of access to key genetic constructs (e.g. certain pest-resistant genes) could limit the future of biotechnology in both the canola and larger agri-food sector. A failure at any one stage in the system, or worse, a simultaneous failure at multiple stages in the system, could bring an end to the short- and medium-term commercial prospects for biotechnology-based agri-food development. Given the level of unease in many markets, even a small environmental or health failure could push biotechnology along the path followed by food irradiation, which, in spite of its demonstrable economic and health benefits, languishes as a vastly underused technology.

Thirdly, unpredictable leadership could reshape what we have today. Lack of leadership in managing the safe and economical transfer of new innovations from the laboratories to the consumer's plate, as discussed above, could jeopardize the entire system. Perhaps more unpredictable is the role of leadership in creating new clusters of research and development. The ease of access to these bodies of knowledge and our ability to use them to innovate will be different. The era of the few individuals who, in Saskatoon, could form the invisible 'college of rapeseed' and work as a team has passed. Today, competitive research is global and no one has a monopoly on new ideas. Therefore, more research on fundamentals and in a cross-cutting team approach is needed. Fundamentally, the presence of commercial and industrial players on the R&D team may need to change. As noted in the canola case, the long-range vision and leadership of a few individuals was vital to the development of both the industry and to the concentration of the activity in Saskatoon. Global society ultimately gained from this new source of healthier oil and generally lower global prices for edible-oil products. As discussed, the critical early efforts were funded and led largely by the public sector. It is no longer clear that leadership of that calibre will be forthcoming under the new conditions that prevail. Given the shorter planning horizon for private capital, it is likely that few private leaders would have the vision or staying power to achieve a similarly elongated development. Furthermore, although the public sector remains important in the global agri-food research effort, much of that effort is either mirroring private, profit-maximizing strategies or is collaborating and partnering with private capital. As a result, public-sector scientists and managers have shortened their planning horizons and also may no longer be able to sustain such an effort. If that is true, society will ultimately lose out, as new innovations will have incrementally smaller social benefits. Having said that, anything can happen. It would only take one dynamic, visionary, determined leader to transform parts or all of the global agricultural biotechnology industry.

Biotechnology Today and Tomorrow

An oft-quoted Chinese curse is 'may you live in interesting times'. The advent of biotechnology in the global agri-food system has certainly created 'interesting times'. From the science perspective, it has rejuvenated interest in the agri-food sector, attracting some of the best and brightest scientists to apply their knowledge and expertise to develop new processes and products. This change caught the attention of the financial markets, and for the first time in decades, entrepreneurs and large companies have entered the industry aggressively with large quantities of capital, radically restructuring the relationships that had evolved slowly over the past century. Perhaps less slowly, but still at a pace unseen before, industrial activity has begun to move, concentrating research at discrete points around the world and causing a perceptible shift in production and trade. In spite of this, the future prosperity of the industry is not certain. Consumers have largely been left out of the system, and they are beginning to show their dissatisfaction through boycotts and political action. The ultimate destiny of this industry remains to be determined.

The fundamental question facing the agri-food world is whether biotechnology (i.e. all those tools that facilitate molecular-based improvements, including but not limited to genetic modification) will transform the industry, or is it 'dead' as Deutsche Bank declared in 1999. If there is one fundamental lesson from this study, it is that the rules and terms of reference for agricultural politics and economics are changing due to innovation. Increasingly, it is necessary to use interdisciplinary approaches to examine the multi-purpose institutions and multi-domain stakeholders that are dominating and influencing the future of the industry (Phillips and Khachatourians, 2001). In essence, the issues of innovation and investment are complex and require the alchemist's skills and magic to find the appropriate mixture of science, society and the market.

Bibliography

ABT Associates of Canada (1996) Survey of New TPR Claimants of the Scientific Research and Experimental Development Income Tax for Revenue Canada, June. Retrieved from the World-wide Web: http://www.fin.gc.ca/toce/1998/resdev_e.html

Ackerlof, G. (1970) The market for lemons: quality, uncertainty and the market mechanism. *Quarterly Journal of Economics* 84, 488–500.

Adolphe, D. (1998) Reflections. *Canola Digest* available at http://www.canola-council.org/about/Digests/Digests.htm

Agriculture and Agrifood Canada (1999) Canadian Grains and Oilseed Outlook. April. Retrieved May 31, 1999 from the World-wide Web: http://www.agr.ca/policy/winn/biweekly/English/gosd/1999/apr99e.htm

Akgüngör, S., Makanda, D.W., Oehmke, J.F., Myers, R.J. and Choe, C.Y. (1996) Dynamic analysis of Kenyan wheat research and rate of return. *Conference on Global Agricultural Policy for the Twenty-First Century, Melbourne, Australia.* 26–28 August, pp. 333–366.

Akino, M. and Hayami, U. (1975) Efficiency and Equity in Public Research: Rice Breeding in Japan's Economic Development. *American Journal of Agricultural Economics* 57, 1–10.

Alberta Agriculture, Food and Rural Development (2000) Varietal performance data – oilseed crops. Retrieved 21 Sept. 2000 from the World Wide-web: http://www.agric.gov.ab.ca/crops/canola/cropsel.html

Alberta Canola Producers' Newsletter (1997) Recording the Research (November). Retrieved from the World-wide Web: http://www.canola-council.org/orgs/acpc/013098.htm#6.

Alston, J. (1991) Research benefits in a multimarket setting: a review. *Review of Marketing and Agricultural Economics* 59(1), 23–52.

Alston, J. and Carter, H. (1994) *Valuing California's Agricultural Research and Extension.* University of California Agricultural Issues Center Publication No. VR-1, Davis.

Alston, J. and Mullen, J. (1992) Economic effects of research into traded goods: the case of Australian wool. *Journal of Agricultural Economics* 43, 268–278.

Alston, J. and Scobie, G. (1983) Distribution of research gains in multistage production systems: comment. *American Journal of Agricultural Economics* 65, 253–256.

Alston, J., Edwards, G. and Freebairn, J. (1988) Market distortions and the benefits from research. *American Journal of Agricultural Economics* 70, 281–288.

Alston, J., Norton, G. and Pardey, P. (1995) *Science Under Scarcity: Principles and Practice of Agricultural Research Evaluation and Priority Setting*. Cornell University Press, Ithaca, New York.

Alston, J., Sexton, R. and Zhang, M. (1997) The effects of imperfect competition on size and distribution of research benefits. *American Journal of Agricultural Economics* 79(November), 1252–1265.

Alston, J., Marra, M., Pardey, P. and Wyatt, T. (1998a) *Research Returns Redux: a Meta-Analysis of the Returns to Agricultural R&D*, Environment and Production Technology Division Discussion Paper No. 38. International Food Policy Research Institute, Washington, DC.

Alston, J., Craig, B. and Pardey, P. (1998b) *Dynamics in the Creation and Depreciation of Knowledge, and the Returns to Research*. GPTD Discussion Paper, No. 35. International Food Policy Research Institute, Washington, DC.

Angus Reid Group Inc. (1999) *Biotechology in Foods*. Angus Reid Group, December.

Appelquist, L. (1972) Historical background. In: Appelquist, L.A. and Ohlson, R. (eds) *Rapeseed: Cultivation, Composition, Processing and Utilization*. Elsevier, Amsterdam, pp. 1–8.

Arrow, K. (ed.) (1988) The balance between industry and agriculture in economic development. *The Basics Proceedings of the 8th World Congress of the IEA*, Vol. 1, Delhi, India. Macmillan Press, London.

Asmussen, E. and Berriot, C. (1993) Le crédit d'impôt recherche, coût et effet incitatif, study for the Ministère de l'Economie et des Finances, April. Canadian Intellectual Property Office. Retrieved from the World-wide Web: http://strategis.ic.gc.ca/ sc_mrksv/cipo/prod_ ser/online/guides_e/pateng/III.html

Audretsch, D. (1998) Agglomeration and the location of innovative activity. *Oxford Review of Economic Policy* 14(2), 18–29.

Ayer, H. and Schuh, G. (1972) Social rates of returns and other aspects of agricultural research: the case of cotton research in Sao Paulo, Brazil. *American Journal of Agricultural Economics* 54, 557–569.

Baldwin, R. and Krugman, P. (1988) Industrial policy and international competition in wide-bodied jet aircraft. In: Baldwin, R. (ed.) *Trade Policy Issues and Empirical Analysis*. University of Chicago Press for the NBER, Chicago.

Baldwin, R. and Krugman, P. (1992) Market access and international competition: A simulation study of 16K random access memories. In: Grossman, G. (ed.) *Imperfect Competition and International Trade*. The MIT Press, Cambridge, Massachusetts.

Baltimore, D. (1982) Priorities in biotechnology. In: *Priorities in Biotechnology Research for International Development: Proceedings of a Workshop*. National Academy Press, Washington, DC.

Barzel, Y. (1989) Economic analysis of property rights. In: Alt, J. and North, D. (eds) *Political Economy of Institutions and Decisions*. Cambridge University Press, Cambridge.

Bell, J.M. (1955) The nutritional value of rapeseed meal: a review. *Canadian Journal of Agriculture Science* 35, 242–251.

Bell, J.M. (1995) Meal and by-product utilization in animal nutrition. In: Kimber, D.S. and McGregor, D.I. (eds) *Brassica Oilseeds: Production and Utilization*. CAB International, Wallingford, UK, pp. 301–337.

Bell, J.M. and Wetter, L. (1974) Utilization of rapeseed meal. In: *The Story of Rapeseed in Western Canada*. Saskatchewan Wheat Pool, Regina, Saskatchewan, Chapter 3.

Blakely, R. and Anderson, W. (1948) The effects of various levels of rapeseed oilcake meal intake diet on the weight of the thyroid glands of turkey poults. *Scientific Agriculture* 28, 393–397.

Boulter, G. (1983) The history and marketing of rapeseed oil in Canada. In: Kramer, J.K.G., Sauer, F.D. and Pigden, W.J. (eds) *High and Low Erucic Acid Rapeseed Oils*. Academic Press Canada, Toronto, Ontario.

Bowditch Group (1999) Retrieved from the World Wide Web.

Bowland, J., Clandinin, D. and Wetter, L. (1965) *Rapeseed Meal for Livestock and Poultry– a Review*. Publication No. 1257. Canadian Department of Agriculture, Ottawa.

Braun, R. (1999) Swiss referendum on genetic engineering, Briefing paper #8. *Australian Biotechnology* 9(1), 38–42.

Brill, W. (1988) Why engineered organisms are safe. *Issues in Science and Technology* 4(3), 44–50.

Buckingham, D. and Phillips, P. (2001) Hot potato, hot potato: regulating products of biotechnology by the international community. *Journal of World Trade Law* (in press).

Buckingham, D., Gray, R., Phillips, P., Roberts, T., Bryce, J., Morris, B., Stovin, D., Isaac, G. and Anderson, B. (1999) *The International Coordination of Regulatory Appoaches to Products of Biotechnology*. Agriculture and Agri-Food Canada.

Bureau, J.-C., Marette, S. and Schiavina, A. (1997) Trade, labels and consumer information: the case of hormone-treated beef. Contributed Paper, *XXII Conference of IAAE, Sacramento, August*.

Busch, L. and Tanaka, K. (1996) Rites of passage: constructing quality in a commodity subsector. *Science, Technology and Human Values* 21(1), 3–27.

Busch, L., Gunter, V., Mentele, F., Tachikawa, M. and Tanaka, K. (1994) Socializing nature: technoscience and the transformation of rapeseed into canola. *Crop Science Society of America* 34(3), 607–614.

Buzza, G. (1995) Plant breeding. In: Kimber, D. and McGregor, D. (eds) *Brassica Oilseeds – Production and Utilization*. CAB International, Wallingford, UK, pp. 153–176.

Calgene (1998) http://calgene.com/oils/index.html, p.4

Canola Council of Canada (1995) *Canada's Canola*. Canola Council of Canada, Winnipeg, Manitoba.

Canola Council of Canada (1998a) Provincial Acreages and Yields. December. Retrieved 31 May 1999 from the World-wide Web: http://www.canola- council.org/stats/acreage productionandyield.htm

Canola Council of Canada (1998b) http://www.canola-council.org/stats/acreageproductionandyield. htm. December.

Canola Council of Canada (1998c) http://www.canola-council.org/manual/2 varint.htm. December.

Canola Council of Canada (1999) *Canola connection* (http://www.canola-council.org/index.shtml).

Canola Research Survey (1997) College of Agriculture, University of Saskatchewan.

Cantley, M. (1995) The regulation of modern biotechnology: a historical and European perspective: a case study in how societies cope with new knowledge in the last quarter of the twentieth century. In: Rehm, H. and Reed, G. (eds) (in cooperation with Puhler, A. and Stadler, P.) *Biotechnology*, 2nd edn. Germany.

Cantley, M. (1998) Seeds of self-destruction: over-regulation is stifling biotechnology in Europe. Personal View, *The Financial Times* December, London.

Carew, R. (2000) Intellectual property rights: implications for the canola sector and publicly funded research. *Canadian Journal of Agricultural Economics* 48, 175–194.

Carr, R. (1995) Canola oil processing. In: Kimber, D. and McGregor, D. (eds) *Brassica Oilseeds: Production and Utilization*. CAB International, Wallingford, UK.

CFIA (Canadian Food Inspection Agency) (1998a) *Canadian Varieties, January 1, 1923 to June 24, 1998*. Special tabulation from the Plant Health and Production Division.

CFIA (Canadian Food Inspection Agency) (1998b) Status of New Plant Varieties applied for Protection under PBR: List of Varieties, by Crop Kind. Retrieved from the Worldwide Web at http://www.cfia-acia.agr.ca/english/plant/pbr/home_e.html

CFIA (Canadian Food Inspection Agency) (1999) *Summary of Experimental Releases*. Special tabulation for *Brassica napus* and *rapa*.

Chaudhary, S., Parmenter, D. and Moloney, M. (1998) Transgenic *Brassica carinata* as a vehicle for the production of recombinant proteins in seeds. *Plant Cell Reports* 17, 195–200.

Chavas, J.-P. and Cox, T. (1992) A nonparametric analysis of the effects of research on agricultural productivity. *American Journal of Agricultural Economics* 74(August), 583–591.

Chen, Z. and McDermot, A. (1997) International comparisons of biotechnology policies. Retrieved from the World-wide Web at http://strategis.ic.gc.ca/SSG/ca00910e.html

Choate, G. and Maser, S. (1992) The impact of asset specificity on single-period contracting. *Journal of Economic Behavior and Organization* 18, 373–389.

Christaller, W. (1933) *Central Places in Southern Germany*. Fischer, Jena.

CIPO (Canadian Intellectual Property Office) (1998) Retrieved from the World-wide Web: http://strategis.ic.gc.ca/sc_mrksv/cipo/prod_ser/online/guides_e/pateng/III.html

Coase, R. (1937) The nature of the firm. *Economica* 4, 386–405.

Coase, R. (1960) The problem of social cost. *The Journal of Law and Economics* 3, 1–44.

Coombs, J. and Campbell, P. (1991) *Biotechnology Worldwide*. CPL Press, Newbury.

CSGA (Canadian Seed Growers Association) (1994) *Regulations and Procedures for Pedigree Seed Crop Production*. Circular 6. CSGA, Ottawa, Canada.

Dahlman, C. (1979) On the problem of externality. *Journal of Law and Economics* 22, 141–162.

Del Vecchio, A. (1996) High laurate canola. *Inform* 7, 230.

Doern, B. (1997) The federal biotechnology regulatory system: a commentary on an institutional work in progress. Retrieved from the World-wide Web at http://strategis.ic.gc.ca/SSG/ca00911e.html

Downey, K. and Harvey, B. (1963) Methods for breeding for oil quality in rape. *Canadian Journal of Plant Science* 43, 271–275.

Downey, K. and Rimmer, S. (1993) Agronomic improvement in oilseed brassicas. *Advances in Agronomy* 50, 1–66.

Downey, K. and Robbelen, G. (1989) Brassica species. In: Downey, K., Robbelen, G. and Ashri, A. (eds) *Oil Crops of the World: their Breeding and Utilization*. McGraw-Hill, New York.

Downey, K., Klassen, A. and Pawlowski, S. (1974) Breeding quality improvements into Canadian Brassica oilseed crops. *Proceedings of the 4th International Rapeseed Congress*, Giessen, Germany, pp. 57–61.

Drucker, P. (1993) *Post Capitalist Society*. Harper Books, New York.

Dupont, J., White, P., Jonston, M., Heggtveity, H., McDonald, B., Crundy, S. and Bonanoma, A. (1989) Food safety and health effects of canola oil. *Journal of The American College of Nutrition* 8, 360–375.

Eggertsson, T. (1994) The economics of institutions in transaction economics. In: Schiavo-Campo, S. (ed.) *Institutional Change and The Public Sector in Transactional Economics*. The World Bank, Washington, DC.

EuropaBio (1997) Biotechnology: Consumer/User Information and Informed Choice for Products of GM and non-GM Sources. Position Document 267, June.

European Commission (1983) *Biotechnology in the Community: Communication from the Commission to the Council*. COM(83) 672 Final, October, Brussels.

European Commission (1986) A Community Framework for the Regulation of Biotechnology: Communication from the Commission to the Council. COM(86) 573 Final, June, Brussels.

European Commission (1988a) *The European Community and the Contained Use of Genetically Modified Micro-Organisms*. DG-XI/A/2, Brussels.

European Commission (1988b) Proposal for a Council Directive on the deliberate release to the environment of genetically modified organisms: COM(88) 160 Final – SYN 131. *Official Journal* C 198. 19–27 July, Brussels.

European Council (1990a) Council Directive of 23 April 1990 on the contained use of genetically modified micro-organisms: 90/219/EEC. *Official Journal* L No.117, May, Brussels.

European Council (1990b) Council Directive of 23 April 1990 on the deliberate release into the environment of genetically modified organisms: 90/220/EEC. *Official Journal* L No.117, May, Brussels.

Evans, K. and Scarisbrick, D. (1994) Integrated insect pest management in oilseed rape crops in Europe. *Crop Protection* 13, 403–412.

Evenson, R. (1967) The contribution of agricultural research to production. *Journal of Farm Economics* 49, 1415–1425.

Evenson, R. (1996) Two blades of grass: research for U.S. Agriculture. In: Antle, J. and Sumner, D. (eds) *The Economics of Agriculture*, Vol. 2, *Papers in Honor of D. Gale Johnson*. University of Chicago Press, Chicago, pp. 171–203.

Evenson, R. (1998) The economics of intellectual property rights for agricultural technology introduction. Paper presented to the NC-208 meeting at CIMMYT, 6 March.

Falck-Zepeda, J.B., Traxler, G. and Nelson, R.G. (2000) Surplus distribution from the introduction of a biotechnology innovation. *American Journal of Agricultural Economics*, 360–369.

Falk, K., Rakow, G. and Downey, K. (1998) The utilization of heterosis for seed yield in hybrid and synthetic cultivars of summer turnip rape (*B. rapa* L.). *Canadian Journal of Plant Science* 78, 383–387.

FAO (Food and Agriculture Organization) (1998) *State of the World's Plant Genetic Resources for Food and Agriculture*. Appendix 1. FAO, Rome.

FDA (Food and Drug Administration, US) (1985) *Direct Food Substances Affirmed as Generally Recognized as Safe; Low Erucic Acid Rapeseed Oil*. Federal Food and Cosmetic Act. 21 CFR Part 184, 1555 (c). Food and Drug Administration, Washington, DC, Amended 9 January.

Finance Canada (1998) The federal system of income tax incentives for scientific research and experimental development: evaluation report prepared by the Department of Finance Canada and Revenue Canada. http://www.fin.gc.ca/resdev/fedsys_e.html

Forhan, M. (1993) Canola breeding goes private. *Canola Guide*, pp. 28–31.

Freebairn, J., Davis, J. and Edwards, G. (1982) Distribution of research gains in multi-stage production systems. *American Journal of Agricultral Economics* 64(1), 39–46.

Fuglie, K., Ballenger, N., Day, K., Klotz, C., Ollinger, M., Reilly, J., Vasavada, U. and Yee, J. (1996) *Agricultural Research and Development: Public and Private Investment Under Alternative Markets and Institutions.* AER-735, USDA.

Fukuyama, F. (1995) *Trust: the Social Virtues and the Creation of Prosperity.* Penguin, London.

Fulton, M. (1997) The economics of intellectual property rights: discussion. *American Journal of Agricultural Economics* 79, 1592–1594.

Fulton, M. and Keyowski, L. (1999) The producer benefits of herbicide-resistant canola. *AgBioForum* 2, 2; retrieved at http://www.agbioforum.missouri.edu/

Gadbow, R. and Richards, T. (eds) (1990) *Intellectual Property Rights – Global Consensus, Global Conflict?* Westview Press, Boulder, Colorado.

Gianessi, L. and Carpenter, J. (1999) *Agricultural Biotechnology: Insect Control Benefits.* National Centre for Food and Agricultural Policy, Washington, DC.

Giannakas, K. (1999) Enforcement issues in GMO segregation systems. Paper presented to the Economics of Quality Control in Agriculture Conference, Saskatoon, 3 December.

Giannakas, K. and Fulton, M. (2000) Market effects of genetic modification under alternative labeling regimes and enforcement scenarios. Paper presented at the 2000 American Agricultural Economics Association Annual Meeting, Tampa, Florida.

Gibbons, M. (1995) *Technology and the Economy: a Review of Some Critical Relationships.* Micro-Economic Policy Analysis, Industry Canada.

Gold, E.R. (2000) Patents in Genes. Paper commissioned by Canadian Biotechnology Advisory Committee, available at http://cbac.gc.ca/

Gray, R. and Malla, S. (2001) The evaluation of the economic and external health benefits from canola research. In: Alston, J., Pardey, P. and Taylor, M. (eds) *Agriculture Science Policy.* IFPRI and Johns Hopkins University Press, Baltimore, Maryland.

Gray, R. and Malla, S. (2001) The evaluation of the economic and external health benefits from canola research. In: Alston, J.M., Pardey, P.G. and Taylor, M. (eds) *Agricultural Science Policy.* IFPRI and Johns Hopkins University Press.

Gray, R., Malla, S. and Phillips, P. (1999) *The Effectiveness of the Research Funding in the Canola Industry.* Saskatchewan Agriculture and Food.

Green, C. (1997) The industrial economics of biotechnology. Retrieved from the World-wide Web at http://strategis.ic.gc.ca/SSG/ca00913e.html

Griffith, G. and Meikle, K. (1993) The impact on the Canadian rapeseed industry of removing EEC import tariffs. *Journal of Policy Modelling* 5, 37.

Griliches, Z. (1958) Research costs and social returns: hybrid corn and related innovations. *Journal of Political Economy* 66, 419–431.

Grossman, G. and Helpman, E. (1991) *Innovation and Growth in the Global Economy.* The MIT Press, London.

Grossman, S. and Hart, O. (1986) The costs and benefits of ownership: a theory of vertical and lateral integration. *Journal of Political Economy* 94, 691–719.

Hall, B. (1993) R&D tax policy during the 1980s: success or failure? *Tax Policy and The Economy* 7.

Hart, F., Vincent, B. and Bubber, J. (1997) Industry perspectives on the UN biosafety discussions: agricultural commodities issues. Discussion Paper Prepared for Industrial Biotechnology Association of Canada, Canadian Seed Trade Association, Canola Council of Canada, Saskatchewan Wheat Pool, Ag-West Biotech.

Harvey, D. (1989) Agricultural research priorities. *Outlook on Agriculture* 18(2), 77–84.

Hayami, Y. and Herdt, R. (1977) Market price effects of technological change on income distribution in semisubsistence agriculture. *American Journal of Agricultural Economics* 59, 245–256.

Hayenga, M. (1998) Structural change in the biotech seed and chemical industrial complex. *AgBioForum* 1(2), 43–55. Retrieved 31 May 1999 from the World-wide Web:

Hedley, H. (1998) Pursuing Regulatory Approvals For Genetically Modified Crops in Europe. Presentation to a seminar held 29 January at Innovation Place in Saskatoon. Retrieved 11 January 2000, from the World-wide Web: http://www.agwest.sk.ca/

Heller, J. (1995) James G. Heller Consulting, June. Quoted in Goudey, J. and Nath, D. (1997) *Canadian Biotech '97: Coming of Age.* Ernst & Young, Toronto.

Hertford, R., Ardila, J., Rocha, A. and Trujillo, C. (1977) Productivity of agricultural research in Colombia. In: Arndt, T. *et al.* (eds) *Resource Allocation and Productivity in National and International Agricultural Research,* pp. 86–123.

Hills, M. and Murphy, D. (1991) Biotechnology of oilseeds. *Biotechnology and Genetic Engineering Reviews* 9, 1–46.

Hoban, T. (1997) Consumer acceptance of biotechnology: an international perspective. *Nature Biotechnology* 15, 232–234.

Hoban, T. (1998a) International acceptance of agricultural biotechnology. *NABC Report 10, Agricultural Biotechnology and Environmental Quality: Gene Escape and Pest Resistance.* NABC, Ithaca, New York.

Hoban, T. (1998b) Trends in consumer attitudes about agricultural biotechnology. *AgBioForum* 1(1); retrieved from the World-wide Web: http://www.agbioforum. missouri.edu/

Hoban, T.J. (2000) Public perceptions of transgenic plants. In: Khachatourians, G.G., McHughen, A., Nip, W.-K., Scorza, R. and Hui, Y.H. (eds) *Transgenic Plants.* Marcel Dekker, New York.

Hodgson, J. (1995) Biotechnology: feeding the world. In: Fransman, M., Junne, G. and Roobeek, A. (eds) *The Biotechnology Revolution?* Blackwell Publishers, Oxford.

Huang, S. and Sexton, R. (1996) Measuring returns to an innovation in an imperfectly competitive market: application to mechanical harvesting of processing tomatoes in Taiwan. *American Journal of Agricultural Economics* 78, 558–571.

Huffman, W. and Evenson, R. (1989) Supply and demand functions for multiproduct U.S. cash grain farms: biases caused by research and other policies. *American Journal of Agricultural Economics* 71, 761–773.

Huffman, W. and Evenson, R. (1992) Contributions of public and private science and technology to U.S. agricultural productivity. *American Journal of Agricultural Economics* 74, 752–756.

Huffman, W. and Evenson, R. (1993) *Science for Agriculture: a Long-Term Perspective.* Iowa State University Press, Ames.

Hui, Y. and Khachatourians, G. (1995) *Food Biotechnology: Microorganisms.* John Wiley & Sons, New York, p. 976.

IBM (2000) Intellectual property network retrieved US patent data from the world-wide web: http://www.patents.ibm.com/

Industry Commission (1995) *Research and Development,* 3 Volumes. Australian Government Publishing Service, Canberra.

Isaac, G. and Phillips, P. (1999) *The Potential Impact of the Biosafety Protocol: The Agricultural Commodities Case.* Final Report for the Biotechnology Unit of the Canadian Food Inspection Agency. CFIA Biotechnology Unit, Ottawa.

Isard, W. (1956) *Location and Space-economy*. MIT Press, Cambridge, Massachusetts.

ISI (Institute for Scientific Investigation) (1997) Citations database, special tabulation of academic publications based on key word search for canola, November.

Jacquemin, A. (1987) *The New Industrial Organization*. MIT Press, Cambridge, Massachusetts.

James, C. (1999) Transgenic crops worldwide: current situation and future outlook. Paper presented at the Conference 'Agricultural Biotechnology in Developing Countries: Towards Optimizing the Benefits for the Poor' organized by ZEF and ISAAA in collaboration with AgrEvo and DSE in Bonn, 15–16 November, 1999. Retrieved November 30, 1999 from the World-wide Web: http://www.zef.de/zef_englisch/f_veranstalt_biotech.htm

Joskow, P. (1987) Contract duration and relationship-specific investments: empirical evidence from coal market. *American Economic Review* 71, 168–185.

Josling, T. (1998) *Agricultural Trade Policy: Completing the Reform*. Institute for International Economics, Washington, DC.

Juska, A. and Busch, L. (1994) The production of knowledge and the production of commodities: The case of rapeseed technoscience. *Rural Sociology* 59(4), 581–597.

Just, R.E. and Hueth, D.L. (1993) Multimarket exploitation: the case of biotechnology and chemicals. *American Journal of Agricultural Economics* 75, 936–945.

Kasschau, K.D. and Carrington, J.C. (1998) A counter defensive strategy of plant viruses: suppression of post transcriptional gene silencing. *Cell* 95, 461–470.

Katz, J., Hicks, D., Sharp, M. and Martin, B. (1995) *The Changing Shape of British Science*. Science Policy Research Unit, University of Sussex, Brighton.

Kenney, M. (1986) *Biotechnology: the University–Industrial Complex*. Yale University Press, New Haven, Connecticut.

Keyowski, L. (1998) The impact of technological innovation on producer returns: the case of transgenic canola. BSc. in Agriculture thesis. University of Saskatchewan, Saskatoon, Canada.

Khachatourians, G.G. (2000) Agriculture and food crops: development, science and society. In: Khachatourians, G.G., McHughen, A., Nip, W.-K., Scorza, R. and Hui, Y.H. (eds) *Transgenic Plants and Crops*. Marcel Dekker, New York.

Khachatourians, G., McHughen, A., Nip, W.-K., Scorza, R. and Hui, Y. (2001) *Transgenic Plants and Crops*. Marcel Dekker, New York.

Kimber, D. and McGregor, D. (1995) The species and their origin, cultivation and world production. In: Kimber, D. and McGregor, D. (eds) *Brassica Oilseeds: Production and Utilization*. CAB International, Wallingford, UK, pp. 1–8.

Klein, B. and Crawford, R.G. (1978) Vertical integration, appropriable rents, and the competitive contracting process. *Journal of Law and Economics* 21, 297–326.

Klein, S. and Rosenberg, N. (1986) An overview of innovation. In: Landau, R. and Rosenberg, N. (eds) *The Positive Sum Strategy: Harnessing Technology for Economic Growth*. National Academy Press, Washington, DC.

Kneen, B. (1992) *The Rape of Canola*. NC Press, Toronto.

Koss, P., Eaton, A. and Curtis, B. (1997) Co-specific investments, hold-up and self-enforcing contracts. *Journal of Economic Behavior and Organization* 32(3), 457–470.

Kozul-Wright, Z. (1995) The role of the firm in the innovation process. UNCTAD Discussion Paper 98, April, p. 3.

KPMG (1997) Executive briefing on the economic opportunities in canola research. Regina, Canada.

Krugman, P. (1998) What's new about the new economic geography? *Oxford Review of Economic Policy* 14(2), 7–17.

Landau, R. and Rosenberg, N. (eds) (1986) *The Positive Sum Strategy: Harnessing Technology for Economic Growth.* National Academy Press, Washington, DC.

Lee, F. and Has, H. (1996) A quantitative assessment of high-knowledge industries versus low-knowledge industries. In: Howitt, P. (ed.) *The Implications of Knowledge Based Growth for Micro-economic Policies.* University of Calgary Press, pp. 39–76.

Leiby, J. and Adams, G. (1991) The returns to agricultural research in Maine: the case of a small northeastern experiment station. *Northeastern Journal of Agricultural and Resource Economics* 20, 1–14.

Lemieux, C. and Wohlgenant, M. (1989) Ex ante evaluation of the economic impact of agricultural biotechnology: the case of porcine somatotropin. *American Journal of Agricultural Economics* 71, 903–914.

Lesser, W. (1994) Global interdependence and the private sector. *NABC Report 6: Agricultural Biotechnology and the Public Good.* NABC, Ithaca, New York.

Lesser, W. (1998a) Managing IPR and the agricultural research process: lessons from university technology transfer offices. Paper presented to the NC-208 meeting at CIMMYT, 5 March.

Lesser, W. (1998b) Intellectual property rights and concentration in agricultural biotechnology. *AgBioForum* 1(2), 56–61; retrieved 1 January, 1999 from the World-wide Web: http://www.agbioforum.missouri.edu

Lucas, R. (1988) On the mechanics of economic development. *Journal of Monetary Economics* 22, 30–42.

Lundvall, B.-A. (1992) *National Systems of Innovation: Towards a Theory of Innovation and Interactive Learning.* Pinter, New York.

Mahoney, J. (1992) The choice of organisational form: vertical financial ownership versus other methods of vertical integration. *Strategic Management Journal* 13, 576.

Majone, G. (1990) *Deregulation vs. Reregulation: Regulatory Reform in Europe and the United States.* Pinter, London.

Majone, G. (1994) *Interdependence vs. Accountability: Non-Majoritarian Institutions and Democratic Government in Europe.* EUI Working Paper SPS 94/3. European University Institute, San Domenico di Fiesole.

Makki, S., Tweeten, L. and Thraeu, C. (1996) Returns to agricultural research: are we assessing right?' *Conference on Global Agricultural Science Policy for the Twenty-First Century, Melbourne, Australia, 26–28 August,* pp. 89–114.

Malecki, E. (1997) *Technology and Economic Development: the Dynamics of Local, Regional and National Competitiveness.* Longman, Toronto.

Malla, S. (1995) The distribution of the economic and health benefits from canola research. MSc. thesis, University of Saskatchewan, Saskatoon, Canada.

Malla, S., Gray, R. and Stephen, A. (1995) The Health Care Cost Savings due to Canola Consumption in Canada. Saskatchewan Agriculture and Food.

Malla, S., Gray, R. and Phillips, P. (1998) *The Evolution of Canola Research: an Example of Success.* Saskatchewan Agriculture and Food.

Manitoba Crop Insurance Corporation (1998) *Variety Summary.* 1993–1997.

Marshall, A. (1980) *Principles of Economics.* Macmillan, London.

Mayer, H. (1997) The economics of transgenic herbicide-tolerant canola. Master's thesis, University of Saskatchewan, Saskatoon.

McAnsh, J. (No date) The First Six Years – 1967–1973. Rapeseed Association of Canada.

McHughen, A. (2000) *Pandora's Picnic Basket: the Potential and Hazards of Genetically Modified Foods*. Oxford University Press, Oxford.

McLeod, A. (ed.) (1974) *The Story of Rapeseed in Western Canada*. Saskatchewan Wheat Pool, Regina.

Meilke, K. (1998) *Trade Liberalization of the International Oilseed Complex*. Economic and Policy Analysis Directorate, Agriculture and Agri-food Canada.

Metcalfe, J. (1995) Technology systems and technology policy in an evolutionary framework. *Cambridge Journal of Economics* 19, 25–46.

Milgrom, P. and Roberts, J. (1992) *Economics of Organization and Management*. Prentice-Hall, Englewood Cliffs, New Jersey.

Moed, H., Burger, W., Frankfort, J. and van Raan, A. (1985) The use of bibliometrics data for the measurement of university research performance. *Research Policy* 23(2), 187–222.

Moloney, M. (1995) The future of molecular farming: An interview. *PBI Bulletin*, August. Retrieved from the World-wide Web: http://www.pbi.nrc.ca/bulletin/aug95/aug95.html#MOLFARM

Moschini, G. and Lapan, H. (1997) Intellectual property rights and the welfare effects of agricultural R&D. *American Journal of Agricultural Economics* 79, 1229–1242.

Moschini, G., Lapan, H. and Sobolevsky, A. (1999) Trading technology as well as final products: Roundup Ready™ soybeans and welfare effects in the soybean complex. *Proceedings of the ICABR Conference on The Shape of the Coming, Agricultural Biotechnology Transformation: Strategic Investment and Policy Approaches from an Economic Perspective, University of Rome, Tor Vergata*.

Mowery, D. and Oxley, J. (1995) Inward technology transfer and competitiveness: the role of national innnovation sytems. *Cambridge Journal of Economics* 19, 67–93.

Mullen, J., Wohlgenant, M. and Farris, D. (1988) Input substitution and the distribution of surplus gains from lower U.S. beef-processing costs. *American Journal of Agricultural Economics* 70, 245–254.

Mulongoy, K. (1997) Different perceptions on the International Biosafety Protocol. *Biotechnology and Development* 31, June.

Murphy, J., Furtan, H. and Schmitz, A. (1993) The gains from agricultural research under distorted trade. *Journal of Public Economics* 51, 161–172.

NABC (1993) *Agricultural Biotechnology: a Public Conversation About Risk*. NABC Report 5. National Agricultural Biotechnology Council, Ithaca, New York.

Nabli, M. and Nugent, J. (1989) The new institutional economics and its applicability to development. *World Development* 17, 1333–1347.

Nagy, J. and Furtan, H. (1977) *The Socio-Economic Costs and Returns from Rapeseed Breeding in Canada*. Department of Agricultural Economics, University of Saskatchewan, Saskatoon.

Nagy, J. and Furtan, H. (1978) Economic costs and returns from crop development research: the case of rapeseed breeding in Canada. *Canadian Journal of Agricultural Economics* 26, 1–14.

Natunola (1998) http://www.ottawa.lifesciences.com/corporate/natunola.htm

Nordfeldt, S., Gellerstedt, N. and Falkmer, S. (1954) Studies with rapeseed oilmeal and its goitrogenic effect in pigs. A nutritional and histopathological study. *Acta Pathologica et Microbiologica Scandinavica* 35, 217–236.

North, D. (1989) Institutions and economic growth: an historical introduction. *World Development* 17, 1319–1331.

North, D. (1991) Institutions. *Journal of Economic Perspectives* 5, 97–112.

NRC (National Research Council) (1992) *From Rapeseed to Canola: the Billion Dollar Success Story*. National Research Council, Saskatoon, Saskatchewan, p. 79.

NRC (1997) Bi-Annual Report. Retrieved from the World-wide Web at http://www.pbi.nrc.ca/96annrpt/bus.html

NRC (1998) Commercialization of plants with novel traits. *PBI Bulletin*, September, retrieved 31 May, 1999 from the World-wide Web: http://www.pbi.nrc.ca/bulletin/pbibintro.html

OECD (1992) *Technology and the Economy: the Key Relationships*. OECD, Paris.

OECD (1996) *The Knowledge Based Economy*. OECD, Paris: Retrieved from the World-wide Web at http://www.oecd.org/dsti/sti/s_t/inte/prod/kbe.htm

OECD (1999) *Policy Brief*. OECD, Paris.

Ohlson, R. (1972) Production of and trade in rapeseed. In: Appelquist, L.A. and Ohlson, R. (eds) *Rapeseed: Cultivation, Composition, Processing and Utilization*. Elsevier, Amsterdam, pp. 9–35.

Olson, M. (1965) *The Logic of Collective Action: Public Goods and the Theory of Groups*. Harvard University Press, London.

OTA (Office of Technology Assessment) US Congress (1984) *Commercial Biotechnology: an International Analysis*. US Government Printing Office, Washington, DC.

OTA (1991) *Biotechnology in a Global Economy*. US Government Printing Office, Washington, DC.

Palast, G. (1999) Outrage Over Monsanto's Underhand Tactics in the EU. *The London Observer* 14 March.

Palmer, C.E. and Keller, W. (2001) Transgenic oilseed Brassicas. In: Khachatourians, G.G., McHughen, A., Nip, W.-K., Scorza, R. and Hui, Y.H. (eds) *Transgenic Plants and Crops*. Marcel Dekker, New York.

Pardey, P. and Craig, B. (1989) Causal relationships between public sector agricultural research expenditures and output. *American Journal of Agricultural Economics* 71, 9–19.

Parkin, I. and Lydiate, D. (2001) The dynamics of plant genome organization. In: Khachatourians, G.G., McHughen, A., Nip, W.-K., Scorza, R. and Hui, Y.H. (eds) *Transgenic Plants and Crops*. Marcel Dekker, New York.

Parmenter, L., Boothe, J., VanRooijen, G., Yeung, E. and Moloney, M. (1995) Production of biologically active hirudin in plant seeds using oleosin partitioning. *Plant Molecular Biology* 29, 1167–1180.

Patel, P. (1995) Localized production of technology for global markets. *Cambridge Journal of Economics* 19, 141–153.

Pearce, D. (ed.) (1996) *The MIT Dictionary of Modern Economics*, 4th edn. MIT Press, Cambridge, Massachusetts.

Pelletier, G., Primard, C., Vedel, F., Chetrit, R. and Renard, M. (1983) Intergeneric cytoplasmic hybridization in Cruciferae by protoplast fusion. *Molecular and General Genetics* 191, 240–250.

Perillat, B. and Phillips, P. (1999) Farmer returns from herbicide tolerant canola: a Saskatchewan case study. University of Saskatchewan Working Paper.

Persley, G. and Lantin, M. (eds) (1999) *Agricultural Biotechnology and the Poor*. CGIAR, Washington, DC.

Peterson, W. (1967) Returns to poultry research in the United States. *Journal of Farm Economics* 49, 656–669.

Phillips, P. (1998) Innovation and restructuring in Canada's canola industry. *Proceedings* of 3rd International Conference on Chain Management in Agribusiness and the Food Industry, May.

Phillips, P. and Foster, H. (2000) Labeling for GM foods: theory and practice. *Proceedings of the ICABR conference, University of Rome 'Tor Vergata', August.*

Phillips, P. and Isaac, G. (1998) GMO labelling: threat or opportunity? *AgBioForum* 1(1) (www.agbioforum.missouri.edu/).

Phillips, P., Smyth, S. and McCormick, C. (1999) An economic assessment of the activities and investments of Ag-West Biotech Inc. and ICAST, 1989–99. (Mimeograph, 31 August.)

Phillips, P.W.B. (2001) Regional systems of innovation as a modern R&D entrepot: the case of the Saskatoon biotechnology cluster. In: Chrisman, J. *et al.* (eds) *Innovation, Entrepreneurship, Family Business and Economic Development: a western Canadian Perspective.* University of Calgary Press.

Phillips, PW.B. and Buckingham, D. (2001) Agricultural biotechnology, the environment and international trade regulation: your place or mine?. In: Stabler, J. *et al.* (eds) *Globalization and New Agricultural Trade Rules for the 21st Century.* Lynne Rynner.

Phillips, P.W.B. and Khachatourians, G.G. (2001) Political and economic consequences. In: Khachatourians, G.G., McHughen, A., Nip, W.-K., Scorza, R. and Hui, Y.H. (eds) *Transgenic Plants and Crops.* Marcel Dekker, New York.

Picciotto, R. (1995) *Putting Institutional Economics to Work: from Participation to Governance.* World Bank Discussion Paper 304.

Pioneer Grain Company Ltd (1999) 1999 Canola Product Line.

Plant, A., VanRooijen, G., Anderson, C. and Moloney, M. (1994) Regulation of an *Arabidopsis* oleosin gene promoter in transgenic *Brassica napus. Plant Molecular Biology* 25, 193–205.

Plunket, M.D. and Gaisford, J.D. (2000) Limiting biotechnology? Information problems and policy responses. *Current Agriculture, Food and Resource Issues* 1, 21–28. Retrieved September 21, 2000 from the World Wide-web: http://www.CAFRI.org

Poirier, Y. (1999) Production of new polymeric compounds in plants. *Current Opinion in Biotechnology* 10, 181–185.

Porter, M. (1990) *The Comparative Advantage of Nations.* Free Press, New York.

Porter, M. (1991) *Canada at the Crossroads – the Reality of a New Competitive Environment.* Study prepared for the Business Council on National Issues and the Government of Canada, October 1991.

Porter, M.E. (1998) *On Competition.* Harvard Business School, Boston, Massachusetts.

Prairie Pools Inc. Various. *Prairie Grain Variety Survey,* 1967–1992.

Putnam, R. (1993) *Making Democracy Work: Civic Traditions in Modern Italy.* Princeton University Press, Princeton, New Jersey.

RAFI (1995) Genetically engineered high-lauric rapeseed (canola): What threat to tropical lauric oil producers? RAFI communique Mar/Apr.

Rakow, G. and Woods, D. (1987) Outcrossing in rape and mustard under Saskatchewan prairie conditions. *Canadian Journal of Plant Science* 67, 147–151.

Raney, J. and Falk, K. (1998) Outcrossing in turnip rape (*Brassica rapa* L.). *11th International Crucifer Genetics Workshop, Montreal, Canada.* Abst. p.36.

Ram's Horn, The. Available on the World-wide Web at: http://www.ramshorn.bc.ca/back.html

Romer, P. (1990) Endogenous technological change. *Journal of Political Economy* 98(5:2), S71–S102.

Romer, P. (1995) Beyond the knowledge worker. *Worldlink,* Jan/Feb; retrieved from http://www~leland.stanford.edu/~promer/wrld_lnk.htm

Rose, F. (1980) Supply shifts and research benefits: comments. *American Journal of Agricultural Economics* 62, 834–837.

Rosenberg, N., Landau, R. and Mowery, D. (1992) *Technology and the Wealth of Nations.* Stanford University Press, Stanford.

Roth, M. (1999) Transitions in agbiotech: economics of strategy and policy. Presentation to the NE-165 Conference in Washington, DC, 24–25 June.

Russell, A. (1990) Biotechnology in the European Community: promotion and regulation. Paper presented at the annual convention of the International Society for Political Psychology, Tel Aviv, Israel.

Sacristan, M. (1982) Resistance responses to *Phoma lingam* of plants regenerated from selected cell and embryogenic cultures of haploid *Brassica napus. Theoretical and Applied Genetics* 61, 193–200.

Santaniello, V., Evenson, R., Zilberman, D. and Carlson, G. (eds) (2000) *Agriculture and Intellectual Property Rights: Economic Institutional and Implementation Issues in Biotechnology.* CAB International, Wallingford, UK.

Saskatchewan Canola Development Commission (1998) Special tabulation.

Saskatchewan Canola Development Commission (2000) Benefits of herbicide-tolerant canola systems vary, study shows. Retrieved 21 Sept. 2000 from the World-wide Web: http://www.canolainfo.org/scdc/html/news.html#anchor15571

Sauer, D. and Kramer, J. (1983) The problems associated with the feeding of high erucic rapeseed oils and some fish oils to experimental animals. In: Kramer, J., Sauer, F. and Pidgen, W. (eds) *High and Low Erucic Acid Rapeseed Oils.* Academic Press, Orlando, Florida.

Schmitz, A. and Seckler, D. (1970) Mechanized agriculture and social welfare: The case of the tomato harvester. *American Journal of Agricultural Economics* 52, 569–578.

Schumpeter, J. (1954) *Capitalism, Socialism, and Democracy.* George Allen and Unwin, London.

Science Watch (1996) *Science Watch* 7(1), 2.

Scobie, G. and Posada, R. (1978) The impact of tecnological change on income distribution: the case of rice in Columbia. *American Journal of Agricultural Economics* 60(1), 85–92.

Shapiro, C. and Varian, H. (1999) *Information Rules: a Strategic Guide to the Network Economy.* Harvard University Press, Boston, Massachusetts.

Smyth, S. and Phillips, P. (2000) Science and regulation: Assessing the impacts of incomplete institutions and information in the global agricultural biotechnology industry. *Proceedings of the ICABR conference,* University of Rome 'Tor Vergata'.

Solow, R. (1956) A contribution to the theory of economic growth. *Quarterly Journal of Economics* 70, 65–94.

Sparks Companies Inc. (1997) The impact of the introduction of roundup ready™ canola on quantifiable environmental conditions. Study contracted by Monsanto Company.

SREDA (1998) Biotechnology Industry Profiles. Mimeograph.

Statistics Canada (1997) Economic impact of canola industry, special tabulations.

Stiglitz, J. (1999) *Wither Reform? Ten Years of the Transition.* World Bank Annual Bank Conference on Development Economics, 28–30 April.

Storey, G. (1998) *Central Marketing Requirements for the Development of the Saskatchewan*

Small Fruit Industry. Department of Agricultural Economics, University of
Saskatchewan.

Stringam, G. and Downey, K. (1978) Effectiveness of isolation distance in turnip rape.
Canadian Journal of Plant Science 58, 427–434.

Swanson, E., Coumans, S., Barsby, T. and Beversdorf, W. (1987) Efficient isolation of
microspores and production of microspore-derived embryos from *Brassica napus.*
Plant Cell Reports 6, 94–97.

Swanson, T. (1997) *Global Action for Biodiversity.* Earthscan Publications Ltd, London.

Tait, J. and Levidow, L. (1992) Proactive and Reactive Approaches to Risk Regulation:
the Case of Biotechnology. *Futures,* April, 219–231.

The Economist (1998) In defense of the demon seed, 13 June, p. 13.

The Financial Times (1999) 15 March.

The Western Producer (2000a) Dairies debate GMO use. 10 February.

The Western Producer (2000b) Test show no GMO residue in animals. 10 February.

The Western Producer (2000c) Triple-resistant canola weeds found in Alta. 10 February.

Thirtle, C. and Bottomley, P. (1988) Is publicly funded agricultural research excessive?
Journal of Agricultural Economics 31, 99–111.

Tirole, J. (1988) *The Theory of Industrial Organization.* MIT Press, Cambridge,
Massachusetts.

Trajtenberg, M. (1990) *Economic Analysis of Product Innovation: the case of CT scanners.*
Harvard University Press, Cambridge, Massachusetts.

Ulrich, A. and Furtan, H. (1985) *An Investigation in the Rates of Return from the Canadian
Crop Breeding Program.* University of Saskatchewan.

Ulrich, A., Furtan, H. and Downey, K. (1984) Biotechnology and rapeseed breeding:
some economic considerations. A Manuscript Report Prepared for the Science
Council of Canada.

Ulrich, A., Furtan, H. and Schmitz, A. (1986) Public and private returns from joint venture research: an example from agriculture. *Quarterly Journal of Economics,*
February, 103–129.

UMI Proquest (1998) Digital dissertations. Searched on http://wwwlib.umi.com/
dissertations/search

UMI Proquest (1999) Digital dissertations Available from http://wwwlib.umi.com/
dissertations/gateway

Unnevehr, L. (1986) Consumer demand for rice grain quality and returns to research
for quality improvements in southeast Asia. *American Journal of Agricultural
Economics* 68, 634–641.

UPOV (1999) *About UPOV.* Retrieved from http://www.upov.int/eng/content.htm

USDA (1997) *International Agriculture and Trade, Europe Update,* 9 July.

USDA (1998) Revealed comparative advantages. Retrieved at http://mann77.mannlib.
cornell.edu/data-sets/trade/93005/Table174-Table187/

Vain, P., Worland, B., Kohli, A., Snape, J., Christou, P., Allen, G. and Thompson,
W. (1999) Matrix attachment regions increase transgene expression levels and
stability in transgenic rice plants and their progeny. *The Plant Journal* 18,
233–242.

Van Rooijen, G. and Moloney, M. (1995) Plant seed oil bodies as carriers for foreign proteins. *Biotechnology* 13, 72–77.

Vernon, R. (1966) International investment and internationl trade in the product cycle.
Quarterly Journal of Economics 80(2).

Von Thunen, J. (1826) *The Isolated State* [English edition]. Pergamon Press, London.

Voon, J. (1991) Measuring research benefits from a reduction of pale, soft and exudative pork in Australia. *Journal of Agricultural Economics* 43, 180–184.

Voon, J. and Edwards, G. (1992) Research payoff from quality improvement: the case of backfat depth in pigs. *American Journal of Agricultural Economics* 74, 564–572.

Warda, J. (1997) *R&D Tax Incentives in OECD Countries: How Canada Compares*. Members' Briefing 190-97, The Conference Board of Canada, January.

Weber, A. (1909) *Theory of Location of Industries*. University of Chicago Press, Chicago.

Wensley, M. (1996) Institutional change and the canola industry in Saskatchewan. University of Saskatchewan mimeograph.

Western Grains Research Foundation (1998) Special Tabulation.

White, W. (1974) Production of rapeseed in Canada. In: *The Story of Rapeseed in Western Canada*. Saskatchewan Wheat Pool, Regina, Saskatchewan, Chapter 1.

Wiegele, T. (1991) *Biotechnology and International Relations: the Political Dimensions*. University of Florida Press, Gainesville.

Williamson, O. (1979) Transaction-cost economics: the governance of contractual relations. *Journal of Law and Economics* 22, 233–261.

Williamson, O. (1981) The economics of organization: transaction cost approach. *American Journal of Sociology* 87, 548–577.

Williamson, O. (1983) Organization form, residual claimants, and corporate control. *Journal of Law and Economics* 26, 675–680.

Williamson, O. (1985) *The Economic Institutions of Capitalism: Firms, Markets, Relational Contracting*. The Free Press, New York.

Woolcock, S. (1998) European and North American Approaches to Regulation: Continued Divergence? In: van Scherpenberg, J. and Thiel, E. (eds) *Towards Rival Regionalism? US and EU Regional Regulatory Regime Building*. Nomos Verlagsgesellschaft, Baden-Baden, pp. 257–276.

Youngs, C. (1974) Uses of rapeseed oil. In: McLeod, A. (ed.) *The Story of Rapeseed in Western Canada*. Saskatchewan Wheat Pool, Regina.

Zatylny, T. (1998) Canola on Biotechnology. Canola Council of Canada Website.

Zenter, R.P. (1982) An economic evaluation of public wheat research expenditures in Canada. PhD thesis, University of Minnesota, St Paul, Minnesota.

Zilberman. D., Yarkin, C. and Heiman, A. (1997) Agricultural biotechnology: economic and international implications. Paper presented at the *XXIII International Conference of Agricultural Economists, 'Food Security, Diversification, and Resource Management: Refocusing the Role of Agriculture'*, Sacramento, 10–16 August.

Zucker, L., Darby, M. and Brewer, M. (1998) Intellectual human capital and the birth of U.S. biotechnology enterprises. *American Journal of Agricultural Economics* 88(1), 290–306.

Index

£55.00

£55.00